Physics Research and Technology

nova science publishers

www.novapublishers.com

Physics Research and Technology

Electromagnetic Waves: Advances in Applications and Research
Manuel B. Hutchinson (Editor)
2022. ISBN: 979-8-88697-254-2 (Softcover)
2022: ISBN: 979-8-88697-265-8 (e-book)

Beyond Special Relativity:
Looking for the Intrinsic Properties of Space-Time
Raul Tozzi and
Riccardo Zancan
2022. ISBN: 979-8-88697-108-8 (Hardcover)
2022. ISBN: 979-8-88697-219-1 (eBook)

Second Harmonic Generation: Pathways of Nonlinear Photonics
Abdel-Baset M. A. Ibrahim, PhD
Pankaj Kumar Choudhury, PhD (Editors)
2022. ISBN: 978-1-68507-888-1 (Hardcover)
2022. ISBN: 979-8-88697-002-9 (eBook)

Advances in Sustainable Materials and Technology
Sandeep Singh, PhD
Er. Shristi Kanoungo, M.E.
Ajay Goyal, PhD (Editors)
2022. ISBN: 978-1-68507-967-3 (Hardcover)
2022. ISBN: 979-8-88697-070-8 (eBook)

Quantum Field Theory and Applications
Natale Palerma (Editor)
2022. ISBN: 978-1-68507-930-7 (Hardcover)
2022. ISBN: 978-1-68507-957-4 (eBook)

More information about this series can be found at
https://novapublishers.com/product-category/series/physics-research-and-technology/

Marian Apostol, PhD

Radiation and Matter

nova science publishers

www.novapublishers.com

DOI: https://doi.org/10.52305/ZRBX8691

Library of Congress Cataloging-in-Publication Data

ISBN: 979-8-88697-246-7

Published by Nova Science Publishers, Inc. † New York

Contents

1 Introduction

The main part of this book relates to the relativistic quantum mechanical motion and quantum-mechanical electromagnetic effects. Elementary electrical charges are pointlike particles with mass. The electromagnetic field produced by these charges is singular at the position of the charges. The self-interaction of the charges should be avoided. Therefore, we should endow the charges with a small, at least infinitesimal, radius. The pointlike charges move in an indefinite manner inside such a surrounding region.

The electromagnetic field propagates with the speed of light in vacuum c ($c = 3 \times 10^{10} cm/s$), which is higher than the velocity of any other body. Consequently, a charge may interact with its own propagating field. This self-interaction is avoided only if the region surrounding the charge has the radius $a = q^2/mc^2$, where q is the charge of the particle and m is the particle mass. This is the classical electromagnetic radius of the charge (particle). For electrons $a \simeq 2.8 \times 10^{-13} cm$ ($q = e = -4.8 \times 10^{-10} esu$, $m \simeq 10^{-27} g$).

A relativistic particle with mass m may have a quantum-mechanical motion, as long as the variations of its mechanical action are comparable to Planck's constant $\hbar$ ($\hbar \simeq 10^{-27} erg \cdot s$). For electrons this motion is described by the Dirac equation. The Dirac equation includes matrices, which mix up the spin states and the charge (particle-antiparticle) states. Consequently, this quantum-mechanical motion is undetermined to some extent. According to the Dirac equation, the undetermined motion proceeds over distances of the order of the Compton wavelength $\lambda_c = \hbar/mc$, with an uncertainty mc in momentum and an uncertainty mc^2 in energy; the time uncertainty is $\hbar/mc^2$. This is the Zitterbewegung (or trembling motion). A relativistic pointlike electron has an indefinite motion inside a region with radius $\lambda_c = \hbar/mc$. For electrons $\lambda_c \simeq 3.8 \times 10^{-11} cm$. The existence of the cutoff length λ_c originates in the negative relativistic energy, *i.e.*

in the existence of the antiparticles. Therefore, these cutoff length, momentum, energy and time should be incorporated in the description of the quantum-mechanical motion of a relativistic particle. In a quantum-mechanical elementary act of interaction a particle cannot receive an energy larger than mc^2, since this is the maximum available energy povided by a disappearing particle; similarly for the maximum momentum mc. No photon can be created with a larger energy (momentum), because this is the maximum energy (momentum) provided by a disappearing particle.

A quantum-mechanical motion can be superposed over the Zitterbewegung. This would be a measurable motion. It is obtained by giving small variations of the Zitterbewegung. For elementary acts of interaction this procedure leads to a boson field with four components, corresponding to spin states and charge states, which satisfies the Klein-Gordon equation. For a charge this equation of motion includes the Pauli interaction term. It should be treated by using the Compton cutoff momentum (length, energy, time). This treatment induces an uncertainty in the physical quantities, which is a substitute for the renormalization technique. This uncertainty should be visible in the experimental data, *e.g.* in the Lamb shift and the anomalous magnetic moment. Quantum-mechanical effects may accompany relativistic motion. Both the Dirac equation and the Klein-Gordon equation can be used for relativistic fields, with these limitations, since the relativistic motion implies a change of the reference frame and both equations are relativistically invariant. However, the Dirac equation includes an indefinite motion.

Any wavefunction can be written as $\psi = Ae^{\frac{i}{\hbar}S}$, where A and S are functions of time and position. For a particle we may write locally $S = S_0 + \frac{\partial S}{\partial t}\delta t + \frac{\partial S}{\partial \boldsymbol{r}}\delta \boldsymbol{r} + ...$, such that S is the classical action with energy $\mathcal{E} = -\frac{\partial S}{\partial t}$ and momentum $\boldsymbol{p} = \frac{\partial S}{\partial \boldsymbol{r}}$. The wavefunction $\psi \sim e^{-\frac{i}{\hbar}\mathcal{E}\delta t + \frac{i}{\hbar}\boldsymbol{p}\delta \boldsymbol{r} + ...}$ is meaningful only if $\delta\mathcal{E}\delta t > \frac{1}{2}\hbar$ and $\delta p \delta l > \frac{1}{2}\hbar$, where δl is the variation of the distance (the factor $1/2$ is half the unit radius of the circle over which the phase varies with π). These are the quantum-mechanical uncertainty relations. For a classical motion the variation of the action is much greater than $\hbar$, the energy and the momentum vary slowly over large durations and distances, the function S can be preserved as the classical action over such large durations and distances, and the uncertainty relations are well satisfied ($\delta\mathcal{E}\delta t \gg \frac{1}{2}\hbar$ and $\delta p\delta l \gg \frac{1}{2}\hbar$). For a quantum-mechanical motion the variations of the action are of the order of $\hbar$, the classical form of the action cannot be preserved over large durations and positions, the energy and the momentum vary fastly and the quantum-mechanical uncertainty relations are satisfied as $\delta\mathcal{E}\delta t \gtrsim \frac{1}{2}\hbar$ and $\delta p\delta l \gtrsim \frac{1}{2}\hbar$. The quantum-mechanical motion is an erratical motion over small distances and short durations without any trajectory.

Over short durations of time and short distances the Schroedinger equation

$$i\hbar\frac{\partial\psi}{\partial t} = -\frac{\hbar^2}{2m}\Delta\psi$$

leads to

$$2m\frac{\delta l^2}{\delta t} = \hbar\frac{\delta(\delta\psi)}{\delta\psi} < \hbar$$

(with all the variations in absolute value), *i.e.* $\delta p\delta l < \frac{1}{2}\hbar$, where $\delta l/\delta t = \delta v$, since l and t are distinct variables. This is an unphysical motion.

Moreover, the Schroedinger equation shows that $\mid \psi \mid^2 = \mid A \mid^2$ is a conserved density of probability of localization of the particle. This gives a meaning to the wavefunction. Also, in the (quasi-) classical limit $\hbar \rightarrow 0$ it recovers the Hamilton-Jacobi equation of motion.

A relativistic particle with mass m has a minimum energy mc^2. Since the relativistic distances δl are related to the relativistic durations δt through the velocity of light in vacuum, $\delta l = c\delta t$, a relativistic particle has also a minimum momentum mc. It follows that the quantum-mechanical uncertainty relations for a relativistic particle read $\frac{1}{2}mc^2\delta t > \frac{1}{2}\hbar$ and $\frac{1}{2}mc\delta > \frac{1}{2}\hbar$, or $\delta t > \hbar/mc^2$ and $\delta l > \hbar/mc$. A quantum-mechanical relativistic motion has a meaning only for distances larger than the Compton wavelength $\lambda_c = \hbar/mc$ and for durations longer than the Compton lifetime $\hbar/mc^2$. The quantum-mechanical motion of a relativistic electron is governed by the Dirac equation

$$i\gamma^\mu\partial_\mu\psi = \frac{mc}{\hbar}\psi \ ,$$

where γ^μ are the Dirac matrices and ψ is a bispinor. First we note that this equation has not a classical limit. Indeed, in the classical limit it becomes $\gamma^\mu p_\mu = mc$, where p_μ is the relativistic momentum; this is an invalid equation. Therefore, the Dirac equation is a purely quantum-mechanical equation. Second, the Dirac equation leads to a probability conservation

$$\frac{\partial}{\partial t}(\psi^*\psi) + c \cdot grad(\psi^* \boldsymbol{\alpha}\psi) = 0 \ ,$$

where $\boldsymbol{\alpha} = \gamma^0\boldsymbol{\gamma}$. This equation shows that the probability density $\psi^*\psi$ propagates with velocity $c\boldsymbol{\alpha}$, which is meaningless. The Dirac equation has no conventional quantum-mechanical meaning. Its applications to the motion in external fields amount in fact to relativistic corrections to the quantum-mechanical motion. Moreover, the projection of the Dirac equation along the direction of variation of the coordinates leads to

$$\frac{mc}{\hbar}\psi = \frac{\delta\psi}{\delta l} < \frac{\psi}{\delta l} \ ,$$

i.e. $\delta l < \hbar/mc$. Therefore, the relativistic quantum-motion over distances smaller than the Compton wavelength (and durations shorter than the Compton lifetime) is meaningelss. This is the Zitterbewegung. The Dirac equation has the structure of an eigenvalue equation

$$\Gamma\psi = (mc^2\gamma^0 + c\boldsymbol{p}\boldsymbol{\gamma})\psi = \mathcal{E}\psi.$$

The eigenvalues of the matrix Γ are $\mathcal{E} = \pm\sqrt{c^2p^2 + m^2c^4}$. It follows that the mean value of the matrix Γ is vanishing, $\overline{\Gamma} = 0$, and its deviation $\left(\overline{\Gamma^2}\right)^{1/2} = \mathcal{E}$ is of the same order of magnitude as its eigenvalues. Therefore, the Zitterbewegung is unphysical and undetermined, while over distances much longer than the Compton wavelength (and durations much longer that the Compton lifetime) the relativistic quantum-mechanical motion has not a conventional quantum-mechanical meaning.

The quantum-mechanical motion of a relativistic electron gets a meaning if the Zitterbewegung moves slowly with wavelenghts longer than the Compton wavelength; this motion is a quantum-field motion. To this end, we give a variation $\delta x^\mu = u^\mu$ to the coordinates x^μ, such that $\delta x_\mu \delta x^\mu = u_\mu u^\mu = u^2$. The solution of this equation $\delta x^\mu = u^\mu = \frac{1}{2} u \gamma^\mu$. The change in the Dirac equation is given by

$$\gamma^\mu \left(\partial_\mu \psi + u^\nu \partial_\nu \partial_\mu \psi\right) = \frac{mc}{i\hbar} \left(\psi + u^\nu \partial_\nu \psi\right) \ ,$$

or

$$\partial^\mu \partial_\mu \psi = -\frac{m^2 c^2}{\hbar^2} \psi \ .$$

This is the Klein-Gordon equation with a four-component boson field (u is absorbed in the original bispinor ψ). The labels of the boson field correspond to the spin states and the charge (particle-antiparticle) states. In the presence of the electromagnetic field with potentials A_μ we need to use the covariant derivative $D_\mu = \partial_\mu - \frac{e}{ic\hbar} A_\mu$, according to $\psi \longrightarrow \psi + u^\mu D_\mu \psi$, where e is the particle charge; we get the Klein-Gordon equation

$$\left(p_\mu - \frac{e}{c} A_\mu\right) \left(p^\mu - \frac{e}{c} A^\mu\right) \psi - \frac{ie\hbar}{2c} \sigma^{\mu\nu} F_{\mu\nu} \psi = m^2 c^2 \psi$$

with electromagnetic field $F_{\mu\nu}$, which includes the Pauli interaction term with $\sigma^{\mu\nu} = \frac{1}{2}[\gamma^\mu, \gamma^\nu]$. The Klein-Gordon equation is solved with the Compton cutoff momentum mc, *i.e.* for wavelengths longer than the Compton wavelength $\lambda_c = \hbar/mc$. Being associated to an undetermined, unphysical motion region, this length is not subject to the relativistic invariance requirements. We recall that the fermionic nature of the 1/2-spin particles is achieved by viewing the bispinor as a field. The boson occupancy number is the number of electrons, the cutoff length λ_c ensures the exclusion principle.

The Klein-Gordon equation has no definite quantum-mechanical probability; for field quanta such a probability is meaningless. Moreover, although the Klein-Gordon equation in the classical limit $\hbar \rightarrow 0$ goes over into the Hamilton-Jacobi equation (quasi-classical motion), the motion of the function S in $\psi = Ae^{\frac{i}{\hbar}S}$ is the motion of a free relativistic field, which is meaningless from a classical standpoint. The Klein-Gordon equation describes a purely quantum-mechanical motion, namely the motion of a quantum field. While the Dirac equation includes both an undefinite motion and a cutoff, the Klein-Gordon equation includes only a cutoff.

1 Introduction

At the beginning there was the quantum of action h. The elementary acts of electromagnetic interaction of the radiation with matter imply a quantum of action h. Radiation consists of waves with wavelengths λ and frequency $\nu = 1/T$, where T is the oscillation period. Therefore, such a wave should have a momentum $p = h/\lambda$ and an energy $\mathcal{E} = h\nu$. Thus, the photon appeared. This quantization of the radiation has nothing to do with any equation of motion. The emission or the absorption of a photon by a particle means a small displacement δx of that particle and a small change δp of its momentum along the direction of motion; these variations should be connected by a relation like $\delta p \delta x \sim h$. Similarly, the variations of the energy and the time should be connected by $\delta\mathcal{E}\delta t \sim h$. In fact, a wave reads $e^{-i2\pi\frac{t}{T}+i2\pi\frac{x}{\lambda}}$, such that the variations are related as $2\pi\delta x\delta p \simeq \frac{1}{2}h$ and $2\pi\delta t\delta\mathcal{E} \simeq \frac{1}{2}h$, or $\delta x\delta p \simeq \frac{1}{2}\hbar$ and $\delta t\delta\mathcal{E} \simeq \frac{1}{2}\hbar$. This way, the momentum can be represented as $p = -i\hbar\frac{\partial}{\partial x}$ and the energy as $\mathcal{E} = i\hbar\frac{\partial}{\partial t}$, such that we get the momentum and the energy by applying these operators to a plane wave.[1] Since $\mathcal{E} = \frac{1}{2m}p^2$ for a non-relativistic particle with mass m, we get the Schroedinger equation $i\hbar\frac{\partial\psi}{\partial t} = -\frac{\hbar^2}{2m}\Delta\psi$ and, of course, $i\hbar\frac{\partial\psi}{\partial t} = -\frac{\hbar^2}{2m}\Delta\psi + V\psi$, where V is an interaction. The Schroedinger equation makes sense in the classical limit $\hbar \to 0$ and shows that the wavefunction ψ is the amplitude of the localization probability. In the quantum-mechanical limit it describes the quantum-mechanical motion, as expected. We note that the quantization was the first, and thereafter the Schroedinger equation. The same quantization rule applied to photons, with $\mathcal{E}^2 = c^2p^2$, leads to the wave equation $\frac{1}{c^2}\frac{\partial^2\psi}{\partial t^2} - \Delta\psi = 0$, where the ψ loses its quantum-mechanical meaning. This is already an early signal that the quantization with standard rules, *i.e.* by replacing $\mathcal{E}$ and $\boldsymbol{p}$ in the relativistic energy-momentum relation with $i\hbar\frac{\partial}{\partial t}$ and $-i\hbar grad$, does not work for photons. Indeed, the photon does not move quantum-mechanically, like a massive particle, by changing the position and the momentum by small variations; it moves like an integral entity, by appearing and disappearing. This is a quantum-field picture, as distinct from the standard quantum-mechanical picture. This is so because the relativistic energy-momentum relation $\mathcal{E}^2 = c^2p^2$ implies both positive

[1]The relation between the uncertainties and the operators is as follows: $p\psi = -i\hbar\frac{\partial\psi}{\partial x} = -i\hbar\frac{\delta\psi}{\delta x}$, and $\mid(\delta\psi, p\psi)\mid = \delta p = \hbar\frac{\mid(\delta\psi,\delta\psi)\mid}{\delta x} > \hbar/\delta x$.

and negative energies, and the negative energies are viewed as the absence (destruction) of a photon; and, of course, the positive energies are viewed as the creation of a photon.

The situation is similar for a massive relativistic particle, with energy given by $\mathcal{E}^2 = c^2p^2 + m^2c^4$. Such a particle has already a momentum mc, so a relativistic pointlike particle moves already in a region with a small dimension of the order $\delta x \simeq \hbar/mc$ (we note the similarity of this relation with the relations $\delta x \sim \lambda$, $p \sim mc$ for photons); of course, its motion inside this region is indefinite. If we wish to quantize its motion, we should quantize the motion of this motion, *i.e.* the motion of this small region; and similarly for the variations of the time. Indeed, the standard replacement of $\mathcal{E}$ and $\boldsymbol{p}$ in the relativistic energy-momentum relation leads to the Klein-Gordon equation $\hbar^2 \frac{\partial^2 \psi}{\partial t^2} - c^2\hbar^2 \Delta\psi + m^2c^4\psi = 0$, which has no meaning: ψ is not an amplitude of probability, and the equation has not a classical limit. It was thought that the way out is the linearization of the relativistic energy-momentum relation, by the Dirac equation. We could say from the beginning that an equivalent linear equation would not resolve the problem, because it is equivalent with the quadratic equation, but, indeed, the Dirac equation brought to light new features, valid for fermions: namely, the existence of the half-integral spin and the explanation of the exclusion principle for fermions. The unphysical motion inside the region with dimensions of the order of the Compton wavelength $\lambda_c = \hbar/mc$ persists; it is known as the Zitterbewegung. The Dirac equation leads to a undetermined velocity of the probability current ($c\boldsymbol{\alpha}$, where $\boldsymbol{\alpha}$ are the Dirac matrices) and has not a classical limit. It is a quantum-field equation for free electrons, which includes an undefinite motion in describing the quantum-mechanical motion, in particular the interaction. This is the reason for the occurrence of the ambiguities in the regularization and the renormalization of the divergences of the quantum electrodynamics (and, in general, the quantum field theory). The quantum-mechanical motion of the quantum fields is the motion of their Compton region. We arrive at its description by using the displacement field of the bosons presented above, and by using the cutoff length of the order of the Compton wavelength (and the other equivalent cutoffs for time, energy, momentum). The results of such computations are inevitably plagued by the uncertainty in the cutoff length. The motion of an indefinite motion cannot be fully definite. We should be content with a fundamental relativistic uncertainty in describing the quantum-mechanical motion.

By 1926 Schroedinger quantized the relativistic energy-momentum equation for a particle with mass m, having in mind the electron, and got what is now called the Klein-Gordon equation. In Coulomb potential this equation does not give the experimentally observed fine structure of the hydrogen atom, because it lacks the spin. The equation was a failure. People continue to think that there should exist a relativistic wave equation, like the non-relativistic Schroedinger equation discovered in 1926. For Dirac, a relativistic wave equation should include the first-order time derivative, as a wave equation, and also, being relativistic, similar first-order spatial derivatives. In 1928 he arrived at the Dirac equation. In electromagnetic field the equation predicts the existence of the spin of the electron in the Klein-Gordon form of the equation. In Coulomb potential the equation gives the correct fine structure of the hydrogen atom, though, rather, in a perturbation theory framework with relativistic corrections. The Dirac equation started to be widely accepted, though many unknowns were related to it. The most troubling inconsistency was the states with negative energy, though the motion of the probability density with an undetermined velocity was no less a big trouble. Dirac noticed that the negative energies are associated with a particle with positive charge, and was told that this particle must be the proton. However, the electron-neutron annihilation rate came out too high, and matter would have been quite unstable. It was shown soon that this positive-charge particle should have the same mass as the electron, so the idea of an antiparticle, the positron, appeared, but still the annihilation rate was there and very high. Then, Dirac assumed the existence of the filled holes in the vacuum, and so the Dirac equation became, from a single-particle equation, a multi-particle equation, which enhanced confusion. It was thought that the idea of a quantum-field of electrons, advanced by Heisenberg by using the Jordan formalism, will improve the things. Pauli was especially satisfied, because the Dirac equation explained this way his exclusion principle. Meanwhile, the positron was discovered, so the Dirac equation gained more credibility. Nevertehless, its meaning continues to escape. Its use both as a wave equation and a quantum-field equation for the fine structure of the hydrogen raises serious problems of interpretation. Moreover, the Foldy-Wouthuysen transformation preserves positive-energy states, separately from the negative-energy ones, though it does so by a canonical transformation, which indicates that the mixing up of the states is still there. The transitions between positive- and negative-energy states ensures the classical Thomson formula of radiation scattering. To complicate the things further, according to Heisenberg, the Dirac equation was viewed as the equation of a classical field, though it has no classical limit. It is safe to say that the meaning of the Dirac equation, as well as of its equivalent Klein-Gordon equation, was never understood.

The application of the Dirac equation to the electromagnetic interaction, *i.e.* to what was called the quantum electrodynamics, is even more troublesome. Suppose that we want to compute the anomalous magnetic moment of the electron. Following Heitler, we arrive at an integral which is ambiguous, both divergent and convergent, depending on the various ways we may estimate it. In one way of viewing it, the integral has the structure of a tensor. Since there is no tensor in such a calculation, we put the integral equal to zero. Such a procedure is called the covariance principle of regularization. It is claimed that the result agrees with the experimental one, though there are several experimental results, which differ between themselves, beyond the experimental errors. Similar problems appear for the Lamb shift, especially visible in Bethe's calculation.

The book includes also chapters on electromagnetism in matter. Matter is both polarizable and magnetizable. The polarization of matter exhibits longitudinal (plasmon) modes and transverse (polaritonic) modes. The latter are responsible for the (Ewald-Oseen) extinction of the incident ray and the occurrence of the refracted ray. The longitudinal modes do not participate in charge-field interaction. The polaritons can be quantized. For them the Minkowski momentum is relevant. For small bodies a resonant dipolar force appears; it is related to the time variation of the Abraham momentum. The surface is special, so surface electromagnetism appears.

2 Electromagnetic Field

2.1 Maxwell equations

E and **H** denote the electric field and the magnetic field in vacuum, respectively, ρ and $\mathbf{j}$ are the densities of electrical charge and electrical current, all real functions of the time t and the position $\mathbf{r}$. The Maxwell equations are

$$\begin{gathered} div\mathbf{E} = 4\pi\rho \ , \ div\mathbf{H} = 0 \ , \\ curl\mathbf{E} = -\tfrac{1}{c}\tfrac{\partial\mathbf{H}}{\partial t} \ , \ curl\mathbf{H} = \tfrac{1}{c}\tfrac{\partial\mathbf{E}}{\partial t} + \tfrac{4\pi}{c}\mathbf{j} \ , \end{gathered} \tag{2.1}$$

where $c \simeq 3 \times 10^{10} cm/s$ is the speed of light. The c in these equations is a universal constant. The first two equations are Gauss's laws, the third equation is Faraday's law, the fourth equation is Maxwell-Ampere's equation. The charges and the currents satisfy the continuity equation

$$\frac{\partial\rho}{\partial t} + div\mathbf{j} = 0 \ . \tag{2.2}$$

By taking the *div* of the *curl* and using the continuity equation, the Gauss laws follow.

The Maxwell equations and the Theory of Electricity and Magnetism (Electromagnetism) have been established in the 18th and 19th centuries and the beginning of the 20th century, by Coulomb, Gauss, Oersted, Ampere, Ohm, Fresnel, Faraday, Maxwell, Heaviside, Kirchhoff, Hertz, Lorentz, Einstein, etc.[1]

[1] J. C. Maxwell, *Lehrbuch der Elektrizitaet und der Magnetismus*, Berlin (1983); H. A. Lorentz, *The Theory of Electrons*, Teubner (Leipzig, 1916); see also L. Landau and E. Lifshitz, *Course of Theoretical Physics*, vol. 2 (*Classical Theory of Fields*), vol. 8 (*Electrodynamics of Continuous Media*), Elsevier (2004).

By using *curl curl* $= -\Delta + grad\,div$ in equations above, we get two equivalent wave equations with sources

$$\begin{aligned} \tfrac{1}{c^2}\tfrac{\partial^2 \boldsymbol{E}}{\partial t^2} - \Delta \boldsymbol{E} = -4\pi grad\rho - \tfrac{4\pi}{c^2}\tfrac{\partial \boldsymbol{j}}{\partial t} \ , \\ \tfrac{1}{c^2}\tfrac{\partial^2 \boldsymbol{H}}{\partial t^2} - \Delta \boldsymbol{H} = \tfrac{4\pi}{c} curl \boldsymbol{j} \ , \end{aligned} \tag{2.3}$$

which show that we have only one unknown field in Maxwell's equations (either $\boldsymbol{E}$, or $\boldsymbol{H}$). According to these equations, the fields are electromagnetic waves propagating with the velocity c. Of course, there exist also static solutions. The solution is provided by a particular field produced by sources plus a superposition of free waves, *i.e.* solutions of the homogeneous equation. Besides the equations, the solution is determined by boundary conditions and initial conditions.

The fields can be obtained from the vector potential protential **A** and the scalar potential Φ,

$$\mathbf{E} = -\frac{1}{c}\frac{\partial \mathbf{A}}{\partial t} - grad\Phi \ , \ \mathbf{H} = curl\mathbf{A} \ , \tag{2.4}$$

defined up to the gauge transformation

$$\mathbf{A} = \mathbf{A}' + grad\chi \ , \ \Phi = \Phi' - \frac{1}{c}\frac{\partial \chi}{\partial t} \ , \tag{2.5}$$

which leaves the fields unchanged; χ is an arbitrary function.

The representation given by equation (2.4) satisfies the homogeneous Maxwell equations (the Faraday equation and the Gauss equation for magnetic field). By using the same representation in the Gauss equation for electric field and the Maxwell-Ampere equation we get[2]

$$\begin{aligned} \tfrac{1}{c^2}\tfrac{\partial^2 \Phi}{\partial t^2} - \Delta\Phi = 4\pi\rho + \tfrac{1}{c}\tfrac{\partial}{\partial t}\left(\tfrac{1}{c}\tfrac{\partial \Phi}{\partial t} + div\boldsymbol{A}\right) \ , \\ \tfrac{1}{c^2}\tfrac{\partial^2 \boldsymbol{A}}{\partial t^2} - \Delta\boldsymbol{A} = \tfrac{4\pi}{c}\boldsymbol{j} - grad\left(\tfrac{1}{c}\tfrac{\partial \Phi}{\partial t} + div\boldsymbol{A}\right) \ , \end{aligned} \tag{2.6}$$

which, by imposing the Lorenz ("Lorentz") gauge

$$\frac{1}{c}\frac{\partial \Phi}{\partial t} + div\boldsymbol{A} = 0 \ , \tag{2.7}$$

[2]Various vectorial identities can be obtained by using $(\mathbf{a}\times\mathbf{b})_i = \varepsilon_{ijk}a_j b_k$ and $(curl\mathbf{a})_i = \varepsilon_{ijk}\partial_j a_k$, where ε_{ijk} is the totally antisymmetric tensor of rank 3 (Levi-Civita symbol, $\varepsilon_{ijk}\varepsilon_{lmk} = \delta_{il}\delta_{jm} - \delta_{im}\delta_{jl}$); $\delta_{ij} = 1$ for $i = j$, $\delta_{ij} = 0$ for $i \neq j$ is the Kronecker symbol.

become wave equations with sources

$$\frac{1}{c^2}\frac{\partial^2\mathbf{A}}{\partial t^2} - \Delta\mathbf{A} = \frac{4\pi}{c}\mathbf{j} \; , \; \frac{1}{c^2}\frac{\partial^2\Phi}{\partial t^2} - \Delta\Phi = 4\pi\rho \; . \tag{2.8}$$

The gauge condition is necessary in order to preserve three independent unknowns (the components of a field) out of the four unknowns introduced by the potentials (three components of $\boldsymbol{A}$ and the scalar potential Φ). Subjected to the Lorenz gauge, the potentials are still defined up to a χ-gauge which satisfies the free wave equation $\frac{1}{c^2}\frac{\partial^2\chi}{\partial t^2} - \Delta\chi = 0$. The two wave equations (2.8) are not independent: they are coupled by the Lorenz gauge (and the continuity equation). It is sufficient to solve one equation, for instance the equation for the vector potential $\mathbf{A}$ and derive the scalar potential by using the Lorenz gauge.

The charges are pointlike particles with mass. A pointlike charge q placed at $\boldsymbol{r}_0(t)$ has a charge density $\rho = q\delta(\boldsymbol{r} - \boldsymbol{r}_0)$; since

$$\dot{\rho} = -q(\boldsymbol{v}grad)\delta(\boldsymbol{r} - \boldsymbol{r}_0) = -qdiv[\boldsymbol{v}\delta(\boldsymbol{r} - \boldsymbol{r}_0)] \; , \tag{2.9}$$

where $\boldsymbol{v} = \dot{\boldsymbol{r}}_0$ is the charge velocity, from the continuity equation we get the current density $\boldsymbol{j} = q\boldsymbol{v}\delta(\boldsymbol{r} - \boldsymbol{r}_0)$.

Under spatial inversion ($\boldsymbol{r} \rightarrow -\boldsymbol{r}$) ρ does not change, while $\boldsymbol{j} \rightarrow -\boldsymbol{j}$ (it is a polar vector); under time reversal ($t \rightarrow -t$) ρ does not change, while $\boldsymbol{j} \rightarrow -\boldsymbol{j}$. Consequently, according to Maxwell equations, under spatial inversion $\boldsymbol{E}$ and $\boldsymbol{A}$ are polar vectors ($\mathbf{E} \rightarrow -\mathbf{E}$, $\boldsymbol{A} \rightarrow -\boldsymbol{A}$), $\boldsymbol{H}$ is an axial vector ($\boldsymbol{H} \rightarrow \boldsymbol{H}$) and Φ does not change. Under temporal inversion $\boldsymbol{E}$ and Φ do not change, while $\boldsymbol{A} \rightarrow -\boldsymbol{A}$ and $\boldsymbol{H} \rightarrow -\boldsymbol{H}$.

2.2 Fourier transformation

Let us consider circular functions written as $e^{i\boldsymbol{kr}}$, where $\boldsymbol{k}$ is a constant vector. Such functions are useful because their spatial derivatives yield algebraic factors; for instance, $\Delta e^{i\boldsymbol{kr}} = -k^2 e^{i\boldsymbol{kr}}$. These functions are called plane waves, because they oscillate in space along the radius r, with plane phase surfaces (*i.e.* they are constant on the surfaces perpendicular to $\boldsymbol{r}$), and with a wavelength $\lambda = 2\pi/k$; $\boldsymbol{k}$ is called wavevector. If we are going to use them as fields, we must take their real part.

Let us consider a parallelipiped with long sides $L_{x,y,z}$ along the x, y, z-axes, and divide these sides in small lengths $a_{x,y,z}$, such that $L_{x,y,z} = a_{x,y,z}N_{x,yz}$, where $N_{x,y,z}$ are large integers. The coordinates are written now as x, y, $z = a_{x,y,z}s_{x,y,z}$, where $s_{x,y,z}$ are integers. Let us assume first that $N_{x,y,z}$ are even integers, so the integers $s_{x,y,z}$ are in the range $-N_{x,y,z}/2 \leq s_{x,y,z} < N_{x,y,z}/2$. Let us define the components of the wavevector as $k_{x,y,z} = \frac{2\pi}{L_{x,y,z}}n_{x,y,z}$, where $n_{x,y,z}$ are integers. Obviously, their range is $-N_{x,y,z}/2 \leq n_{x,y,z} < N_{x,y,z}/2$, such that the phase variation is limited to a multiple of 2π. This natural restriction is called the cyclic boundary condition. We compute the summation

$$\begin{gathered}\sum\nolimits_{\boldsymbol{k}} e^{i\boldsymbol{k}\boldsymbol{r}} = \sum\nolimits_{n_x,n_y,n_z} e^{i\frac{2\pi}{N_{x,y,z}a}n_{x,y,z}a_{x,y,z}s_{x,y,z}} = \\ = \left(\sum\nolimits_{n_x} e^{i\frac{2\pi}{N_x}n_x s_x}\right)\left(\sum\nolimits_{n_y} e^{i\frac{2\pi}{N_y}n_y s_y}\right)\left(\sum\nolimits_{n_z} e^{i\frac{2\pi}{N_z}n_z s_z}\right) \ ;\end{gathered} \tag{2.10}$$

it is sufficient to compute only one of these summations,

$$\begin{gathered}\sum\nolimits_{n_x=-N_x/2}^{n_x=N_x/2-1} e^{i\frac{2\pi}{N_x}n_x s_x} = \frac{e^{-i\pi s_x} - e^{i\pi s_x}}{1-e^{i\frac{2\pi}{N_x}s_x}} = \\ = e^{-i\pi s_x}\frac{1-e^{i2\pi s_x}}{1-e^{i\frac{2\pi}{N_x}s_x}} \ ;\end{gathered} \tag{2.11}$$

the result is zero, except for $s_x = 0$, where it is N_x. Therefore,

$$\sum_{\boldsymbol{k}} e^{i\boldsymbol{k}\boldsymbol{r}} = N\delta_{\boldsymbol{r},0} \ , \tag{2.12}$$

where $N = N_x N_y N_z$. Since the points $\boldsymbol{k}$ are sufficiently dense, we may pass to the integral in equation (2.12), which becomes

$$\frac{V}{(2\pi)^3}\int d\boldsymbol{k}e^{i\boldsymbol{k}\boldsymbol{r}} = N\delta_{\boldsymbol{r},0} \ , \tag{2.13}$$

or

$$\frac{1}{(2\pi)^3}\int d\boldsymbol{k}e^{i\boldsymbol{k}\boldsymbol{r}} = \frac{1}{a_x a_y a_z}\delta_{x,0}\delta_{y,0}\delta_{z,0} = \delta(\boldsymbol{r}) \ , \tag{2.14}$$

where $V = L_x L_y L_z$ is the volume of the parallelipiped. The integration is peformed in the domain $-\pi/a_x < k_x < \pi/a_x$, etc; since $a_{x,y,z} \to 0$, this is the whole $\boldsymbol{k}$-space. Also, a convenient representation of the δ-function is

$$\frac{1}{V}\sum_{\boldsymbol{k}} e^{i\boldsymbol{kr}} = \delta(\boldsymbol{r}) \ . \tag{2.15}$$

Let us do the summation

$$\begin{aligned} \sum_{s_x=-N_x/2}^{s_x=N_x/2-1} e^{i\frac{2\pi}{N_x}n_x s_x} &= \frac{e^{-i\pi n_x} - e^{i\pi n_x}}{1-e^{i\frac{2\pi}{N_x}s_x}} = \\ &= e^{-i\pi n_x}\frac{1-e^{i2\pi n_x}}{1-e^{i\frac{2\pi}{N_x}nx}} = N_x\delta_{n_x,0} \ ; \end{aligned} \tag{2.16}$$

therefore,

$$\sum_{\boldsymbol{r}} e^{i\boldsymbol{kr}} = \frac{1}{a_x a_y a_z}\int d\boldsymbol{r} e^{i\boldsymbol{kr}} = N\delta_{\boldsymbol{k},0} \ , \tag{2.17}$$

or

$$\int d\boldsymbol{r} e^{i\boldsymbol{kr}} = V\delta_{\boldsymbol{k},0} = (2\pi)^3\delta(\boldsymbol{k}) \ . \tag{2.18}$$

It is easy to see that the same results are obtained for odd integers $N_{x,y,z}$. In passing from $a_{x,y,z}$ to $a_{x,y,z} \to 0$ an error of the order of the area of the surface of the parallelipiped occurs, which in the limit $V \to \infty$ (for reasonable shapes of the region with volume V) may be neglected; such that the above formulae do not depend on the shape of the region with volume V in this limit.

The representations of the δ-function given above allow the expansion of any function $f(\boldsymbol{r})$ in Fourier series (integrals), according to

$$f(\boldsymbol{r}) = \frac{1}{V}\sum_{\boldsymbol{k}} f_{\boldsymbol{k}} e^{i\boldsymbol{kr}} \ ; \tag{2.19}$$

indeed, from this equation we get immediately

$$\int d\boldsymbol{r} f(\boldsymbol{r}) e^{-i\boldsymbol{kr}} = f_{\boldsymbol{k}} \ ; \tag{2.20}$$

various other conventions can be used, regarding the integrals, the discrete summations, the change of the factor $1/V$, etc. Errors with this representation occur at discontinuities. The functions are repeated periodically over the volume V and over the region $(-\pi/a_x, \pi/a_x)$, etc; for $a_x \to 0$ this range is infinite. For real functions $f^*_{-\boldsymbol{k}} = f_{\boldsymbol{k}}$.

A similar Fourier representation holds for time-dependent functions. By a similar procedure, we get

$$\delta(t) = \frac{1}{2\pi} \int d\omega e^{i\omega t} \ , \ \delta(\omega) = \frac{1}{2\pi} \int dt e^{-i\omega t} \tag{2.21}$$

and we have for any function $f(t)$ the Fourier representation

$$f(t) = \frac{1}{2\pi} \int d\omega f(\omega) e^{-i\omega t} \ , \ f(\omega) = \frac{1}{2\pi} \int dt f(t) e^{i\omega t} \ . \tag{2.22}$$

The functions $e^{i\omega t}$ oscillate with the period $T = 2\pi/\omega$; they are called "monochromatic" waves. ω is angular frequency (or, simply, frequency) and $\nu = 1/T$ is frequency.

2.3 Solutions

Let us consider the static equation

$$\Delta\Phi = -4\pi\rho \tag{2.23}$$

for the scalar potential (Poisson equation). We look for solutions vanishing at infinity. If we knew the solution G of the equation

$$\Delta G = -4\pi\delta(\boldsymbol{r}) \ , \tag{2.24}$$

corresponding to a pointlike charge $q = 1$ placed at $\boldsymbol{r}_0 = 0$, the solution Φ of equation (2.23) is

$$\Phi(\boldsymbol{r}) = \int d\boldsymbol{r}' \rho(\boldsymbol{r}') G(\boldsymbol{r} - \boldsymbol{r}') \ . \tag{2.25}$$

Noteworthy, $\delta(\boldsymbol{r})$ plays the role of a boundary condition at $\boldsymbol{r} = 0$. This is the Green theorem, and G is a Green function. The Fourier transform of equation (2.24) gives $G_{\boldsymbol{k}} = 4\pi/k^2$, and

$$G(\boldsymbol{r}) = \tfrac{1}{V} \textstyle\sum_{\boldsymbol{k}} \tfrac{4\pi}{k^2} e^{i\boldsymbol{kr}} = \tfrac{1}{\pi} \int_0^\infty dk \tfrac{e^{ikr} - e^{-ikr}}{ikr} \ . \tag{2.26}$$

We change k into $-k$ in the second integral, and get a singular integral for $k=0$. We put $k \longrightarrow k+i0^+$and get

$$G(r)=\frac{1}{r} \ . \tag{2.27}$$

This is the Coulomb potential of a pointlike charge $q=1$. We note that

$$\Delta\frac{1}{r}=-4\pi\delta(\boldsymbol{r}) \ . \tag{2.28}$$

According to equation (2.25), for an arbitrary charge density the potential is

$$\Phi(\boldsymbol{r})=\int d\boldsymbol{r}'\frac{\rho(\boldsymbol{r}')}{\mid\boldsymbol{r}-\boldsymbol{r}'\mid} \ . \tag{2.29}$$

For a pointlike charge q placed at $\boldsymbol{r}_0$ the static potential is $\Phi(\boldsymbol{r}) = q/\mid\boldsymbol{r}-\boldsymbol{r}_0\mid$. Similarly, the vector potential of a static current density. *i.e.* the solution of the equation

$$\Delta\boldsymbol{A}=-\frac{4\pi}{c}\boldsymbol{j} \tag{2.30}$$

($div\boldsymbol{j}=0$) is given by the Biot-Savart (Ampere) law

$$\boldsymbol{A}(\boldsymbol{r})=\frac{1}{c}\int d\boldsymbol{r}'\frac{\boldsymbol{j}(\boldsymbol{r}')}{\mid\boldsymbol{r}-\boldsymbol{r}'\mid} \ . \tag{2.31}$$

For a pointlike charge moving with the constant velocity $\boldsymbol{v}$ the vector potential is $\boldsymbol{A}(\boldsymbol{r})=q\boldsymbol{v}/c\mid\boldsymbol{r}-\boldsymbol{r}_0\mid$.
Let us seek a solution of the wave equation

$$\frac{1}{c^2}\frac{\partial^2\Phi}{\partial t^2}-\Delta\Phi=4\pi\rho \tag{2.32}$$

for the scalar potential Φ. We look for a potential Φ which is vanishing at infinity and, also, is causal, *i.e.* it vanishes for $t<0$. Φ is given by

$$\Phi(t,\boldsymbol{r})=\int dt'd\boldsymbol{r}'\rho(t',\boldsymbol{r}')G(t-t',\boldsymbol{r}-\boldsymbol{r}') \ , \tag{2.33}$$

where the Green function G satisfies the equation

$$\frac{1}{c^2}\frac{\partial^2 G}{\partial t^2}-\Delta G=4\pi\delta(t)\delta(\boldsymbol{r}) \ ; \tag{2.34}$$

by Fourier transforming, it leads to

$$G(t,\boldsymbol{r}) = -\frac{4\pi c^2}{(2\pi)^4}\int d\omega d\boldsymbol{k}\frac{e^{-i\omega t+i\boldsymbol{kr}}}{\omega^2-c^2k^2} \ . \tag{2.35}$$

In order to get a vanishing G for $t<0$, the ω-integration must be performed in the upper half-plane, so the poles $\omega=\pm ck$ $(k\neq 0)$ must lie in the lower half-plane, *i.e.* $\omega=\pm ck-i\varepsilon$, $\varepsilon\rightarrow 0^+$; on integration they give $-2\pi i$. The replacement $\omega\rightarrow\omega+i\varepsilon$ ensures the causal condition. Therefore, for $t>0$ we get

$$\begin{aligned} G(t,\boldsymbol{r}) &= -\tfrac{4\pi c^2}{(2\pi)^4}\int d\omega d\boldsymbol{k}\tfrac{e^{-i\omega t+i\boldsymbol{kr}}}{(\omega-ck+i\varepsilon)(\omega+ck+i\varepsilon)} = \\ &= \tfrac{4\pi^2 ci}{(2\pi)^4}\int d\boldsymbol{k}\left(\tfrac{e^{-ickt+i\boldsymbol{kr}}}{k}-\tfrac{e^{ickt+i\boldsymbol{kr}}}{k}\right) = \\ &= \tfrac{\delta(t-r/c)}{r} \end{aligned} \tag{2.36}$$

and

$$\Phi(t,\boldsymbol{r}) = \int d\boldsymbol{r}'\frac{\rho(\boldsymbol{r}',t-\mid\boldsymbol{r}-\boldsymbol{r}'\mid/c)}{\mid\boldsymbol{r}-\boldsymbol{r}'\mid} \ . \tag{2.37}$$

A similar representation holds for the vector potential

$$\boldsymbol{A}(t,\boldsymbol{r}) = \frac{1}{c}\int d\boldsymbol{r}'\frac{\boldsymbol{j}(\boldsymbol{r}',t-\mid\boldsymbol{r}-\boldsymbol{r}'\mid/c)}{\mid\boldsymbol{r}-\boldsymbol{r}'\mid} \ . \tag{2.38}$$

The same results are obtained by using this procedure for the Fourier transform of the wave equation

$$(\omega^2/c^2-k^2)\Phi(\omega,\boldsymbol{k}) = -4\pi\rho(\omega,\boldsymbol{k}) \tag{2.39}$$

and introducing $\rho(\omega,\boldsymbol{k})=\int d\boldsymbol{r}'\rho(\omega,\boldsymbol{r}')e^{-i\boldsymbol{kr}'}$. It is worth noting that the superposition of monochromatic plane waves $e^{-i\omega t+i\boldsymbol{kr}}$ with arbitrary ω and $\boldsymbol{k}$, resulting from the wave equation, reduces to the contribution of the plane waves with $k=\mid\omega\mid/c$, *i.e.* $\omega^2=c^2k^2$. These latter waves satisfy the homogeneous wave equation $(\frac{1}{c^2}\frac{\partial^2}{\partial t^2}-\Delta)\Phi(t,\boldsymbol{r})=0$. We call these waves free (plane) waves, or "real" waves, while the plane waves with $\omega^2\neq c^2k^2$ are called "virtual" waves. The coefficients of the free waves are determined by sources, as if the sources are boundary conditions. The limiting process of reaching the boundary represented by an instantaneous pointlike source ($\sim \delta(t)\delta(\boldsymbol{r})$) is obviously $ct > r \rightarrow 0$.[3] The singularity of the potentials at the position of a pointlike charge and current indicates that a pointlike particle should be viewed as an infinitesimal sphere. If we integrate the Gauss equation $div\boldsymbol{E} = 4\pi\delta(\boldsymbol{r})$ over an infinitesimal sphere with radius $r \rightarrow 0$ enclosing the point $\boldsymbol{r} = 0$, we get the surface integral $4\pi r^2 E = 4\pi$, or $4\pi r^2(\frac{1}{r^2}) = 4\pi$, since $\boldsymbol{E} = -grad(1/r)$, valid for any $r \neq 0$, but not for $r = 0$. Similarly, by using the Stokes theorem in the Maxwell-Ampere equation $curl\boldsymbol{H} = \frac{4\pi}{c}\boldsymbol{j}$ for a uniform, linear current intensity $I = \oint d\boldsymbol{s}\boldsymbol{j}$ passing through a surface $\boldsymbol{S}$ with the infinitesimal element $d\boldsymbol{s}$, we get a magnetic field $H = 2I/cr$, where r is the distance from the current to the observation point and $\boldsymbol{H}$ is tangent to the circles with radius r, in the rotation sense of a right-hand screw which advances along the current (Ampere's law). Apart from $H \rightarrow \infty$ for $r \rightarrow 0$, we need also a finite cross-section for the infinitesimal charges.

According to equations (2.37) and (2.38), the field at the moment of time t is produced by the sources at the retarded moment of time $t- \mid \boldsymbol{r} - \boldsymbol{r}' \mid /c$, the retardation time $\mid \boldsymbol{r} - \boldsymbol{r}' \mid /c$ being the time needed by the fields to propagate from the source placed at $\boldsymbol{r}'$ to the observation point placed at $\boldsymbol{r}$, with the finite velocity c. These are the retarded Kirchhoff potentials. Similar formulae are valid for the fields $\boldsymbol{E}$ an $\boldsymbol{H}$. Writing $\boldsymbol{r}' - \boldsymbol{r} = \boldsymbol{R}$ and

$$\begin{aligned} \Phi(t,\boldsymbol{r}) &= \int d\boldsymbol{R}\frac{\rho(\boldsymbol{r}+\boldsymbol{R},t-R/c)}{R} \ , \\ \boldsymbol{A}(t,\boldsymbol{r}) &= \frac{1}{c}\int d\boldsymbol{R}\frac{\boldsymbol{j}(\boldsymbol{r}+\boldsymbol{R},t-R/c)}{R} \ , \end{aligned} \tag{2.40}$$

the Lorenz gauge $\frac{1}{c}\frac{\partial\Phi}{\partial t} + div\boldsymbol{A} = 0$ is verified immediately by the continuity equation $\frac{\partial\rho}{\partial t} + div\boldsymbol{j} = 0$. The Green function given by equation (2.36) is called spherical wave. Also, its monochromatic component $e^{-i\omega r/c}/r$ is called spherical wave.

[3]M. Apostol, "On unphysical terms in the elastic Hertz potentials", Acta Mech. **228** 2733 (2017).

2.4 Free waves

Let us consider again the wave equation

$$\frac{1}{c^2}\frac{\partial^2 G}{\partial t^2} - \Delta G = 4\pi\delta(t)\delta(\boldsymbol{r}) \ . \tag{2.41}$$

Obviously, the solution G is a function of r, so it is convenient to use spherical coordinates. For $r > 0$ this equation becomes the free-wave equation

$$\frac{1}{c^2}\frac{\partial^2 G}{\partial t^2} - \Delta G = 0 \ , \tag{2.42}$$

or

$$\frac{1}{c^2}\frac{\partial^2 G}{\partial t^2} - \frac{1}{r^2}\frac{\partial}{\partial r}\left(r^2\frac{\partial G}{\partial r}\right) = 0 \ ; \tag{2.43}$$

by using the substitution $G = \chi/r$, this equation becomes the wave equation in one dimension

$$\frac{1}{c^2}\ddot{\chi} - \chi'' = 0 \ . \tag{2.44}$$

We choose the retarded solution $\chi(t - r/c)$, which leads to $G = \chi(t - r/c)/r$. In order to satisfy the boundary condition in equation (2.41) we should approach the point $r = 0$ for finite times; for $r \to 0$ we neglect the time dependence in equation (2.41), which reduces to $\Delta G = -4\pi\delta(t)\delta(\boldsymbol{r})$. The solution of this equation is $G = \delta(t)/r$, such that we get

$$G(t, r) = \frac{\delta(t - r/c)}{r} \tag{2.45}$$

for equation (2.41). This is the spherical wave given by equation (2.36). We can see that the source acts as a boundary condition for free waves. We can check that $G(t, r)$ given by equation (2.45) is a solution of the free-wave equation (2.42) for $r > 0$. Outside the source the solutions of the wave equation are free waves.

Let us consider now the solution of the Fourier-transformed wave equation (2.39)

$$\Phi(\omega, \boldsymbol{k}) = -\frac{4\pi\rho(\omega, \boldsymbol{k})}{\omega^2/c^2 - k^2} \tag{2.46}$$

(for $k \neq 0$), which, according to the causality principle, is written as

$$\Phi(\omega, \boldsymbol{k}) = -\frac{4\pi c^2 \rho(\omega, \boldsymbol{k})}{(\omega - ck + i\varepsilon)(\omega + ck + i\varepsilon)} \quad . \tag{2.47}$$

The integral

$$\int d\omega \Phi(\omega, \boldsymbol{k}) f(\omega) = \frac{4\pi^2 ic}{k} \left[\rho(ck, \boldsymbol{k}) f(ck) - \rho(-ck, \boldsymbol{k}) f(-ck)\right] \tag{2.48}$$

for any function $f(\omega)$ is given by the contribution of the two poles $\omega = \pm ck + i\varepsilon$ ($\varepsilon \rightarrow 0^+$). It is worth noting that the causality principle removes the time-reversal invariance, such that there is no definite parity for the functions ρ and f anymore. By using the same procedure, the solution

$$\Phi_0(\omega, \boldsymbol{k}) = C(\omega)\delta(\omega^2 - c^2k^2) \tag{2.49}$$

of the free-wave equation

$$\left(\omega^2 - c^2k^2\right) \Phi_0(\omega, \boldsymbol{k}) = 0 \tag{2.50}$$

becomes

$$\Phi_0(\omega, \boldsymbol{k}) = \frac{C_1(\omega, \boldsymbol{k})}{2ck} \delta(\omega - ck + i\varepsilon) + \frac{C_2(\omega, \boldsymbol{k})}{2ck} \delta(\omega + ck + i\varepsilon) \tag{2.51}$$

in the limit $\varepsilon \rightarrow 0^+$, where $C_{1,2}(\omega, \boldsymbol{k})$ are some undetermined functions of ω and $\boldsymbol{k}$.[4] The integral in equation (2.48) is now

$$\begin{aligned} &\int d\omega \Phi_0(\omega, \boldsymbol{k}) f(\omega) = \\ &= \frac{1}{2ck} \left[C_1(ck, \boldsymbol{k}) f(ck) + C_2(-ck, \boldsymbol{k}) f(-ck)\right] \quad . \end{aligned} \tag{2.52}$$

By comparing equations (2.48) and (2.52), we can see that the relevant coefficients $C_{1,2}(\pm ck)$ are determined by

$$C_1(ck) = 8\pi^2 ic^2 \rho(ck, \boldsymbol{k}) \ , \ C_2(-ck) = -8\pi^2 ic^2 \rho(-ck, \boldsymbol{k}) \tag{2.53}$$

irrespective of the arbitrary function $f(\omega)$. It follows that the solution of the wave equation is represented by free-wave solutions with coefficients determined by sources.

[4] E. C. G. Stueckelberg, "Relativistisch invariante Stoerungstheorie des Diracschen Elektrons", Ann. Phys. **21** (5. Folge) 367 (1934); "An unambiguous method of avoiding divergence difficulties in quantum theory", Nature **153** 143 (1944); D. Rivier and E. C. G. Stueckelberg, "A propos de divergences en theorie de champs quantifie", Helv. Phys. Acta **23**, Suppl. 3, 236 (1950).

2.5 Transverse waves

The free waves satisfy the homogeneous Maxwell equations

$$div\mathbf{E}=0\ ,\ div\mathbf{H}=0\ ,$$
$$curl\mathbf{E}=-\tfrac{1}{c}\tfrac{\partial\mathbf{H}}{\partial t}\ ,\ curl\mathbf{H}=\tfrac{1}{c}\tfrac{\partial\mathbf{E}}{\partial t}\ . \tag{2.54}$$

These equations show that $\boldsymbol{E}$ and $\boldsymbol{H}$ are perpendicular to one another. If we expand the fields in plane waves $e^{i\boldsymbol{kr}}$, the Fourier transforms $\boldsymbol{E_k}$ and $\boldsymbol{H_k}$ are perpendicular to each other and both are perpendicular to the wavevector $\boldsymbol{k}$. These fields are transverse waves. The magnetic field is always transverse, irrespective of the presence of the currents, due to the Gauss equation $div\boldsymbol{H}=0$. The presence of the currents determines the boundary condition for the magnetic field. If electrical charges are present, we have also a longitudinal electric field, with boundary conditions determined by charges.

For free fields the potentials satisfy the wave equations

$$\frac{1}{c^2}\frac{\partial^2\mathbf{A}}{\partial t^2}-\Delta\mathbf{A}=0\ ,\ \frac{1}{c^2}\frac{\partial^2\Phi}{\partial t^2}-\Delta\Phi=0 \tag{2.55}$$

and the Lorenz gauge

$$\frac{1}{c}\frac{\partial\Phi}{\partial t}+div\boldsymbol{A}=0\ . \tag{2.56}$$

From the second equation (2.55) we get

$$\frac{1}{c^2}\frac{\partial^2}{\partial t^2}\left(-\frac{1}{c}\frac{\partial\Phi}{\partial t}\right)-\Delta\left(-\frac{1}{c}\frac{\partial\Phi}{\partial t}\right)=0\ , \tag{2.57}$$

such that we may take $\Phi'=-\frac{1}{c}\frac{\partial\Phi}{\partial t}$ as a new scalar potential. Indeed, the electric field is then

$$\boldsymbol{E}=-\tfrac{1}{c}\tfrac{\partial\boldsymbol{A}}{\partial t}-grad\left(-\tfrac{1}{c}\tfrac{\partial\Phi}{\partial t}\right)=$$
$$=-\tfrac{1}{c}\tfrac{\partial}{\partial t}\left(\boldsymbol{A}-grad\Phi\right)=-\tfrac{1}{c}\tfrac{\partial\boldsymbol{A}'}{\partial t}\ , \tag{2.58}$$

the magnetic field is $\boldsymbol{H}=curl\boldsymbol{A}=curl\boldsymbol{A}'$ and the Lorenz gauge becomes

$$-\frac{1}{c^2}\frac{\partial^2\Phi}{\partial t^2}+div\boldsymbol{A}'+\Delta\Phi=div\boldsymbol{A}'=0\ , \tag{2.59}$$

where $\boldsymbol{A}' = \boldsymbol{A} - grad\Phi$. This is a gauge transformation. The vector potential $\boldsymbol{A}'$ is a transverse potential. The new gauge condition $div\boldsymbol{A}' = 0$ is the Coulomb gauge. We can see that the longitudinal degrees of freedom are removed from equations; they are superfluous. The transverse (free) electromagnetic waves is a pure radiation field.

Far away from sources, within a limited spatial region, we may neglect the spatial variation of the fields, and view them as superpositions of free waves, of the form $e^{-i\omega(t-r/c)}$, *i.e.* $e^{-i\omega_k t + i\boldsymbol{kr}}$, where $\omega_k = ck$, with arbitrary coefficients (for arbitrary sources). The space inversion and time reversal symmetries of the Maxwell equations are determined by charges and currents. In an empty space, without charges and currents, these symmetries are not necessary. In addition, the time reversal symmetry is removed by the causality principle. In this way the free field becomes an independent entity. For a transverse field included in the volume V we use the Fourier expansion

$$\boldsymbol{A} = \frac{1}{\sqrt{V}} \sum_{\boldsymbol{k}} \alpha_k \boldsymbol{e_k} \left(a_{\boldsymbol{k}} e^{-i\omega_k t + i\boldsymbol{kr} + i\gamma_{\boldsymbol{k}}} + a^*_{\boldsymbol{k}} e^{i\omega_k t - i\boldsymbol{kr} - i\gamma_{\boldsymbol{k}}} \right) , \quad (2.60)$$

where α_k and $a_{\boldsymbol{k}}$ are undetermined coefficients, $\omega_k = ck$, $\boldsymbol{e_k}$ are polarization unit vectors ($\boldsymbol{e_k e_k} = 1$) and $\gamma_{\boldsymbol{k}}$ are undetermined phases. The polarization vector $\boldsymbol{e_k}$ is perpendicular to $\boldsymbol{k}$ ($\boldsymbol{e_k k} = 0$), and $\boldsymbol{e_{-k}} = -\boldsymbol{e_k}$, $\gamma_{-\boldsymbol{k}} = -\gamma_k$; the polarization vector is a two-dimensional vector, the field has two independent polarizations.

2.6 Energy, force and momentum

From the Faraday equation and the Maxwell-Ampere equation (equations (2.1)) multiplied by $\mathbf{H}$ and $\boldsymbol{E}$, respectively, we get

$$\frac{1}{8\pi}\frac{\partial}{\partial t}(E^2 + H^2) + \frac{c}{4\pi} div(\mathbf{E} \times \mathbf{H}) + \mathbf{jE} = 0 \ . \quad (2.61)$$

This equation suggests that $u = (E^2 + H^2)/8\pi$ is a density of electromagnetic energy,

$$\mathbf{S} = \frac{c}{4\pi} \mathbf{E} \times \mathbf{H} \quad (2.62)$$

is an energy flux density (*i.e.* energy per unit time per unit cross-sectional area) and $\mathbf{jE}$ is the rate of change in time of the mechanical work done by the field on the charges, or by charges on the field. When the field energy u decreases, part of it is lost through the surrounding surface and part of it is transferred by mechanical work to charges. When the field energy u increases, part of it comes through the surrounding surface and part of it comes by mechanical work done by charges upon the field. **S** is called the Poynting vector. For a charge density ρ the current density is $\mathbf{j} = \rho\mathbf{v}$, where $\mathbf{v}$ is the velocity of the charge; therefore, $\rho\mathbf{E}$ is a force density $\mathbf{f}$ (of electromagnetic origin), and $\mathbf{jE} = \rho\boldsymbol{v}\boldsymbol{E} = \boldsymbol{v}\mathbf{f}$ is indeed the rate of change in time of the density of a mechanical (kinetic) energy E_m under the action of that force (if w is the mechanical work, then $dw/dt = \mathbf{f}\boldsymbol{v}$). Denoting

$$U = \frac{1}{8\pi}\int d\mathbf{r}(E^2 + H^2) \tag{2.63}$$

we get

$$\frac{\partial}{\partial t}(U + E_m) + \int d\mathbf{s}\mathbf{S} = 0 \ , \tag{2.64}$$

where the integration is performed over the surface **s** enclosing the volume where the fields and charges are placed. Equations (2.61) and (2.64) tell that the volume electromagnetic energy (U) of the electromagnetic fields, the electromagnetic part of the mechanical energy (E_m) of the charges (*i.e.* that part which is changed by the action of electromagnetic forces) and the flux of the electromagnetic energy through the surface enclosing the volume (*i.e.* the surface integral in equation (2.64) multiplied by time) are all balanced against each other. These equations represent the energy conservation. For fields vanishing at infinity ($\mathbf{S} = 0$) the energy $U + E_m$ is conserved (as for any closed motion). u is the pressure of the electromagnetic field. It is worth noting that these equations may include both an external field (which satisfies the homogeneous Maxwell equations) and the field generated by the charges and the currents.

According to Maxwell equations, the charges and the currents generate an electromagnetic field; we can see that they do that by mechanical work. Also, we see that the electromagnetic field acts by a force upon charges and currents. Of course, the self-force, *i.e.* the force acting upon a charge by that charge itself should be avoided (similarly, the force acting upon a field on behalf of that very field should be avoided).

We may view the charges and the currents, on one side, and the field, on the other side, as distinct entities, interacting by the energy density $\int dt\boldsymbol{jE}$. Making use of $\boldsymbol{E} = -\frac{1}{c}\frac{\partial \boldsymbol{A}}{\partial t} - grad\Phi$, we get the interaction energy

$$V = \int d\boldsymbol{r}\left(\rho\Phi - \frac{1}{c}\boldsymbol{jA}\right) . \tag{2.65}$$

Indeed, $\int dt\boldsymbol{j}\left(-\frac{1}{c}\frac{\partial \boldsymbol{A}}{\partial t}\right) = -\frac{1}{c}\boldsymbol{jA}$ (for fixed $\boldsymbol{j}$); as regards $-\int dt\boldsymbol{j}grad\Phi$, we note that $\boldsymbol{j} = \rho\boldsymbol{v} = -\rho\boldsymbol{v}'$ for a pointlike charge density $\rho = q\delta(\boldsymbol{r}-\boldsymbol{r}_0)$, where $\boldsymbol{v} = d\boldsymbol{r}_0/dt$ and $\boldsymbol{v}' = d\boldsymbol{r}/dt$, such that

$$-\int dt\,(\boldsymbol{j}grad\Phi) = \int dt\rho\frac{d\boldsymbol{r}}{dt}grad\Phi = \int dt\rho\frac{d\Phi}{dt} = \rho\Phi \tag{2.66}$$

It is worth noting that this interaction is not applicable to a self-interaction: it requires the charges and the currents, on one side, and the field, on the other side, to be viewed as distinct entities.

Let us form the combination $\rho\mathbf{E}+\mathbf{j}\times\mathbf{H}/c$ from the Maxwell equations (2.1). We get

$$\begin{gathered}4\pi\rho\mathbf{E} + \tfrac{4\pi}{c}\mathbf{j}\times\mathbf{H} = \\ = \mathbf{E}div\mathbf{E} + \left(curl\mathbf{H} - \tfrac{1}{c}\tfrac{\partial\mathbf{E}}{\partial t}\right)\times\mathbf{H} = \\ = \mathbf{E}div\mathbf{E} + \mathbf{H}div\mathbf{H} - (\mathbf{E}\times curl\mathbf{E} + \mathbf{H}\times curl\mathbf{H}) - \\ -\tfrac{1}{c}\tfrac{\partial}{\partial t}(\mathbf{E}\times\mathbf{H}) ;\end{gathered} \tag{2.67}$$

here we write for components $(\mathbf{E}div\mathbf{E})_i = \partial_j(E_iE_j) - (\mathbf{E}grad)E_i$ and use $grad(E^2) = 2(\mathbf{E}grad)\mathbf{E} + \mathbf{2E}\times curl\mathbf{E}$. Equations (2.67) can be cast in the form

$$\mathbf{f} = \mathbf{t} - \frac{\partial \mathbf{g}}{\partial t} , \tag{2.68}$$

where

$$\mathbf{f} = \rho\mathbf{E} + \frac{1}{c}\mathbf{j}\times\mathbf{H} \tag{2.69}$$

is the density of the Lorentz force (acting upon charges and currents on behalf of the electromagnetic field), $t_i = \partial_j\sigma_{ij}$,

$$\sigma_{ij} = \frac{1}{4\pi}\left[E_iE_j + H_iH_j - \frac{1}{2}\delta_{ij}(E^2+H^2)\right] , \tag{2.70}$$

are the stress forces (σ_{ij} is the electromagnetic stress tensor) and

$$\mathbf{g} = \frac{1}{4\pi c}\mathbf{E} \times \mathbf{H} = \frac{1}{c^2}\mathbf{S} \tag{2.71}$$

is the density of the electromagnetic momentum. We can see that the electromagnetic field acts upon charges and currents with the Lorentz force which contains $\rho\mathbf{E}$; the Lorentz term $\mathbf{j} \times \mathbf{H}/c$ does not change the motion energy and does not do mechanical work (since $\mathbf{v}(\rho\mathbf{v} \times \mathbf{H})/c = 0$).

Equation (2.68) is the momentum conservation. The Lorentz force $\boldsymbol{f}$ acting upon charges and currents arises from the field stress force $\boldsymbol{t}$ minus the time variation of the field momentum $\boldsymbol{g}$. Also, the Lorentz force $-\boldsymbol{f}$ force acting upon the field is the time variation of the field momentum $\boldsymbol{g}$ minus the field stress force $\boldsymbol{t}$. We note that the stress forces are surface forces, *i.e.*

$$\int d\mathbf{r} f_i = \int ds_j \sigma_{ij} - \frac{\partial}{\partial t}\int d\mathbf{r} g_i \ . \tag{2.72}$$

The tensor σ_{ij} is the flux of forces (*i.e.* forces divided by the cross-sectional area), or the flow of momentum, *i.e.* the rate of change in time of the field momentum per unit cross-sectional area. There is a state of stress in the field, as in an elastic body, suggesting an aether. In contrast to elastic bodies, the aether has only transverse waves in the far-field region. For a pure radiation field, *i.e.* a free field, the rate of change in time of the electromagnetic momentum is compensated by the stress forces. For fields vanishing at infinity (a pulse for instance), the stress forces are zero, and the electromagnetic momentum is conserved. Similarly, the electromagnetic energy is then conserved. The existence of energy and momentum for radiation field indicates that the radiation field may be viewed as an independent physical entity. Let us assume that charges and currents act by a force $-\boldsymbol{f}$ and generate an electromagnetic field, *i.e.* they give to the field a change of momentum $\partial\boldsymbol{g}/\partial t$ and an additional stress force $-\boldsymbol{t}$; according to the law of action and reaction, and the momentum conservation, the newly created field acts upon the charges which generated it by an equal and opposite force $\boldsymbol{f}$. We can see that the field generated by charges and currents reacts back upon these charges and currents with a reaction force.

It is tempting, by the equations above, to give charges and currents a full electromagnetic nature.[5] Indeed, from equations (2.61) and (2.68) we may write

$$e_m = -u - \int dt \cdot div\mathbf{S} \ ,$$
$$\mathbf{p} = \int \mathbf{f} dt = \int dt \cdot \mathbf{t} - \mathbf{g} \ , \tag{2.73}$$

where e_m is the mechanical energy density of charges and currents and $\mathbf{p}$ denotes the density of their momentum. Then, it may appear that the energy and the momentum of particles (charges) are completely of electromagnetic nature, arising from their own electromagnetic field. It is worth noting that the integration over time makes these expressions relativistically non-invariant, and an additional space integration wouldn't do.

We multiply equation (2.68) by $\mathbf{r}$:

$$\mathbf{rf} = \partial_j(x_i\sigma_{ij}) - \mathbf{u} - \frac{\partial}{\partial t}(\mathbf{rg}) \tag{2.74}$$

($\sigma_{ii} = -u$); this is the electromagnetic virial theorem.[6] Let us multiply equation (2.68) for the k-th component by $\varepsilon_{ijk}x_j$:

$$\varepsilon_{ijk}x_j f_k = \varepsilon_{ijk}x_j\partial_l(\sigma_{kl}) - \varepsilon_{ijk}x_j\frac{\partial}{\partial t}g_k =$$
$$= \partial_l(\varepsilon_{ijk}x_j\sigma_{kl}) - \varepsilon_{ijk}x_j\frac{\partial}{\partial t}g_k \ , \tag{2.75}$$

because σ_{kl} is a symmetric tensor. We recognize in equation (2.75) the torque of the force on the left, the rate of change in time of the angular momentum and the flux of the torque (flow of the angular momentum) on the right. This is the law of conservation of the angular momentum for the electromagnetic field.

[5] M. Abraham, *Theorie der Elektrizitaet*, Leipzig (1923) (ed. by R. Becker, Leipzig (1933)).

[6] For a mechanical particle $\mathbf{rf} = d(\mathbf{rp})/dt - \mathbf{vp} = d(\mathbf{rp})/dt - 2T$, where T is the kinetic energy; for $\mathbf{f} = const \cdot \mathbf{r}/r^3$, as for Coulomb force, $\mathbf{rf} = V$, where V is the potential energy. We get $d(\mathbf{rp})/dt = 2T + V$, whose average over a finite (bound) motion leads to $2T + V = 0$.

2.7 Longitudinal field

Let us represent the scalar potential in the volume V as the Fourier expansion

$$\Phi(t,\boldsymbol{r}) = \frac{1}{\sqrt{V}}\sum_{\boldsymbol{k}}\Phi_{\boldsymbol{k}}(t)e^{i\boldsymbol{k}\boldsymbol{r}} \ , \tag{2.76}$$

where $\Phi^*_{-\boldsymbol{k}} = \Phi_{\boldsymbol{k}}$. Similarly, the vector potential is represented as

$$\boldsymbol{A}(t,\boldsymbol{r}) = \frac{1}{\sqrt{V}}\sum_{\boldsymbol{k}}\boldsymbol{A}_{\boldsymbol{k}}(t)e^{i\boldsymbol{k}\boldsymbol{r}} + \frac{1}{\sqrt{V}}\sum_{\boldsymbol{k}}\frac{i\boldsymbol{k}}{ck^2}\dot{\Phi}_{\boldsymbol{k}}(t)e^{i\boldsymbol{k}\boldsymbol{r}} \ , \tag{2.77}$$

where $\boldsymbol{k}\boldsymbol{A}_{\boldsymbol{k}} = 0$. The first term in equation (2.77) is the transverse part of the vector potential, while the second term is its longitudinal part. The Lorenz gauge is satisfied by this representation. The electric field is given by

$$\begin{gathered}\boldsymbol{E} = \tfrac{1}{\sqrt{V}}\sum_{\boldsymbol{k}}\left(-\tfrac{1}{c}\dot{\boldsymbol{A}}_{\boldsymbol{k}}\right)e^{i\boldsymbol{k}\boldsymbol{r}}+ \\ +\tfrac{1}{\sqrt{V}}\sum_{\boldsymbol{k}}\tfrac{i\boldsymbol{k}}{k^2}\left(-\tfrac{1}{c^2}\ddot{\Phi}_{\boldsymbol{k}} - k^2\Phi_{\boldsymbol{k}}\right)e^{i\boldsymbol{k}\boldsymbol{r}} = \\ = \tfrac{1}{\sqrt{V}}\sum_{\boldsymbol{k}}\left(-\tfrac{1}{c}\dot{\boldsymbol{A}}_{\boldsymbol{k}}\right)e^{i\boldsymbol{k}\boldsymbol{r}} - \tfrac{1}{\sqrt{V}}\sum_{\boldsymbol{k}}\tfrac{4\pi i\boldsymbol{k}}{k^2}\rho_{\boldsymbol{k}}e^{i\boldsymbol{k}\boldsymbol{r}} \ ,\end{gathered} \tag{2.78}$$

where $\rho_{\boldsymbol{k}}$ is the Fourier component of the charge density; it has both a transverse part and a longitudinal part. The magnetic field is

$$\boldsymbol{H} = \frac{1}{\sqrt{V}}\sum_{\boldsymbol{k}} i\boldsymbol{k}\times\boldsymbol{A}_{\boldsymbol{k}}e^{i\boldsymbol{k}\boldsymbol{r}} \ ; \tag{2.79}$$

it is a transverse field. The energy of the electromagnetic field can now be written as

$$\begin{gathered}U = \tfrac{1}{8\pi}\int d\boldsymbol{r}\,(E^2+H^2) = \\ = \tfrac{1}{8\pi}\sum_{\boldsymbol{k}}\left(\tfrac{1}{c^2}\dot{\boldsymbol{A}}_{\boldsymbol{k}}\dot{\boldsymbol{A}}_{-\boldsymbol{k}} + k^2\boldsymbol{A}_{\boldsymbol{k}}\boldsymbol{A}_{-\boldsymbol{k}}\right) + \\ +\sum_{\boldsymbol{k}}\tfrac{2\pi}{k^2}\rho_{\boldsymbol{k}}\rho_{-\boldsymbol{k}} \ .\end{gathered} \tag{2.80}$$

We recognize in the last row of equation (2.80) the Coulomb energy (with $4\pi/\sqrt{V}k^2$ the Fourier transform of the Coulomb potential). We have replaced $\frac{1}{c^2}\ddot{\Phi}_{\boldsymbol{k}} + k^2\Phi_{\boldsymbol{k}}$ in equation (2.78) by $4\pi\rho_{\boldsymbol{k}}$, as if the potential Φ is generated by a charge distribution. Consequently, the Coulomb energy in equation (2.80) arises from a self-interaction.

Let us now compute the interaction energy $V = \int dtd\boldsymbol{r}\boldsymbol{j}\boldsymbol{E}$, where $\boldsymbol{j}$ is the current density, by using the electric field given by equation (2.78). We get

$$V = \sum_{\boldsymbol{k}} \int dt \left(-\tfrac{1}{c}\boldsymbol{j}_{\boldsymbol{k}}\dot{\boldsymbol{A}}_{-\boldsymbol{k}}\right) + \sum_{\boldsymbol{k}} \int dt \tfrac{4\pi i\boldsymbol{k}\boldsymbol{j}_{\boldsymbol{k}}}{k^2}\rho_{-\boldsymbol{k}} \; . \tag{2.81}$$

Since $\sum_{\boldsymbol{k}} \left(-\frac{1}{c}\boldsymbol{j}_{\boldsymbol{k}}\dot{\boldsymbol{A}}_{-\boldsymbol{k}}\right)\Delta t = \sum_{\boldsymbol{k}} \left(-\frac{1}{c}\boldsymbol{j}_{\boldsymbol{k}}\Delta\boldsymbol{A}_{-\boldsymbol{k}}\right)$ is a mechanical work done in time Δt, it follows that $\sum_{\boldsymbol{k}} \left(-\frac{1}{c}\boldsymbol{j}_{\boldsymbol{k}}\boldsymbol{A}_{-\boldsymbol{k}}\right)$ is an interaction energy; therefore,

$$\begin{aligned} V &= \sum_{\boldsymbol{k}} \left(-\tfrac{1}{c}\boldsymbol{j}_{\boldsymbol{k}}\boldsymbol{A}_{-\boldsymbol{k}}\right) + \sum_{\boldsymbol{k}} \int dt \tfrac{4\pi i\boldsymbol{k}\boldsymbol{j}_{\boldsymbol{k}}}{k^2}\rho_{-\boldsymbol{k}} = \\ &= \sum_{\boldsymbol{k}} \left(-\tfrac{1}{c}\boldsymbol{j}_{\boldsymbol{k}}\boldsymbol{A}_{-\boldsymbol{k}}\right) - \sum_{\boldsymbol{k}} \int dt \tfrac{4\pi}{k^2}\dot{\rho}_{\boldsymbol{k}}\rho_{-\boldsymbol{k}} \; . \end{aligned} \tag{2.82}$$

In the second summation in equation (2.82) we change $\boldsymbol{k}$ into $-\boldsymbol{k}$ and write $\frac{1}{2}\left(\dot{\rho}_{\boldsymbol{k}}\rho_{-\boldsymbol{k}} + \dot{\rho}_{-\boldsymbol{k}}\rho_{\boldsymbol{k}}\right) = \frac{1}{2}\frac{d}{dt}\left(\dot{\rho}_{\boldsymbol{k}}\rho_{-\boldsymbol{k}}\right)$. Finally, we get the interaction

$$V = \sum_{\boldsymbol{k}} \left(-\frac{1}{c}\boldsymbol{j}_{\boldsymbol{k}}\boldsymbol{A}_{-\boldsymbol{k}}\right) - \sum_{\boldsymbol{k}} \frac{2\pi}{k^2}\rho_{\boldsymbol{k}}\rho_{-\boldsymbol{k}} \; . \tag{2.83}$$

The Coulomb interaction in equation (2.83) is a self-interaction. We can see that this interaction cancels out the Coulomb energy in equation (2.80). The longitudinal degrees of freedom do not participate in the dynamics of the charges, the currents and the electromagnetic field. This amounts to view the field as a free radiation field, with a vanishing scalar potential, interacting with currents; although the longitudinal part of the field remains in the electric field (equation (2.78)). We are left with the transverse field energy

$$U = \frac{1}{8\pi}\sum_{\boldsymbol{k}} \left(\frac{1}{c^2}\dot{\boldsymbol{A}}_{\boldsymbol{k}}\dot{\boldsymbol{A}}_{-\boldsymbol{k}} + k^2\boldsymbol{A}_{\boldsymbol{k}}\boldsymbol{A}_{-\boldsymbol{k}}\right) \tag{2.84}$$

and the transverse part of the interaction

$$V = \sum_{\boldsymbol{k}} \left(-\frac{1}{c} \boldsymbol{j}_{\boldsymbol{k}} \boldsymbol{A}_{-\boldsymbol{k}} \right) . \tag{2.85}$$

It is worth noting that this interaction is not a self-interaction. We can see that the field is a set of harmonic oscillators with eigenfrequencies $\pm\omega_k$, $\omega_k = ck$, with momenta $\boldsymbol{P}_{\boldsymbol{k}} = \frac{1}{4\pi c^2}\dot{\boldsymbol{A}}_{\boldsymbol{k}}$ and coordinates $\boldsymbol{A}_{\boldsymbol{k}}$, whose equation of motion is the Maxwell-Ampere equation

$$\frac{1}{c^2}\ddot{\boldsymbol{A}}_{\boldsymbol{k}} + k^2 \boldsymbol{A}_{\boldsymbol{k}} = \frac{4\pi}{c}\boldsymbol{j}_{\boldsymbol{k}} . \tag{2.86}$$

Noteworthy, $\boldsymbol{j}_{\boldsymbol{k}}$ in this equation is the transverse part of the current. Since $\boldsymbol{A}^*_{-\boldsymbol{k}} = \boldsymbol{A}_{\boldsymbol{k}}$, we may write $\boldsymbol{A}_{\boldsymbol{k}} = \alpha_k \boldsymbol{e}_{\boldsymbol{k}} a_{\boldsymbol{k}}(t) + \alpha_k \boldsymbol{e}_{-\boldsymbol{k}} a^*_{-\boldsymbol{k}}(t)$, where α_k and $a_{\boldsymbol{k}}$ are undetermined coefficients and $\boldsymbol{e}_{\boldsymbol{k}}$ is the polarization vector ($\boldsymbol{e}_{\boldsymbol{k}}\boldsymbol{e}_{\boldsymbol{k}} = 1$, $\boldsymbol{e}_{\boldsymbol{k}}\boldsymbol{k} = 0$). For free fields $a_{\boldsymbol{k}}(t) = a_{\boldsymbol{k}} e^{-i\omega_k t}$, $\omega_k = ck$, because $\boldsymbol{A}_{\boldsymbol{k}}$ should satisfy the free wave equation $\frac{1}{c^2}\ddot{\boldsymbol{A}}_{\boldsymbol{k}} + k^2 \boldsymbol{A}_{\boldsymbol{k}} = 0$. This is equation (2.60) (with the phases $\gamma_{\boldsymbol{k}}$ included in the coefficients). The free field energy becomes

$$U = \sum_{\boldsymbol{k}} \frac{\omega_k^2 \alpha_k^2}{4\pi c^2} \left(a^*_{\boldsymbol{k}} a_{\boldsymbol{k}} + a_{\boldsymbol{k}} a^*_{\boldsymbol{k}} \right) . \tag{2.87}$$

Similarly, the total momentum of the free field (equation (2.71)) is

$$\begin{aligned} \boldsymbol{G} &= \tfrac{1}{4\pi c} \int d\boldsymbol{r} \boldsymbol{E} \times \boldsymbol{H} = \tfrac{1}{c^2} \int d\boldsymbol{r} \boldsymbol{S} = \\ &= \textstyle\sum_{\boldsymbol{k}} \frac{\omega_k \alpha_k^2}{4\pi c^2} \boldsymbol{k} \left(a^*_{\boldsymbol{k}} a_{\boldsymbol{k}} + a_{\boldsymbol{k}} a^*_{\boldsymbol{k}} \right) \end{aligned} \tag{2.88}$$

and the total angular momentum is vanishing. Also, we note that the volume integral of the *div* of the Fourier transform of any vector is vanishing, due to the cyclic boundary conditions (*e.g.*, $div\boldsymbol{S}$).

2.8 Static field. Self-interaction

In the immediate vicinity of a pointlike charge q the field is given by the Coulomb potential $\Phi = q/r$. There exists only the electric field

$\boldsymbol{E} = -grad\Phi$. The field energy

$$U = \frac{1}{8\pi}\int d\boldsymbol{r}E^2 = -\frac{1}{8\pi}\int d\boldsymbol{r}\boldsymbol{E}grad\Phi =$$
$$= \frac{1}{8\pi}\int d\boldsymbol{r}div\boldsymbol{E}\cdot\Phi = \frac{1}{2}\int d\boldsymbol{r}\rho\Phi \tag{2.89}$$

arises from a self-interaction $-U$. Both are infinite, and can be written with Fourier components as

$$U = \sum_{\boldsymbol{k}} \frac{2\pi}{k^2}\rho_{\boldsymbol{k}}\rho_{-\boldsymbol{k}} \ , \tag{2.90}$$

where $\rho_{\boldsymbol{k}} = q/\sqrt{V}$ is the Fourier transform of the charge density $\rho = q\delta(\boldsymbol{r})$ (equation (2.80)), *i.e.*

$$U = \frac{2\pi q^2}{(2\pi)^3}\int d\boldsymbol{k}\frac{1}{k^2} \ ; \tag{2.91}$$

A cutoff k_0 makes this energy finite, $U = q^2k_0/\pi$. If the charge is uniformly distributed in a sphere of radius a, the potential is $\Phi = \frac{q}{2a^3}(3a^2 - r^2)$ and equation (2.89) gives $U = 3q^2/5a$. The cutoff is $k_0 = 3\pi/5a$. If we limit ourselves to the external region $r > a$, the energy is $U = q^2/2a$ and $k_0 = \pi/2$. The total energy (self-energy+self-interaction) is zero.

The stress tensor (equation (2.70)) is

$$\sigma_{ij} = \frac{q^2}{4\pi}\left(\frac{x_ix_j}{r^6} - \frac{1}{2r^4}\delta_{ij}\right) \tag{2.92}$$

(where x_i are the coordinates of the position $\boldsymbol{r}$). The density of the stress forces $t_i = \partial_j\sigma_{ij}$ is vanishing, in contrast to the density of the Lorentz force $\boldsymbol{f} = \rho\boldsymbol{E} = q\delta(\boldsymbol{r})\boldsymbol{r}/r^3$, which is not definite. This shows again that we need to view the pointlike charges as infinitesimal spheres.

In its immediate vicinity a pointlike charge carries with it the Coulomb (longitudinal) field, with a vanishing total energy (self-energy+self-interaction), which acts with a vanishing force upon the charge.

2.9 Waves. Self-interaction

A charge in motion generates an electromagnetic field propagating with velocity c. The charge moves in this field, with a velocity smaller than c, so it interacts with its own field, as a consequence of the retardation. This is a self-interaction. Since we are interested in the field near the charge, we may leave aside the scalar potential, whose Coulomb field gives no interaction. The transverse vector potential is generated according to the Maxwell-Ampere equation

$$\frac{1}{c^2}\frac{\partial^2 \boldsymbol{A}}{\partial t^2} - \Delta \boldsymbol{A} = \frac{4\pi}{c}\boldsymbol{j} \ , \tag{2.93}$$

where $\boldsymbol{j}$ is the transverse part of the current density. Let us assume a pointlike charge q with trajectory $\boldsymbol{r}_0(t)$, governed by an external force; its charge density is $\rho = q\delta(\boldsymbol{r} - \boldsymbol{r}_0)$ and its current density is $\boldsymbol{j} = q\boldsymbol{v}\delta(\boldsymbol{r} - \boldsymbol{r}_0)$, where $\boldsymbol{v} = \dot{\boldsymbol{r}}_0$. Equation (2.93) becomes

$$\frac{1}{c^2}\frac{\partial^2 \boldsymbol{A}}{\partial t^2} - \Delta \boldsymbol{A} = \frac{4\pi q}{c}\boldsymbol{v}\delta(\boldsymbol{r} - \boldsymbol{r}_0) \ . \tag{2.94}$$

The field $\boldsymbol{A}$ acts upon the charge. Consequently, the charge with mass m acquires an additional velocity $\boldsymbol{v}'$, whose equation of motion is

$$m\frac{d\boldsymbol{v}'}{dt} = q\boldsymbol{E}_0 + \frac{1}{c}(\boldsymbol{v} + \boldsymbol{v}') \times \boldsymbol{H}_0 \ , \tag{2.95}$$

where $\boldsymbol{E} = -\frac{1}{c}\frac{\partial \boldsymbol{A}}{\partial t}$ is the electric field and $\boldsymbol{H}_0 = curl\boldsymbol{A}$ is the magnetic field. The suffix 0 indicates that the fields are taken at position $\boldsymbol{r}_0$. This is the Lorentz force. Since we are interested in the region close to the charge, we may neglect the second term on the right in equation (2.95), and get

$$\frac{d\boldsymbol{v}'}{dt} = -\frac{q}{mc}\frac{d\boldsymbol{A}_0}{dt} \tag{2.96}$$

and $\boldsymbol{v}' = -\frac{q}{mc}\boldsymbol{A}_0$. It follows that the self-interaction produces a reaction current density

$$\boldsymbol{j}' = q\boldsymbol{v}'\delta(\boldsymbol{r} - \boldsymbol{r}_0) = -\frac{q^2}{mc}\boldsymbol{A}\delta(\boldsymbol{r} - \boldsymbol{r}_0) \ , \tag{2.97}$$

which determines a field given by

$$\frac{1}{c^2}\frac{\partial^2 \boldsymbol{A}'}{\partial t^2} - \Delta \boldsymbol{A}' = -\frac{4\pi q^2}{mc^2}\boldsymbol{A}\delta(\boldsymbol{r}-\boldsymbol{r}_0) \ . \tag{2.98}$$

For $\boldsymbol{r}$ close to $\boldsymbol{r}_0$ we may limit ourselves to the static part of this field, which is $\boldsymbol{A}' = -\frac{q^2}{mc^2}\frac{\boldsymbol{A}_0}{|\boldsymbol{r}-\boldsymbol{r}_0|}$. The total field at the position of the charge is $\boldsymbol{A}_0 - \frac{q^2}{mc^2}\lim_{\boldsymbol{r}\to\boldsymbol{r}_0}\frac{\boldsymbol{A}_0}{|\boldsymbol{r}-\boldsymbol{r}_0|}$, which shows that the charge must be viewed as having an undetermined position inside a region with dimension of the order

$$a = \frac{q^2}{mc^2} \ ; \tag{2.99}$$

the field at the position of the charge is practically zero. The assumption of the self-interaction is meaningless for pointlike charges. If we admit that the charge motion has a finite dimension of the order a, the retarded self-interaction can be included, providing we limit ourselves to distances and wavelengths larger than $\simeq a$, wavevectors smaller than $\simeq 1/a$ and frequencies smaller than $\simeq c/a$. In addition, the self-energy of the charge is of the order $q^2/a = mc^2$. It is worth noting that the interaction density $-\frac{1}{c}\boldsymbol{j}\boldsymbol{A}$ is preserved in the case the self-interaction is included, because the field is generated at a retarded moment of time in comparison with the time of the current, so the current and the field act like independent entities (and the factor 1/2 expected for instantaneous self-interaction does not appear). The self-interaction gives rise to a current density $\boldsymbol{j} = -\frac{q^2}{mc}\boldsymbol{A}\delta(\boldsymbol{r}-\boldsymbol{r}_0)$ and the field moves according to equation

$$\frac{1}{c^2}\frac{\partial^2 \boldsymbol{A}}{\partial t^2} - \Delta \boldsymbol{A} = -\frac{4\pi q^2}{mc^2}\boldsymbol{A}\delta(\boldsymbol{r}-\boldsymbol{r}_0) \ . \tag{2.100}$$

Inside the region with radius a we may replace $\delta(\boldsymbol{r}-\boldsymbol{r}_0)$ by $1/4\pi a^3$, so, inside this region, the radiation frequency ω is given by $\omega^2 = c^2k^2 + c^2/a^2$, while outside this region the frequency is given by $\omega^2 = c^2k^2$. Thus indeed, we are limited by a frequency cutoff of the order c/a.

The fields are limited by a cutoff $E_s = H_s = q/a^2 = m^2c^4/q^3$ (or $qE_sa = mc^2$, $m\omega^2 = q\omega H_s/c$, with $\omega = c/a$; for a quantum-mechanical motion such a field is called Schwinger limit for fields). These are the limits of the (classical) Electromagnetism. a given by equation (2.99) is called the classical electromagnetic radius of the charge. A similar conclusion is reached by a quantum-mechanical motion of the charge, with a replaced by the Compton wavelength $\lambda_c = \hbar/mc$, where $\hbar \simeq 10^{-27} erg \cdot s$ is Planck's constant. The ratio a/λ_c is $a/\lambda_c = q^2/\hbar c$. For electrons $q = -e = -4.8 \times 10^{-10} esu$, $a \simeq 2.8 \times 10^{-13} cm^{-1}$ ($c/a \simeq 10^{23} s^{-1}$, $m \simeq 10^{-27} g$, $mc^2 \simeq 0.5 MeV$, $1eV = 1.6 \times 10^{-12} erg$), $E_s \simeq 6 \times 10^{15} esu$ ($1esu = 3 \times 10^4 V/m$), $\lambda_c \simeq 3.9 \times 10^{-11} cm^{-1}$ and $\alpha = e^2/\hbar c \simeq \frac{1}{137}$ is the fine-structure constant. Being associated with an indefinite motion region the cutoff length is not subject to the relativistic invariance requirement.

2.10 Radiation field

Let us assume that the charge and current distribution is confined to a limited region with dimension l. At large distances r from the distribution we may omit $\boldsymbol{r}'$ in the Kirchhoff formulae in comparison with $\boldsymbol{r}$, such that we get

$$\Phi = \frac{q}{r} + \ldots \; , \; \boldsymbol{A} = \frac{\dot{\boldsymbol{d}}}{cr} \mid_{t-r/c} + \ldots \; , \tag{2.101}$$

where $\boldsymbol{d} = q\boldsymbol{r}_0$ is the dipole moment of the charge placed at $\boldsymbol{r}_0$. Since $div\boldsymbol{A} = -\ddot{\boldsymbol{d}}\boldsymbol{r}/c^2r^2 + \ldots = -\frac{1}{c}\frac{\partial}{\partial t}\frac{\dot{\boldsymbol{d}}\boldsymbol{r}}{cr^2} + \ldots$, the scalar potential has an additional contribution:

$$\Phi = \frac{q}{r} + \frac{\dot{\boldsymbol{d}}\boldsymbol{r}}{cr^2} \mid_{t-r/c} + \ldots = \frac{q}{r} - div\frac{\boldsymbol{d}}{r} \mid_{t-r/c} + \ldots \; . \tag{2.102}$$

We note that in the limit of large distances r the derivatives arise only from the retarded time $t - r/c$. Also, we note that the static Coulomb potential is present in this limit. This is the dipolar approximation. Higher-order terms in powers of $\boldsymbol{r}'$ in the Kirchhoff formulae give quadrupole terms and magnetic dipole terms.

Within the dipolar approximation the fields are given by

$$\begin{aligned} \boldsymbol{E} &= -\frac{\ddot{\boldsymbol{d}}}{c^2r} + grad\, div\frac{\boldsymbol{d}}{r} - grad\frac{q}{r} = \\ &= -\Delta\frac{\boldsymbol{d}}{r} + grad\, div\frac{\boldsymbol{d}}{r} - grad\frac{q}{r} = curl\, curl\frac{\boldsymbol{d}}{r} - grad\frac{q}{r} \end{aligned} \tag{2.103}$$

and

$$\boldsymbol{H} = curl\frac{\dot{\boldsymbol{d}}}{cr} \ , \tag{2.104}$$

at the retarded time $t - r/c$. The last term in equation (2.103) may be neglected.

A finite motion confined in a region of dimension l changes over a characteristic time T, such that this motion generates waves with period of the order of T and wavelengths of the order $\lambda = cT = \frac{c}{v}l$, where v is the charge velocity. Since $v < c$, it follows $\lambda > l$. Therefore, wavelengths shorter than the dimension of the charge and current distribution are meaningless. The fields vary slowly over the region of the charge and current distribution. Indeed, the main contribution of the Fourier components $e^{i\boldsymbol{kr}}$ comes from $kr < 1$, *i.e.* $l < \lambda$. If we preserve in the above formulae the leading contributions for $l \ll \lambda \simeq r$, we are in the near-field region. It is worth noting that the condition $l \ll \lambda$ amounts to $v \ll c$. In fact, for a finite motion we have always $v \ll c$. If we limit ourselves to the leading contributions for $l, \lambda \ll r$, we are in the far-field region (wave zone). This is the radiation field. It is given by

$$\begin{gathered} \boldsymbol{E} = \frac{\boldsymbol{r}\times(\boldsymbol{r}\times\ddot{\boldsymbol{d}})}{c^2r^3} = -\frac{\boldsymbol{r}}{r}\times\boldsymbol{H} \ , \\ \boldsymbol{H} = -\frac{\boldsymbol{r}\times\ddot{\boldsymbol{d}}}{c^2r^2} \ . \end{gathered} \tag{2.105}$$

We can see that the radiation field is transverse. The Poynting vector is

$$\boldsymbol{S} = \frac{c}{4\pi}\boldsymbol{E}\times\boldsymbol{H} = \frac{c\boldsymbol{r}}{4\pi r}H^2 = \frac{\ddot{d}^2}{4\pi c^3r^2}\sin^2\theta\cdot\frac{\boldsymbol{r}}{r} \ , \tag{2.106}$$

where θ is the angle between $\ddot{\boldsymbol{d}}$ and $\boldsymbol{r}$; $\boldsymbol{S}ds = cr^2H^2do/4\pi$ is the radiation intensity through the area $ds = r^2do$, corresponding to the solid angle do. The total intensity is

$$I = \frac{2\ddot{d}^2}{3c^3} \ ; \tag{2.107}$$

this is the total energy radiated by the dipole per unit time.

2.11 Damping force

The momentum carried per unit time and per unit area is $c\boldsymbol{g}$, where $\boldsymbol{g} = \boldsymbol{S}/c^2$ is the momentum density of the field; therefore, the total radiated momentum per unit time is $\oint d\boldsymbol{s}c\boldsymbol{g}$, where the integration is performed over a surface placed at large distances. In this case equation (2.106) is valid, and we can see that the total radiated momentum is zero, because $\boldsymbol{S} \sim \boldsymbol{r}$. It follows that the total momentum of the field does not change ($\Delta \boldsymbol{G} = 0$) and equation (2.68) reduces to $\int d\boldsymbol{r}\boldsymbol{f} = \int d\boldsymbol{r}\boldsymbol{t}$, where $t_i = \partial_j \sigma_{ij}$ are the components of the stress forces. It is easy to see that the i-th component of the total stress force

$$\int d\boldsymbol{r}t_i = \frac{1}{4\pi}\oint \left[E_i \boldsymbol{E} d\boldsymbol{s} + H_i \boldsymbol{H} d\boldsymbol{s} - \frac{1}{2}(E^2 + H^2)ds_i\right] \qquad (2.108)$$

is zero for the radiation field given by equations (2.105). Therefore, the Lorentz force acting between the dipole and its own field, *i.e.* the self-force, is zero (as shown above).

However, the energy radiated per unit time by a charge q can be written as

$$\dot{E} = I = \frac{2q^2\dot{v}^2}{3c^3} = \frac{2q^2}{3c^3}\frac{d}{dt}(\boldsymbol{v}\dot{\boldsymbol{v}}) - \frac{2q^2}{3c^3}\boldsymbol{v}\ddot{\boldsymbol{v}} \qquad (2.109)$$

and the first term on the right in this equation, averaged over a finite motion, is zero. It follows that a force

$$\boldsymbol{f} = \frac{2q^2\ddot{\boldsymbol{v}}}{3c^3} \qquad (2.110)$$

is exerted by the field upon the charge. This is the Lorentz damping force, or the Lorentz frictional force. It is a reaction self-force. Besides the energy loss given by equation (2.109) the dipole suffers also a loss of angular momentum, produced by the damping force (and no loss of momentum).

For a finite motion with a characteristic frequency ω_0 (*e.g.*, a harmonic oscillator, or a rotation) $\ddot{\boldsymbol{v}} = -\omega_0^2\boldsymbol{v}$ and the damping force is $\boldsymbol{f} = -\frac{2q^2}{3c^3}\omega_0^2\boldsymbol{v}$. The equation of motion $m\dot{\boldsymbol{v}} = \boldsymbol{f}$ for a charge with mass m leads to a very slow damping $v \sim e^{-\frac{2a\omega_0^2}{3c}t}$, where $a = q^2/mc^2$ is the classical radius of the charge. This damping decreases the frequency of the emitted radiation, both for positive and negative frequencies, which amounts to both larger and lower frequencies for a positive central frequency; this is the natural line breadth of the emitted radiation.

For higher velocities (in an infinite motion) the damping force (and the equation of motion) should acquire a relativistic form. The relativistic motion of the charge is damped.

2.12 Line breadth

A pointlike charge q with mass m, oscillating with frequency ω_0 along the z-axis, obeys the equation of motion

$$\ddot{z} = -\omega_0^2 z - \gamma \dot{z} \ , \tag{2.111}$$

where

$$\gamma = \frac{2q^2\omega_0^2}{3c^3m} = \frac{2\omega_0^2}{3c}a \ll \omega_0 \tag{2.112}$$

is the damping coefficient ($a = q^2/mc^2$ is the classical radius). The solution of this equation is $z \sim e^{-\gamma t/2}e^{-i\omega_0 t}$; γ^{-1} is the (very long) lifetime of the damped oscillator. The field radiated by the dipole is also damped, E, $H \sim \ddot{z} \sim e^{-\gamma t/2}e^{-i\omega_0 t}$, and has a (long) lifetime. Its Fourier transform goes like

$$E(\omega) \sim \int_0 dt e^{i\omega t} e^{-i\omega_0 t} e^{-\gamma t/2} = \frac{1}{i(\omega_0 - \omega) + \gamma/2} \ ; \tag{2.113}$$

the total energy of the emitted radiation

$$U \sim \int E^2(t)dt = \frac{1}{2\pi}\int d\omega \left|E(\omega)\right|^2 \tag{2.114}$$

gives a spectral intensity (energy)

$$I(\omega) \sim \left|E(\omega)\right|^2 = I_0 \frac{1}{2\pi}\frac{\gamma}{(\omega_0 - \omega)^2 + \gamma^2/4} \ , \tag{2.115}$$

where I_0 is the total intensity ($\omega > 0$). The characteristic function in equation (2.115) is the Lorentz function, or dispersion function. At its half height $\omega = \omega_0 \pm \gamma/2$, so γ is also the breadth $\Delta\omega$ of the dispersion function (the reciprocal of the lifetime). It corresponds to a change

$$\Delta\lambda = 2\pi\frac{c\gamma}{\omega_0^2} = \frac{4\pi}{3}a \simeq 10^{-12}cm = 10^{-4}\text{Å} \tag{2.116}$$

in the wavelength (for electrons). This is the natural breadth of the spectroscopic line.[7]

2.13 Scattering of radiation

Let a monochromatic wave with the electric field $E_0 \cos\omega t$ fall upon a point charge q with mass m at rest. The equation of motion is

$$m\ddot{z} = qE_0 \cos\omega t \ , \tag{2.117}$$

whose solution is $z = -(qE_0/m\omega^2)\cos\omega t$. This dipole $d = qz$ generates a secondary radiation with the same frequency: the radiation is scattered. The energy per unit area scattered in the solid angle do per unit time is $d\dot{E} = Ido$, where I is the intensity per unit area; it is given by

$$I = |\mathbf{S}| = \frac{\ddot{d}^2}{4\pi c^3 r^2}\sin^2\theta = \frac{q^4}{4\pi m^2 c^3 r^2}E_0^2\cos^2\omega t \cdot \sin^2\theta \ , \tag{2.118}$$

where $\mathbf{S}$ is the Poynting vector and θ is the angle between $\mathbf{E}_0$ and $\boldsymbol{r}$ (equation (2.106)). The intensity per unit area of the incident radiation is $I_0 = (cE_0^2/4\pi)\cos^2\omega t$, so we can write

$$I = \frac{q^4}{m^2c^4}\sin^2\theta\left(\frac{I_0}{r^2}\right) \ . \tag{2.119}$$

The quantity

$$\frac{q^4}{m^2c^4}\sin^2\theta = a^2\sin^2\theta \ , \tag{2.120}$$

[7] W. Heitler, *The Quantum Theory of Radiation*, Oxford University Press, London (1954).

where $a = q^2/mc^2$ is the classical radius of the charge, is the scattering cross-section (actually $d\sigma/do = a^2 \sin^2\theta$ is the differential cross-section). We can see that it is of the order of the square of the classical electromagnetic radius of the charge. If the original wave is unpolarized, *e.g.* $\mathbf{E}_0 \sim (\cos\varphi', \sin\varphi', 0)$ (and propagates along the z-direction), then we average over the polarizations and get

$$\overline{\sin^2\theta} = \frac{1}{2}(1+\cos^2\Theta) \ , \tag{2.121}$$

where Θ is the angle of scattering (the angle between $\mathbf{r}$ - the propagation direction of the scattered radiation - and the propagation direction z of the incident radiation). The total scattering cross-section is

$$\sigma = \frac{8\pi}{3}a^2 \ ; \tag{2.122}$$

for electrons $\sigma = 6.65 \times 10^{-25} cm^2$ is a universal constant; it is called the Thomson cross-section.

Let us consider the scattering of radiation by a linear harmonic oscillator. With usual notations, the equation of motion is

$$\ddot{\mathbf{r}} + \omega_0^2\mathbf{r} + \gamma\dot{\mathbf{r}} = \frac{q}{m}\mathbf{E}_0 e^{-i\omega t} \ , \tag{2.123}$$

where the damping is included (equation (2.112)). The solution of this equation is

$$\mathbf{r} = \frac{q}{m}\mathbf{E}_0 \frac{1}{\omega_0^2 - \omega^2 - i\omega\gamma} e^{-i\omega t} \tag{2.124}$$

and the intensity (per unit area) is given by the real part of the solution,

$$I = \frac{q^4 E_0^2}{4\pi m^2 c^3 r^2} \sin^2\theta \frac{\omega^4}{(\omega_0^2 - \omega^2)^2 + \omega^2\gamma^2} \cos^2(\omega t - \delta) \ , \tag{2.125}$$

where

$$\tan\delta = \frac{\gamma\omega}{\omega_0^2 - \omega^2} \ . \tag{2.126}$$

This is the Rayleigh scattering. We note that the intensity of the scattered radiation depends on the incident frequency, the maximum scattering occurring for the resonance $\omega = \omega_0$. The air molecules have a resonance frequency in the blue region, hence the blue colour of the sky. Also, we note that there appears a dephasing δ in the scattered radiation (which is maximal at resonance). Introducing the average intensity of the primary radiation $cE_0^2/8\pi$ and the classical radius a we get the differential cross-section

$$\frac{d\sigma}{do} = a^2 \sin^2\theta \frac{\omega^4}{(\omega_0^2 - \omega^2)^2 + \omega^2\gamma^2} \tag{2.127}$$

and the total cross-section

$$\sigma = \frac{8\pi}{3} a^2 \frac{\omega^4}{(\omega_0^2 - \omega^2)^2 + \omega^2\gamma^2} \ . \tag{2.128}$$

For ω_0, $\gamma = 0$ equation (2.120) for free charges is recovered. Close to the resonance, this formula becomes

$$\sigma = \frac{2\pi}{3} a^2 \frac{\omega_0^2}{(\omega_0 - \omega)^2 + \gamma^2/4} \ . \tag{2.129}$$

This is the fluorescence resonance.

2.14 Absorption of radiation

With the notations used above, a harmonic oscillator subjected to a monochromatic wave,

$$\ddot{\mathbf{r}} + \omega_0^2 \mathbf{r} = \frac{q}{m} \mathbf{E}_0 \cos(\omega t + \delta_\omega) \ , \tag{2.130}$$

exchanges energy with the wave. The general solution of equation (2.130) is a particular solution (frequency ω) plus the solution of the homogeneous equation (frequency ω_0). Let us assume that the interaction begins at $t = 0$. Then, the solution is

$$\begin{aligned} \mathbf{r} = \tfrac{q}{m} \mathbf{E}_0 \tfrac{1}{\omega_0^2 - \omega^2} \left[\cos(\omega t + \delta_\omega) - \cos(\omega_0 t + \delta_\omega)\right] + \\ + \mathbf{A} \cos(\omega_0 t + \theta) \ , \end{aligned} \tag{2.131}$$

where $\mathbf{A}$ and θ are the amplitude and the phase of the free oscillator. The interaction energy per unit time (mechanical work) is given by

$$e_m = q\dot{\mathbf{r}}\mathbf{E}_0 \cos(\omega t + \delta_\omega) \ , \qquad (2.132)$$

the total interaction energy in the time τ is

$$\begin{aligned} \int_0^\tau dt e_m = \tfrac{q^2}{m} E_0^2 \tfrac{\omega_0}{\omega_0^2 - \omega^2} \int_0^\tau dt \sin(\omega_0 t + \delta_\omega) \cos(\omega t + \delta_\omega) - \\ - q\mathbf{A}\mathbf{E}_0 \omega_0 \int_0^\tau dt \sin(\omega_0 t + \theta) \cos(\omega t + \delta_\omega) \end{aligned} \qquad (2.133)$$

(the term with $\cos(\omega t + \delta_\omega)$ in equation (2.131) disappears by integration). This integral may have either sign, depending on the phases δ_ω and θ (the oscillator may give energy to the wave, which is an induced emission of radiation). We assume that the phases δ_ω are randomly distributed (as for incoherent radiaton), and take the average: the second term on the right of equation (2.133) disappears. We get

$$\int_0^\tau dt \overline{e}_m = \frac{q^2}{2m} E_0^2 \frac{\omega_0}{\omega_0^2 - \omega^2} \frac{1 - \cos(\omega_0 - \omega)\tau}{\omega_0 - \omega} \ , \qquad (2.134)$$

which is positive: this is the absorbed energy in the time τ. The intensity (per unit area) of the incident wave is

$$I_0(\omega) = \frac{c}{4\pi} E_0^2 \overline{\cos^2(\omega t + \delta_\omega)} = \frac{c}{8\pi} E_0^2 \ , \qquad (2.135)$$

so we may write for the absorbed energy

$$\Delta E(\omega) = I_0(\omega) \frac{4\pi q^2}{mc} \frac{\omega_0}{\omega_0^2 - \omega^2} \frac{1 - \cos(\omega_0 - \omega)\tau}{\omega_0 - \omega} \ . \qquad (2.136)$$

For large τ the function in equation (2.136) approaches $\tau\delta(\omega_0 - \omega)$. The total energy absorbed per unit time from all the frequencies around ω_0 is

$$\Delta E = I_0(\omega_0) \frac{2\pi^2 q^2}{mc} \tau \ , \qquad (2.137)$$

since

$$\int_{-\infty}^{+\infty} dx \frac{1 - \cos x}{x^2} = \pi \ , \qquad (2.138)$$

with $x = (\omega_0 - \omega)\tau$. The total energy absorbed per unit time is $\Delta E_{ab} \simeq I_0 a c\tau$, while the total energy emitted per unit time is $\Delta \dot{E}_{em} \sim I_0 a^2$ (equation (2.129)). We can see that $\Delta \dot{E}_{em} \ll \Delta E_{ab}$, since $a \ll c\tau$; the difference is spent for the kinetic energy of the charge.

2.15 Sub-wavelength field

We assume an oscillating dipole with the charge and current densities

$$\rho = q\delta(\mathbf{r}-\mathbf{r}_0) \ , \ \mathbf{j} = q\mathbf{v}\delta(\mathbf{r}-\mathbf{r}_0) \ , \ \mathbf{r}_0 = \mathbf{a}\cos\omega t \ , \tag{2.139}$$

where $\mathbf{v} = d\mathbf{r}_0/dt$ is the velocity and ω denotes the frequency; the oscillation amplitude a is much smaller than the distances of interest ($a \ll r$, dipole approximation). In addition, $c/\omega \gg a$, *i.e.* the wavelength is much longer than the amplitude of oscillation. From the Kirchhoff formula the vector potential is

$$\mathbf{A} \simeq \frac{q\mathbf{v}(t-r/c)}{cr} \ . \tag{2.140}$$

The scalar potential

$$\Phi \simeq \frac{q\mathbf{v}\mathbf{r}}{cr^2} + \frac{q\mathbf{r}_0\mathbf{r}}{r^3} \tag{2.141}$$

is obtained, most conveniently, from the gauge condition, where we neglect the Coulomb static contribution q/r (the velocity $\mathbf{v}$ and the position $\mathbf{r}_0$ are taken at the retarded time $t - r/c$). The fields can be decomposed in sub-wavelength fields (s) and far fields (f):

$$\begin{gathered}
\mathbf{E}^s = -\frac{q\mathbf{v}}{cr^2} + \frac{2q(\mathbf{v}\mathbf{r})\mathbf{r}}{cr^4} - \frac{q\mathbf{r}_0}{r^3} + \frac{3q(\mathbf{r}_0\mathbf{r})\mathbf{r}}{r^5} \ , \ \mathbf{H}^s = \frac{q\mathbf{v}\times\mathbf{r}}{cr^3} \ , \\
\mathbf{E}^f = -\frac{q\dot{\mathbf{v}}}{c^2r} + \frac{q(\dot{\mathbf{v}}\mathbf{r})\mathbf{r}}{c^2r^3} = \frac{q[(\dot{\mathbf{v}}\times\mathbf{r})\times\mathbf{r}]}{c^2r^3} = \mathbf{H}^f \times \frac{\mathbf{r}}{r} \ , \\
\mathbf{H}^f = \frac{q\dot{\mathbf{v}}\times\mathbf{r}}{c^2r^2} = \frac{\mathbf{r}}{r} \times \mathbf{E}^f \ .
\end{gathered} \tag{2.142}$$

the far fields go at infinity like $1/r$, while the sub-wavelength fields go faster than $1/r$. In the near-field region, where $\omega r/c \simeq 1$, all the fields in the above equations contribute. For $\omega r/c \ll 1$, *i.e.* in the sub-wavelength region (distances much smaller than the wavelength), the main contribution is brought by the sub-wavelength fields.[8]

Let us consider a capacitor, consisting of two infinite plane-parallel plates placed at $z = \pm d/2$, each with the surface charge desnity q_s,

[8] M. Apostol, "Of longitudinal fields in Electromagnetism", J. Theor. Phys. **265** (2017).

which oscillates in time; the charge and current densities are

$$\begin{aligned} \rho &= q_s[\delta(z+d/2)-\delta(z-d/2)]\cos\omega t \ , \\ j_z &= \omega q_s[\theta(z+d/2)-\theta(z-d/2)]\sin\omega t \ , \end{aligned} \tag{2.143}$$

From the Kirchhoff formulae the scalar potential can be written as

$$\Phi = F(z+d/2) - F(z-d/2) \ , \tag{2.144}$$

where

$$F(z) = q_s Re\left\{ e^{i\omega t}\int d\mathbf{r}^{'} \frac{e^{-i\frac{\omega}{c}\sqrt{(\mathbf{r}-\mathbf{r}^{'})^2+z^2}}}{\sqrt{(\mathbf{r}-\mathbf{r}^{'})^2+z^2}} \right\} \ . \tag{2.145}$$

The integration over $\mathbf{r}^{'}$ is extended from $r^{'}=0$ to $r^{'}=\infty$; the integral in equation (2.145) can then be effected immediately, with the result

$$F(z) = \frac{2\pi c}{\omega} q_s \sin\omega(t- \mid z \mid /c) \ ; \tag{2.146}$$

we get

$$\Phi \simeq -\frac{4\pi c}{\omega} q_s \cos\omega(t-d/2c)\sin\omega z/c \ , \ \mid z \mid < d/2 \tag{2.147}$$

and

$$\begin{aligned} \Phi &\simeq -\tfrac{4\pi c}{\omega} q_s sgn(z)\cos\omega(t- \mid z \mid /c)\cdot \\ &\cdot\sin\omega d/2c \ , \ \mid z \mid > d/2 \ . \end{aligned} \tag{2.148}$$

In the sub-wavelength region for $\omega d/c \ll 1$ the scalar potential reduces to

$$\Phi \simeq -4\pi q_s z\cos\omega t \ , \ \mid z \mid < d/2 \tag{2.149}$$

and zero outside the capacitor; from the gauge condition we can see that the vector potential may be neglected in this case (and the magnetic field). Therefore, we get inside the capacitor a uniform (longitudinal) oscillating electric field; $E_z = 4\pi q_s \cos\omega t$, as expected. It is worth noting that the fields are zero for $\omega d/c > 1$, (for $F(z)$ the vector potential is $A_z = 2\pi(cq_s/\omega)sgn(z)\sin\omega(t- \mid z \mid /c) = sgn(z)F(z)$, $z \neq 0$, and $E_z = -(1/c)\dot{A}_z - \partial F/\partial z = 0$, $\mathbf{H}=0$).

We consider now an infinitely long circular solenoid with radius a and a uniform, oscillating surface current density $j_\varphi = j_s\delta(r-a)\cos\omega t$; j_φ is the φ-component (tangential component) of the current density in cylindrical coordinates. We can see that $div\mathbf{j} = 0$ and, by continuity equation, the charge density ρ is zero. Therefore, we have only the vector potential $\mathbf{A}$. It is easy to see, from Kirchhoff formula, that $div\mathbf{A} = 0$; it follows that $\mathbf{A} \simeq (-y,\, x,\, 0)f(r)$, where $f(r) \to const$ for $r \to 0$ and $f(r) \to 0$ for $r \to \infty$; the magnetic field (H_z) is uniform near the axis of the solenoid and is vanishing for large distances from the axis in the exterior region. Indeed, for the Kirchhoff formulae it is necessary to use the cartesian coordinates, *i.e.*

$$\mathbf{j} = j_s(-\sin\varphi,\, \cos\varphi,\, 0)\delta(r-a)\cos\omega t\ ; \tag{2.150}$$

the vector potential is given by

$$\mathbf{A} = \tfrac{2aj_s}{c}(-\sin\varphi,\, \cos\varphi,\, 0)\cdot$$

$$\cdot Re\left\{e^{-i\omega t}\int_0^\varphi d\varphi'\cos\varphi'\int dz\frac{e^{i\frac{\omega}{c}\sqrt{a^2+r^2-2ar\cos\varphi'+z^2}}}{\sqrt{a^2+r^2-2ar\cos\varphi'+z^2}}\right\}\ ; \tag{2.151}$$

the integral with respect to z is the Hankel function,

$$\int dz\frac{e^{i\frac{\omega}{c}\sqrt{r^2+z^2}}}{\sqrt{r^2+z^2}} = i\pi H_0^{(1)}(\omega r/c) =$$

$$= \begin{cases} i\sqrt{\frac{2\pi}{\omega r/c}}e^{i(\omega r/c-\pi/4)}\ , & \omega r/c \gg 1\ , \\ -2\ln(\omega r/c)\ , & \omega r/c \ll 1\ ; \end{cases} \tag{2.152}$$

we get

$$\boldsymbol{A} = \pi\sqrt{2\pi}\tfrac{aj_s}{cr}\frac{\omega a/c}{\sqrt{\omega r/c}}(-y,\, x\, 0)\cdot$$

$$\cos[\omega(t-r/c)-\pi/4]\ ,\ r > a\ ,\ \omega r/c \gg 1 \tag{2.153}$$

and

$$\boldsymbol{A} = \frac{2\pi j_s}{c}\frac{a^2}{a^2+r^2}(-y,\, x\, 0)\cos\omega t\ ,\ r < a\ ,\ \omega a/c \ll 1\ . \tag{2.154}$$

The magnetic field inside the solenoid is $H_z = (4\pi a^2 j_s/c)/(a^2+r^2)$. For $\omega a/c > 1$ there exists a non-uniform oscillating magnetic field.

3 Relativity

3.1 Electromagnetic field

Let us introduce the notation $x^\mu = (ct, x^i)$ for a contravariant time-position vector, where $\mu = 0, 1, 2, 3$ and $i = 1, 2, 3$, and the covariant notation $x_\mu = (ct, -x_i)$ for the same vector, where c is the speed of light in vacuum (also, we use the notation $x^1 = x$, $x^2 = y$, $x^3 = z$, labels $i, j, ... = 0, 1, 2, 3$, as well as $\alpha, \beta, ... = 1, 2, 3$). We pass from one notation to another by the metric tensor $g_{\mu\nu}$, or $g^{\mu\nu}$, g^ν_μ which is diagonal with elements $(1, -1, -1, -1)$; *e.g.*, $x^\mu = g^{\mu\nu} x_\nu$. We write the scalar product of two vectors A and B as $A^\mu B_\mu = A^0 B_0 - A^i B_i$, where $A^i B_i$ stands for the usual notation$\boldsymbol{AB}$, $A^i B_i = \boldsymbol{AB}$. We introduce the covariant notation $\partial_\mu = \frac{\partial}{\partial x^\mu} = (\frac{\partial}{c\partial t}, \frac{\partial}{\partial x^i})$ and the contravariant notation $\partial^\mu = \frac{\partial}{\partial x_\mu} = (\frac{\partial}{c\partial t}, -\frac{\partial}{\partial x_i})$, such that $\partial_\mu \partial^\mu = \frac{1}{c^2}\frac{\partial^2}{\partial t^2} - \Delta$.

Let us define the electromagnetic potential as $A^\mu = (\Phi, A^i)$, where Φ is the scalar potential and $\boldsymbol{A}$ is the vector potential; then the Lorenz gauge reads

$$\partial_\mu A^\mu = 0 \tag{3.1}$$

and the electric field $\boldsymbol{E}$ and the magnetic field $\boldsymbol{H}$ are given by $F_{\mu\nu} = \partial_\mu A_\nu - \partial_\nu A_\mu$. The gauge transformation is $A_\mu \to A_\mu - \partial_\mu f$, where f is an arbitrary function. The Gauss law $div\boldsymbol{H} = 0$ and the Faraday law $curl\boldsymbol{E} = -\frac{1}{c}\frac{\partial \boldsymbol{H}}{\partial t}$ read

$$\partial_\mu F_{\nu\rho} + \partial_\nu F_{\rho\mu} + \partial_\rho F_{\mu\nu} = 0 \ . \tag{3.2}$$

We define the current density vector $j^\mu = (c\rho, j^i)$, where ρ is the charge density and $\boldsymbol{j}$ is the current density, which obeys the continuity equation $\partial_\mu j^\mu = 0$; we note that $j^\mu = \rho \frac{dx^\mu}{dt}$. Then, the Gauss law

$div\boldsymbol{E} = 4\pi\rho$ and the Maxwell-Ampere equation $curl\boldsymbol{H} = \frac{1}{c}\frac{\partial \boldsymbol{E}}{\partial t} + \frac{4\pi}{c}\boldsymbol{j}$ read

$$\partial_\mu F^{\mu\nu} = \frac{4\pi}{c} j^\nu \tag{3.3}$$

and the potential satisfies the wave equation

$$\partial_\mu \partial^\mu A^\nu = \frac{4\pi}{c} j^\nu \ , \tag{3.4}$$

providing the Lorenz gauge is satisfied.

An energy-momentum tensor $T^{\mu\nu}$ can be defined for the electromagnetic field, which includes the energy density $u = \frac{1}{8\pi}(E^2 + H^2)$, the Poynting vector $\boldsymbol{S}$ divided by c and the stress tensor σ_{ij}. Equation $\partial_\mu T^{\mu\nu} = 0$ is the energy conservation and the momentum conservation for the field, as well as the conservation of the angular momentum ($T^{\mu\nu}$ is a symmetric tensor).

3.2 Lorentz transformation

The equations of the electromagnetic field written above are invariant under a transformation

$$x'^\mu = \alpha^\mu_\nu x^\nu \ , \tag{3.5}$$

where the constant matrix α^μ_ν leaves invariant the quantity

$$s^2 = c^2 t^2 - r^2 = x^\mu x_\nu \ , \tag{3.6}$$

or $ds^2 = dx^\mu dx_\mu$. This is a rotation, with the scalar product $x^\mu x_\mu$, given by an orthogonal matrix $\alpha^\mu_\nu \alpha^\nu_\rho = \delta^\mu_\rho$, with the determinant equal to unity. It leaves the volume element $cdtd\boldsymbol{r}$ unchanged. Under this transformation any vector, tensor changes according to equation (3.5). The condition given by equation (3.6) shows that this transformation is a change of the reference frame with a constant velocity $\boldsymbol{v}$. In the rest frame a particle has $s^2 = c^2\tau^2$, such that

$$s^2 = c^2 t^2 - v^2 t^2 = c^2 \tau^2 \ , \tag{3.7}$$

whence the proper time of the particle

$$\tau = \sqrt{1 - v^2/c^2} t \ ; \tag{3.8}$$

in its own frame the particle has a slower time. This is time dilation. For the electromagnetic field, which propagates with velocity $v = c$, the proper time is zero. We can see that all the velocities should be smaller than c.

For a frame moving with velocity v along the x-axis this transformation is

$$x = \frac{x' + vt'}{\sqrt{1 - v^2/c^2}} \ , \ t = \frac{t' + \frac{v}{c^2}x'}{\sqrt{1 - v^2/c^2}} \ ; \tag{3.9}$$

all the primed quantities are in the moving frame. The length in the moving frame $l' = \sqrt{1 - v^2/c^2} l$ is shorter than the length l in the rest frame; in motion, the length contracts. This is Lorentz's contraction. The invariance of the equations of motion of the electromagnetic field under Lorentz transformations ensures the principle of relativity of the motion of the field. In the non-relativistic limit $v \ll c$ the Lorentz transformation becomes the Galilei transformation.

3.3 Relativistic mechanics

Newton's equation of motion is not invariant under the Lorentz transformation. In order to satisfy the principle of relativity (with Lorentz transformations) Newton's equation of motion needs to be modified;[1] the modification affects the mechanical motion with great velocities, close to the speed of light c. Moreover, the extension of this principle of relativity requires that any other fundamental interaction propagates with the speed of light c in vacuum; c becomes, indeed, a universal constant.

First, we note that the velocity should be defined as dx^μ/ds, *i.e.* $dx^\mu/d\tau$. We introduce the relativistic velocity

$$u^\mu = \left(\frac{1}{\sqrt{1 - v^2/c^2}}, \frac{\boldsymbol{v}/c}{\sqrt{1 - v^2/c^2}} \right) \ ; \tag{3.10}$$

[1] A. Einstein, "Zur Elektrodynamik bewegter Koerper", Ann. Physik **17** 891 (1905); see also, W. Pauli, *Theory of Relativity*, Pergamon Press, London (1958) (*Relativitaetstheorie*, Vieweg, Teubner, Leipzig (1921)).

Newton's law of motion would be

$$mc\frac{d}{dt}\frac{\boldsymbol{v}/c}{\sqrt{1-v^2/c^2}}=\boldsymbol{f} \ , \tag{3.11}$$

for a particle with mass m subjected to a force $\boldsymbol{f}$. In addition, if we multiply this equation by $\boldsymbol{v}$, we get

$$mc^2\frac{d}{dt}\frac{1}{\sqrt{1-v^2/c^2}}=\boldsymbol{vf} \ , \tag{3.12}$$

so we may write

$$mc\frac{d}{dt}\frac{1}{\sqrt{1-v^2/c^2}}=\boldsymbol{vf}/c=f^0 \ . \tag{3.13}$$

The equations of motion become

$$mc\frac{du^\mu}{dt}=f^\mu \ , \tag{3.14}$$

or

$$mc\frac{du^\mu}{ds}=\frac{f^\mu}{\sqrt{1-v^2/c^2}}=F^\mu \ , \tag{3.15}$$

where the force $(f^0,\boldsymbol{f})$ is $(\boldsymbol{vf}/c,\boldsymbol{f})$ and the relativistic force is $F^\mu = f^\mu/\sqrt{1-v^2/c^2}$; in addition, the momentum $\boldsymbol{p}$ and the energy $\mathcal{E}$ of the particle are

$$\boldsymbol{p}=\frac{m\boldsymbol{v}}{\sqrt{1-v^2/c^2}}=mc\boldsymbol{u} \ , \ \mathcal{E}=\frac{mc^2}{\sqrt{1-v^2/c^2}}=mc^2u^0 \ , \tag{3.16}$$

such that $p^\mu=(\mathcal{E}/c,\boldsymbol{p})$ is a vector; and $d\mathcal{E}/dt=\boldsymbol{vf}=f^0/c$, *i.e.* the change of energy is equal to the mechanical work done by the force per unit time. Also, $d\mathcal{E}/dt=\boldsymbol{vf}=\boldsymbol{v}d\boldsymbol{p}/dt$. From equations (3.16) we get

$$\mathcal{E}^2=m^2c^4+p^2c^2 \tag{3.17}$$

and, in the non-relativistic limit,

$$\mathcal{E}=mc^2+\frac{p^2}{2m}+...=mc^2+\frac{1}{2}mv^2+... \ , \ \boldsymbol{p}=m\boldsymbol{v}+\frac{m\boldsymbol{v}}{2c^2}v^2+... \ . \tag{3.18}$$

We can see that the particle has a rest energy mc^2.

Equation (3.11) shows that the motion proceeds as if the mass of the particle increases with increasing velocity. The integration of this equation leads to

$$v^2/c^2 = \frac{\left(\int^t \boldsymbol{f} dt'\right)^2}{m^2c^2 + \left(\int^t \boldsymbol{f} dt'\right)^2} , \tag{3.19}$$

which shows that the velocity may increase in time up to the speed of light c; while its energy may increase indefinitely.

The Lorentz force for a pointlike charge q which moves with velocity $\boldsymbol{v}$ is $\boldsymbol{f} = q\boldsymbol{E} + \frac{q}{c}\boldsymbol{v} \times \boldsymbol{H}$, where $\boldsymbol{E} = -\frac{1}{c}\frac{\partial \boldsymbol{A}}{\partial t} - grad\Phi$ is the electric field and $\boldsymbol{H} = curl\boldsymbol{A}$ is the magnetic field. If we express this force with relativistic notations, we get $f_i = \sqrt{1 - v^2/c^2}\frac{q}{c}F_{i\nu}u^\nu$; similarly, $f_0 = \boldsymbol{v}\boldsymbol{f}/c = \sqrt{1 - v^2/c^2}F_{0\nu}u^\nu$, where $F_{\mu\nu}$ is the tensor of the electromagnetic field. Therefore, we get the equation of motion of the charge in the electromagnetic field

$$mc\frac{du_\mu}{ds} = \frac{q}{c}F_{\mu\nu}u^\nu ; \tag{3.20}$$

we note that the relativistic force is $F_\mu = f_\mu/\sqrt{1 - v^2/c^2} = \frac{q}{c}F_{\mu\nu}u^\nu$. We may view a continuous charge distribution with density ρ, such that the force becomes

$$\begin{aligned} F_\mu &= \tfrac{\rho}{c}F_{\mu\nu}u^\nu = \tfrac{\rho}{c}F_{\mu\nu}\tfrac{dx^\mu}{ds} = \\ &= \tfrac{\rho}{c}F_{\mu\nu}\tfrac{dx^\mu}{dt}\tfrac{dt}{ds} = \tfrac{1}{c}F_{\mu\nu}j^\nu\tfrac{dt}{ds} , \end{aligned} \tag{3.21}$$

according to the definition of the current j^ν; equation of motion (3.20) becomes

$$\mu c\frac{du_\mu}{dt} = \frac{1}{c}F_{\mu\nu}j^\nu , \tag{3.22}$$

where μ is the mass density. Also, for a continuous mass distribution we may define a mass current $\mu\frac{dx^\mu}{dt}$, besides the momentum density $\mu c u^\mu = \mu c\frac{dx^\mu}{ds}$. Then, the energy and the momentum conservation of the mass distribution is a continuity equation for an energy-momentum tensor $T^{\mu\nu}$ of matter. The conservation of the angular momentum is ensured by the symmetry of this tensor. The conservation of the total energy-momentum tensor of the field and matter reduces to the field equations with sources (3.3) and the equation of motion (3.20).

3.4 Lagrangians. Hamilton-Jacobi equation

The lagrangian of the field is

$$L_f = -\frac{1}{16\pi}\int dV F_{\mu\nu}F^{\mu\nu} = \frac{1}{8\pi}\int dV \left(E^2 - H^2\right) \ , \tag{3.23}$$

where dV is the element of volume. The lagrangian of the field-matter interaction is

$$L_{fm} = -\frac{1}{c}\int dV A_\mu j^\mu \ . \tag{3.24}$$

The variation of the action $S_f + S_{fm} = \int dt(L_f + L_{fm})$ with respect to the potential A_μ in $F_{\mu\nu} = \partial_\mu A_\nu - \partial_\nu A_\mu$ leads to the field equations $\partial_\mu F^{\mu\nu} = \frac{4\pi}{c} j^\nu$ (equation (3.3)).

The lagrangian of the field can also be written as

$$\begin{aligned} L_f = \tfrac{1}{8\pi}\int dV\{\left(\tfrac{1}{c}\tfrac{\partial \boldsymbol{A}_l}{\partial t} + grad\Phi\right)^2 + \\ + \left(\tfrac{1}{c}\tfrac{\partial \boldsymbol{A}_t}{\partial t}\right)^2 - (curl\boldsymbol{A}_t)^2\} \ , \end{aligned} \tag{3.25}$$

where $\boldsymbol{A}_{l,t}$ is the longitudinal/transverse part of the vector potential and Φ is the scalar potential. $\frac{1}{c}\partial\boldsymbol{A}_{l,t}/\partial t$ is a field velocity, such that we may define a field momentum density $\boldsymbol{p}_{l,t} = \partial\mathcal{L}_f/\partial(\dot{\boldsymbol{A}}/c)$, where $\mathcal{L}_f$ is the lagrangian density (divided by $1/4\pi$). We get

$$\boldsymbol{p}_l = \frac{1}{c}\dot{\boldsymbol{A}}_l + grad\Phi \ ; \tag{3.26}$$

we see that the velocity cannot be determined by the momentum $\boldsymbol{p}_l$, because of the scalar potential. A canonical hamiltonian does not exist for the longitudinal field. In fact, the longitudinal degrees of freedom are removed by the interaction. By contrast, a hamiltonian

$$H_t = \frac{1}{4\pi}\int dV\left(\frac{1}{2}p_t^2 + \frac{1}{2}(curl\boldsymbol{A}_t)^2\right) \tag{3.27}$$

exists for the transverse degrees of freedom, where $\boldsymbol{p}_t = \dot{\boldsymbol{A}}_t/c$. Its equation of motion leads to the wave equation for $\boldsymbol{A}_t$, providing the Lorenz gauge is satisfied. We can see that we can have a field hamiltonian only if the Lorenz gauge is satisfied.

The action of the particle with mass m is

$$S_m = -mc\int ds = -mc^2\int dt\sqrt{1-v^2/c^2} \tag{3.28}$$

and its lagrangian is

$$L_m = -mc^2\sqrt{1-v^2/c^2} \; . \tag{3.29}$$

For a pointlike charge q with mass m the interaction lagrangian (equation (3.24)) becomes

$$L_{fm} = -\frac{q}{c}A_\mu\frac{dx^\mu}{dt} = -q\Phi + \frac{q}{c}\boldsymbol{v}\boldsymbol{A} \; ; \tag{3.30}$$

the interaction action is

$$S_{fm} = -\frac{q}{c}\int A_\mu dx^\mu \; . \tag{3.31}$$

The variation of the action $S_m + S_{fm}$ with respect to the coordinates x^μ leads to the equation of motion $mc\frac{du_\mu}{ds} = \frac{q}{c}F_{\mu\nu}u^\nu$ (equation (3.20)). The momentum is defined by

$$P_\mu = -\frac{\partial(S_m + S_{fm})}{\partial x^\mu} = mcu_\mu + \frac{q}{c}A_\mu \; , \tag{3.32}$$

where $p_\mu = mcu_\mu$ is the mechanical momentum. In particular, $P_0 = \mathcal{E}/c = (\mathcal{E}_{kin} + q\Phi)/c$, where $E_{kin} = mc^2u_0 = mc^2/\sqrt{1-v^2/c^2}$ and $\boldsymbol{P} = m\boldsymbol{v}/\sqrt{1-v^2/c^2} + \frac{q}{c}\boldsymbol{A}$. Since $u_\mu u^\mu = 1$, we get the Hamilton-Jacobi equation

$$(P_\mu - \frac{q}{c}A_\mu)(P^\mu - \frac{q}{c}A^\mu) = m^2c^2 \; , \tag{3.33}$$

or

$$(\mathcal{E}/c - q\Phi/c)^2 - (\boldsymbol{P} - \frac{q}{c}\boldsymbol{A})^2 = m^2c^2 \; , \tag{3.34}$$

or

$$\frac{1}{c^2}\left(\frac{\partial S}{\partial t}+q\Phi\right)^2-\left(gradS-\frac{q}{c}\boldsymbol{A}\right)^2-m^2c^2=0\ . \tag{3.35}$$

In the non-relativistic limit

$$\mathcal{E}=mc^2+\frac{1}{2m}(\boldsymbol{P}-\frac{q}{c}\boldsymbol{A})^2+q\Phi\ , \tag{3.36}$$

which corresponds to $S=S'-mc^2t$. Indeed,

$$\frac{\partial S'}{\partial t}+q\Phi=-\frac{1}{2m}(\boldsymbol{P}-\frac{q}{c}\boldsymbol{A})^2+\frac{1}{2mc^2}\left(\frac{\partial S'}{\partial t}+q\Phi\right)^2\ , \tag{3.37}$$

or

$$\mathcal{E}=\frac{1}{2m}(\boldsymbol{P}-\frac{q}{c}\boldsymbol{A})^2+q\Phi-\frac{1}{2mc^2}\left(\frac{\partial S'}{\partial t}+q\Phi\right)^2\ , \tag{3.38}$$

which reduces to

$$\mathcal{E}=\frac{1}{2m}P^2-\frac{q}{mc}\boldsymbol{P}\boldsymbol{A}+q\Phi+O(1/c^2)...\ . \tag{3.39}$$

3.5 Lienard-Wiechert potentials

A point charge q moving along a trajectory $\mathbf{r}_0(t)$ generates a vector potential

$$\mathbf{A}(t,\mathbf{r})=\frac{q}{c}\int d\mathbf{r}'\frac{\mathbf{v}(t')}{|\mathbf{r}-\mathbf{r}'|}\delta(\mathbf{r}'-\mathbf{r}_0(t'))\ , \tag{3.40}$$

where

$$t'=t-\frac{|\mathbf{r}-\mathbf{r}'|}{c} \tag{3.41}$$

and $\mathbf{v}=\dot{\mathbf{r}}_0$ is the particle velocity. We rewrite this equation as

$$\mathbf{A}(t,\mathbf{r})=\frac{q}{c}\int dt'd\mathbf{r}'\frac{\mathbf{v}(t')}{|\mathbf{r}-\mathbf{r}'|}\delta(\mathbf{r}'-\mathbf{r}_0(t'))\delta(t'-t+|\mathbf{r}-\mathbf{r}'|/c)\ , \tag{3.42}$$

which leads to

$$\mathbf{A}(t,\mathbf{r})=\frac{q}{c}\int dt'\frac{\mathbf{v}(t')}{|\mathbf{r}-\mathbf{r}_0|}\delta(t'-t+|\mathbf{r}-\mathbf{r}_0|/c)\ ; \tag{3.43}$$

by using

$$\frac{d}{dt'}(t' - t + |\mathbf{r} - \mathbf{r}_0| /c) = 1 - \boldsymbol{v}(\boldsymbol{r} - \boldsymbol{r}_0)/c\,|\mathbf{r} - \mathbf{r}_0| \quad , \tag{3.44}$$

we get

$$\mathbf{A}(t, \mathbf{r}) = \frac{q\mathbf{v}/c}{R - \mathbf{vR}/cR} \,|_{t-R/c} \quad , \tag{3.45}$$

where $\mathbf{R} = \mathbf{r} - \mathbf{r}_0$. Similarly, the scalar potential is given by

$$\Phi(t, \mathbf{r}) = \frac{q}{R - \mathbf{vR}/cR} \,|_{t-R/c} \quad . \tag{3.46}$$

These are the Lienard-Wiechert potentials. We can check that the vector $A^\mu = (\Phi,\, \mathbf{A})$ is given by

$$A^\mu = q\frac{u^\mu}{R_\nu u^\nu} \quad , \tag{3.47}$$

where

$$u^\mu = \frac{1}{\sqrt{1 - v^2/c^2}}(1,\ \mathbf{v}/c) \tag{3.48}$$

is the relativistic velocity and $R_\mu = (c(t - t'),\, -(\mathbf{r} - \mathbf{r}_0))$.
In order to compute the fields we use $s = R - \mathbf{vR}/c$ and

$$\begin{gathered} \tfrac{\partial t'}{\partial t} = \tfrac{R}{s} \ , \ grad t' = -\tfrac{\mathbf{R}}{cs} \ , \ curl \mathbf{v} = grad t' \times \dot{\mathbf{v}} \ , \\ \tfrac{\partial s}{\partial t'} = -\tfrac{\mathbf{Rv}}{R} + \tfrac{v^2}{c} - \tfrac{\mathbf{R}\dot{\mathbf{v}}}{c} \ , \\ grad s = \tfrac{\mathbf{R}}{R} - \tfrac{\mathbf{v}}{c} + \tfrac{\mathbf{R}}{cs}\left(\tfrac{\mathbf{Rv}}{R} - \tfrac{v^2}{c} + \tfrac{\mathbf{R}\dot{\mathbf{v}}}{c}\right) \ . \end{gathered} \tag{3.49}$$

The fields are given by

$$\begin{gathered} \mathbf{E} = q\tfrac{1-v^2/c^2}{s^3}(\mathbf{R} - \tfrac{\mathbf{v}}{c}R) + \\ +q\tfrac{1}{c^2 s^3}\mathbf{R} \times \left[(\mathbf{R} - \tfrac{\mathbf{v}}{c}R) \times \dot{\mathbf{v}}\right] \ , \ \mathbf{H} = \tfrac{1}{R}\mathbf{R} \times \mathbf{E} \ . \end{gathered} \tag{3.50}$$

E and **H** are everywhere perpendicular to each other. The first term on the right in equations (3.50) goes like $\sim q/R^2$ for $R \to \infty$, while

the second term goes like $q\dot{\mathbf{v}}/c^2R$; the latter gives a non-vanishing energy flux through a surface placed at infinity, and depends on particle acceleration: it is transverse (perpendicular to $\mathbf{R}$).
For a uniform motion $\mathbf{v} = const$

$$\mathbf{R}' = \mathbf{R} - \frac{\mathbf{v}}{c}R = \mathbf{R} - \mathbf{v}(t - t') \tag{3.51}$$

is the instantaneous position vector (at the time t); introducing the angle θ between $\mathbf{v}$ and $\mathbf{R}'$ we get easily

$$\mathbf{E} = q\frac{1 - v^2/c^2}{\left(1 - \frac{v^2}{c^2}\sin^2\theta\right)^{3/2}}\frac{\mathbf{R}'}{R'^3} ; \tag{3.52}$$

we can see that the field is contracted in the forward direction, especially for relativistic velocities (it is concentrated mainly on the direction perpendicular to the direction of motion). The magnetic field is $\mathbf{H} = \mathbf{v} \times \mathbf{E}/c$. For $v/c \ll 1$ the magnetic fied is $H = q\boldsymbol{v} \times \boldsymbol{R}/cR^3$.

3.6 Damping in radiation

We consider, with usual notation, the motion of a pointlike charge q with mass m in an electromagnetic field generated by the vector potential $A_z = A_0 \cos\frac{\omega}{c}(ct - x) = A$; the equations of motion are

$$mc\frac{du^\mu}{ds} = \frac{q}{c}F^{\mu\nu}u_\nu , \tag{3.53}$$

where u^μ is the relativistic velocity. The field $F_{\mu\nu} = \partial_\mu A_\nu - \partial_\nu A_\mu$ ($\mathbf{E} = -(1/c)\partial\mathbf{A}/\partial t$, $\mathbf{H} = curl\mathbf{A}$) is given by

$$F^{\mu\nu} = \begin{pmatrix} 0 & 0 & 0 & -E \\ 0 & 0 & 0 & -E \\ 0 & 0 & 0 & 0 \\ E & E & 0 & 0 \end{pmatrix} ,$$
$$F_{\mu\nu} = \begin{pmatrix} 0 & 0 & 0 & E \\ 0 & 0 & 0 & -E \\ 0 & 0 & 0 & 0 \\ -E & E & 0 & 0 \end{pmatrix} , \tag{3.54}$$

where $E = E_z = \frac{\omega}{c} A_0 \sin\frac{\omega}{c}(ct - x)$ (and $H_y = -E$). The length element is $ds = cdt/\gamma$, $\gamma = 1/\sqrt{1 - v^2/c^2}$. We introduce the notation $s = ct - x$. Consequently, the equations of motion are re-written as

$$\begin{gathered} mc\frac{du^0}{ds} = \frac{qE}{c}u^3 \ , \ mc\frac{du^1}{ds} = \frac{qE}{c}u^3 \ , \\ mc\frac{du^3}{ds} = \frac{qE}{c}(u^0 - u^1) \end{gathered} \tag{3.55}$$

($u^2 = 0$); we get immediately $u^0 - u^1 = const = 1$, for a charge initially at rest in the origin ($u^0 = \gamma$, $\mathbf{u} = \gamma\mathbf{v}/c$); it follows

$$\begin{gathered} u^1 = \frac{q^2 A^2}{2m^2c^4} \ , \ u^0 = \gamma = 1 + \frac{q^2A^2}{2m^2c^4} \ , \\ u^3 = -\frac{q}{mc^2}A \end{gathered} \tag{3.56}$$

(we can check $u_i u^i = 1$).[2] The velocities are given by

$$v_x = c\frac{q^2A^2/2m^2c^4}{1 + q^2A^2/2m^2c^4} \ , \ v_z = -c\frac{qA/mc^2}{1 + q^2A^2/2m^2c^4} \tag{3.57}$$

and the energy is

$$\mathcal{E} = mc^2\gamma = mc^2 + \frac{q^2A^2}{2mc^2} \ . \tag{3.58}$$

We can see that for high fields the charge reaches the ultra-relativistic limit along the direction of propagation of the radiation field ($v_x \rightarrow c$, $v_z \rightarrow 0$). From $u^\mu = dx^\mu/ds$ we get the coordinates

$$\begin{gathered} x = \frac{q^2A_0^2/4m^2c^3}{1+q^2A_0^2/4m^2c^4}\left[t + \frac{1}{2\omega}\sin 2\frac{\omega}{c}(ct - x)\right] \ , \\ z = -\frac{qA_0}{mc\omega}\sin\frac{\omega}{c}(ct - x) \ . \end{gathered} \tag{3.59}$$

[2] C. Miron and M. Apostol, "On the motion of an electric charge in high-intensity electromagnetic radiation", Roum. J. Phys. **62** 117 (2017).

The reaction force is given by[3]

$$g^{\mu} = \frac{2q^3}{3mc^3}\frac{\partial F^{\mu\nu}}{\partial x^{\rho}}u_{\nu}u^{\rho} - \tag{3.60}$$
$$\frac{2q^4}{3m^2c^5}\left[F^{\mu\nu}F_{\rho\nu}u^{\rho} - (F_{\nu\rho}u^{\rho})(F^{\nu\sigma}u_{\sigma})u^{\mu}\right] \ .$$

This force reduces to $\boldsymbol{f}/c$ in the non-relativistic limit, where $\boldsymbol{f} = 2q^2\ddot{\boldsymbol{v}}/3c^3$ is the Lorentz damping force; in addition, $g^{\mu}u_{\mu} = 0$, an identity valid for any force, according to the equation of motion and $u_{\mu}u^{\mu} = 1$. The integral $\int dsg^{\mu}$ is the change ΔP^{μ} of the field momentum, due to the relativistic second term in equation (3.60) (for the Lorentz damping force $\Delta P^{\mu} = 0$). It must be much smaller than $q^2/c\gamma a^2$, where $a = q^2/mc^2$ is the classical electromagnetic radius of the charge; this condition, which indicates the limits of applicability of the Electromagnetism, does not imply always that the damping (reaction) force is smaller than the electromagnetic (Lorentz) force; in the limit $v \rightarrow c$ the damping may be higher than the driving force.

By making use of the field given by equations (3.54) we get the reaction force

$$g^0 = \frac{2q^3}{3mc^3}(u^0 - u^1)\left\{\frac{\omega^2}{c^2}Au^3 + \frac{qE^2}{mc^2}\left[1 - (u^0 - u^1)u^0\right]\right\} \ ,$$

$$g^1 = \frac{2q^3}{3mc^3}(u^0 - u^1)\left\{\frac{\omega^2}{c^2}Au^3 + \frac{qE^2}{mc^2}\left[1 - (u^0 - u^1)u^1\right]\right\} \ , \tag{3.61}$$

$$g^3 = \frac{2q^3}{3mc^3}(u^0 - u^1)^2\left(\frac{\omega^2}{c^2}A - \frac{qE^2}{mc^2}u^3\right)$$

(we can check $u_i g^i = 0$). By inserting the velocity given by equations

[3]M. Abraham, *Theorie der Elektrizitaet*, Leipzig (1908); M. v. Laue, "Die Wellenstrahlung einer bewegten Punktladung", Ann. Physik **28** 436 (1909);. L. Landau and E. Lifshitz, *Course of Theoretical Physics*, vol. 2, *The Classical Theory of Fields*, Elsevier (2004).

(3.56) we get[4]

$$g^0 = -\frac{2a^2\omega^2A^2}{3c^3}\left(1 + a\frac{E^2}{2m\omega^2}\right) \; ,$$

$$g^1 = -\frac{2a^2\omega^2}{3c^3}\left[A^2 - \frac{c^2E^2}{\omega^2}\left(1 - \frac{q^2A^2}{2m^2c^4}\right)\right] \; , \tag{3.62}$$

$$g^3 = \frac{2aqA\omega^2}{3c^3}\left(1 - a\frac{E^2}{m\omega^2}\right) \; ,$$

or, in the limit of high-intensity fields,

$$g^0 = g^1 \simeq -\frac{a^3A^2E^2}{3mc^3} \; , \; g^3 \simeq -\frac{2a^2qAE^2}{3mc^3} \; ; \tag{3.63}$$

worth noting, it is a damping force. We consider an electron in a high-intensity field $E = 10^{10}esu$, corresponding to a laser radiation with intensity $I = 10^{22}w/cm^2$ ($I = cE^2/4\pi$), focalized in a pulse of dimension $d = 10\mu m$. The vector potential is $A_0 = cE_0/\omega = 10^{-5}E_0 = 10^5 esu$ for the optical frequency $\omega \simeq 3 \times 10^{15}s^{-1}$ ($\lambda = 0.5\mu m$); the corresponding energy for an electron is $qA_0 = 4.8 \times 10^{-5}erg \simeq 30MeV$. This energy is much higher than the rest energy of the electron $mc^2 = 0.5MeV$. The electron is accelerated along the direction of propagation of the radiation up to energies of the order $1GeV$ (equation (3.58)). Under these conditions the damping force is $g^0 = g^1 \simeq 10^{-12}g/s$, $g^3 \simeq 10^{-13}g/s$ is much smaller than the Lorentz force ($qEu^3/c \simeq 10^{-8}g/s$, $qE/c \simeq 10^{-10}g/s$; $\gamma \simeq 10^4$) and the validity condition $g^\mu \ll q^2/c\gamma a^2$ is satisfied ($q^2/c\gamma a^2 \simeq 10^{-8}g/s$).

We may conclude that in high-intensity radiation field the motion of the charge becomes ultra-relativistic and the damping force has no measurable effect on the motion.

[4]M. Apostol, "Damping (reaction) force for a charge in high-intensity radiation field", J. Theor. Phys. **262** (2017).

3.7 Relativistic effects

3.7.1 Addition of velocities

By taking the differentials in equations (3.9) and get the velocities in the moving frame

$$v_x^{'} = \frac{v_x - v}{1 - \frac{v}{c^2} v_x} \ ,$$
$$v_y^{'} = \sqrt{1-\beta^2} \frac{v_y}{1 - \frac{v}{c^2} v_x} \ , \ v_z^{'} = \sqrt{1-\beta^2} \frac{v_z}{1 - \frac{v}{c^2} v_x} \ , \tag{3.64}$$

where $\beta = v/c$. We can see that the law of composition of the velocities is not commutative, any two velocities less than c combine to a velocity less than c, two parallel velocities c give one velocity c, etc.

3.7.2 Aberration of light

Let us assume that we have a source of light in the frame K, *i.e.* we assume

$$v_x^2 + v_y^2 + v_z^2 = v_0^2 = c^2 \ ; \tag{3.65}$$

we introduce the angles α and $\alpha^{'}$ through $v_x = v_0 \cos\alpha = c \cos\alpha$ and $v_x^{'} = v^{'} \cos\alpha^{'}$, where $v^{'2} = v_x^{'2} + v_y^{'2} + v_z^{'2}$. From equations (3.64) we have

$$v_x^{'} = c\frac{\cos\alpha - \beta}{1 - \beta\cos\alpha} \ ,$$
$$v_y^{'} = \sqrt{1-\beta^2} \frac{v_y}{1 - \beta\cos\alpha} \ , \ v_z^{'} = \sqrt{1-\beta^2} \frac{v_z}{1 - \beta\cos\alpha} \tag{3.66}$$

and

$$v^{'} = c\frac{\left[(\cos\alpha - \beta)^2 + (1-\beta^2)\sin^2\alpha\right]^{1/2}}{1 - \beta\cos\alpha} = c \tag{3.67}$$

as expected. In addition,

$$v^{'} \cos\alpha^{'} = c\cos\alpha^{'} = v_x^{'} = c\frac{\cos\alpha - \beta}{1 - \beta\cos\alpha} \ , \tag{3.68}$$

or

$$\cos\alpha^{'} = \frac{\cos\alpha - \beta}{1 - \beta\cos\alpha} \ , \ \cos\alpha = \frac{\cos\alpha^{'} + \beta}{1 + \beta\cos\alpha^{'}} \tag{3.69}$$

and

$$\sin\alpha' = \sqrt{1-\beta^2}\frac{\sin\alpha}{1-\beta\cos\alpha} \ . \qquad (3.70)$$

This is the aberration of light. Equations (3.69) and (3.70) give the deviation of a ray of light on passing from one frame to another. On moving towards the source $\cos\alpha < 0$, $\cos\alpha' < 0$ and $\cos\alpha' < \cos\alpha$, according to equation (3.69). In this region the observer in the frame K' sees the source deviated to the forward direction. This is called the "forward beaming". On moving away from the source $\cos\alpha > 0$, but $\cos\alpha'$ can be either positive or negative; for $\cos\alpha > \beta$ the ray of light is deviated backward. For $\cos\alpha < \beta$ there is a shadow cone where the light does not reach the observer any longer. This can be easily seen for $\alpha = -\pi/2$.

On moving near the source we see an enhanced luminosity, as a result of the decrease of the solid angle in the moving frame. Indeed, the solid angles are given by

$$\frac{do'}{do} = \frac{d(\cos\alpha')}{d(\cos\alpha)} = \frac{1-\beta^2}{(1-\beta\cos\alpha)^2} \ . \qquad (3.71)$$

For a longitudinal beaming $\cos\alpha = -1$, we get $do'/do = (1-\beta)/(1+\beta)$, which, in the ultra-relativistic limit, becomes $do'/do \simeq 1/4\gamma^2$, where $\gamma = 1/\sqrt{1-\beta^2}$. The volume is contracted by the square root of the ratio of the two solid angles $V'/V = \sqrt{do'/do} = \sqrt{1-\beta^2}/(1-\beta\cos\alpha)$.

The amplitude A of a plane wave is transformed according to

$$A' = A\frac{1-\beta\cos\alpha}{\sqrt{1-\beta^2}} \ , \qquad (3.72)$$

such that its energy $\mathcal{E} = \frac{1}{2}A^2V$ is transformed as

$$\mathcal{E}' = \mathcal{E}\frac{1-\beta\cos\alpha}{\sqrt{1-\beta^2}} \ . \qquad (3.73)$$

3.7.3 Doppler effect

The phase $\omega t - \mathbf{kr}$ of a plane wave must be invariant on changing the frame; therefore, we have

$$\begin{gathered}\omega^{'}\left[t^{'} - \frac{1}{c}(x^{'}\cos\alpha^{'} + y^{'}\sin\alpha^{'})\right] = \\ = \omega\left[t - \frac{1}{c}(x\cos\alpha + y\sin\alpha)\right] \ ;\end{gathered} \tag{3.74}$$

making use of the Lorentz transformations we get

$$\omega^{'} = \omega\frac{1-\beta\cos\alpha}{\sqrt{1-\beta^2}} \tag{3.75}$$

and equations (3.69) and (3.70) of the aberration of light. Equation (3.75) is the Doppler effect: the frequency decreases on moving away from the source and increases on coming towards the source. In addition, there is a transverse Doppler effect for $\cos\alpha = 0$. Comparing equations (3.73) and (3.75) we get

$$\frac{\mathcal{E}^{'}}{\omega^{'}} = \frac{\mathcal{E}}{\omega} \ . \tag{3.76}$$

3.7.4 Fresnel drag

Let light be propagated in a medium which moves with velocity v in the same direction as light. The velocity of light in the medium is c/n, where n is the refraction index of the medium. We are interested in the velocity $v_x = V$ of light in the laboratory frame. Equation (3.64) gives

$$V = \frac{c/n + v}{1 + v/cn} \simeq \frac{c}{n} + v(1 - 1/n^2) \ . \tag{3.77}$$

The refraction index corresponds to the wavelength in the moving medium; from equation (3.75) we have

$$\omega^{'} = \omega\sqrt{\frac{1-\beta}{1+\beta}} \simeq \omega(1-\beta) \tag{3.78}$$

and the wavelengths

$$\lambda^{'} \simeq \lambda(1 + vn/c) \; ; \tag{3.79}$$

we have

$$\frac{c}{n(\lambda')} = \frac{c}{n} - \frac{c}{n^2}\frac{dn}{d\lambda}\frac{\lambda vn}{c} = \frac{c}{n} - \frac{\lambda v}{n}\frac{dn}{d\lambda} \; ; \tag{3.80}$$

inserting this correction into equation (3.77) we get

$$V \simeq \frac{c}{n} + v\left(1 - 1/n^2 - \frac{\lambda}{n}\frac{dn}{d\lambda}\right) \; ; \tag{3.81}$$

a moving medium drags the light along.

3.7.5 Spots, beaming and focusing

Let S be a source of light in motion. For short distances it emits a parallel (paraxial) beam, which does not exhibit aberration. For longer distances, the dispersion gives a spot $\Delta S^{'}$, which should be compared with the spot ΔS, corresponding to the source at rest. From equation (3.70) we have

$$\sin\alpha^{'} = \sqrt{1-\beta^2}\frac{\sin\alpha}{1+\beta\cos\alpha} \tag{3.82}$$

and

$$\sin\alpha^{'} = \sqrt{1-\beta^2}\frac{\sin\alpha}{1+\beta} = \sqrt{\frac{1-\beta}{1+\beta}}\sin\alpha \tag{3.83}$$

for small values of α and $\alpha^{'}$ (the sign of the velocity is reversed in comparison with equations (3.69) and (3.70)). The ratio of the two spots is given by

$$\frac{\Delta S^{'}}{\Delta S} = \frac{\alpha^{'2}}{\alpha^2} = \frac{1-\beta}{1+\beta} \; ; \tag{3.84}$$

the light beams in the forward direction; in the ultra-relativistic limit $\Delta S^{'}/\Delta S \simeq 1/2\gamma^2$. Therefore, the flux of a laser beam (energy per cross-sectional area) increases by a factor $2\gamma^2$, as a result of the light beaming. At the same time, the pulse duration of the laser decreases by a factor γ, so the laser intensity (flow of energy, *i.e.* energy per unit cross-sectional area and per unit time) increases by a factor $2\gamma^3$.

A pulsed, focused laser beam can be viewed as a succession of "slices" of light flying through the geometrical-optics frame of focalized light, the thickness of the slices being dictated by the pulse duration. Visualized on a screen moving towards the focus F, the spot appears of shorter transverse dimensions, due to the aberration of light. According to the laws of the geometrical optics, we deduce that the transverse dimensions of the focus F are shorter, in keeping with the paradigm that we cannot focalize light over distances shorter than the wavelength (the wavelength gets shorter on moving towards the focus). Similarly, on moving away from the focus, the wavelength gets longer and the spot gets wider, in accordance with a wider focus; again, we cannot focalize light on distances shorter than the wavelength.

3.8 Charge in an electromagnetic plane wave

Let a pointlike charge q with mass m be acted by an electromagnetic plane wave given by the vector potential

$$\mathbf{A} = (0,\ a\cos(\omega t - kx),\ 0)\ ; \tag{3.85}$$

the electric field is

$$\mathbf{E} = -\frac{1}{c}\frac{\partial \mathbf{A}}{\partial t} = (0,\ \frac{\omega}{c}a\sin(\omega t - kx),\ 0) \tag{3.86}$$

and the magnetic field is

$$\mathbf{H} = curl\mathbf{A} = (0,\ 0,\ ka\sin(\omega t - kx))\ ; \tag{3.87}$$

a is the amplitude of the vector potential, ω is the frequency and $k = \omega/c$ is the wavevector of the wave. The equation of motion reads

$$\frac{d}{dt}\mathbf{p} = \frac{d}{dt}\frac{m\mathbf{v}}{\sqrt{1 - v^2/c^2}} = q\mathbf{E} + \frac{q}{c}\mathbf{v}\times\mathbf{H}\ ; \tag{3.88}$$

first we note that $d/dt = \partial/\partial t + \mathbf{v}\partial/\partial\mathbf{r}$; the last term here has no effect on $\mathbf{p}$, since $\mathbf{v}$ does not depend on $\mathbf{r}$; next, we have $\mathbf{E} = -(1/c)(\partial\mathbf{A}/\partial t)$, such that equation (3.88) becomes

$$\frac{\partial}{\partial t}\left(\frac{m\mathbf{v}}{\sqrt{1-v^2/c^2}}+\frac{q}{c}\mathbf{A}\right)= \\ =\frac{q}{c}\mathbf{v}\times\mathbf{H}=\frac{q\omega}{c^2}a(v_y,\,-v_x,\,0)\sin(\omega t-kx)\ ; \tag{3.89}$$

or

$$\frac{\partial}{\partial t}\frac{mv_x}{\sqrt{1-v^2/c^2}}=\frac{q\omega}{c^2}av_y\sin(\omega t-kx)\ , \\ \frac{\partial}{\partial t}\left(\frac{mv_y}{\sqrt{1-v^2/c^2}}+\frac{q}{c}a\cos(\omega t-kx)\right)= \\ =-\frac{q\omega}{c^2}av_x\sin(\omega t-kx)\ , \\ \frac{\partial}{\partial t}\frac{mv_z}{\sqrt{1-v^2/c^2}}=0\ ; \tag{3.90}$$

we can see that the motion is in the xy-plane (we take $v_z=0$ at the initial moment); the above equations can also be written as

$$\frac{\partial}{\partial t}\frac{\beta_x}{\sqrt{1-\beta^2}}=\frac{q\omega}{mc^2}a\beta_y\sin(\omega t-kx)\ , \\ \frac{\partial}{\partial t}\frac{\beta_y}{\sqrt{1-\beta^2}}=\frac{q\omega}{mc^2}a\sin(\omega t-kx)-\frac{q\omega}{mc^2}a\beta_x\sin(\omega t-kx)\ , \tag{3.91}$$

where we have introduced $\boldsymbol{\beta}=\mathbf{v}/c$; with $\beta_\pm=v_x\pm iv_y$ we get

$$\frac{\partial}{\partial t}\frac{\beta_+}{\sqrt{1-\beta^2}}=\frac{iq\omega}{mc^2}a(1-\beta_+)\sin(\omega t-kx) \tag{3.92}$$

(and $\beta_-=\beta_+^*$).

Usually, the parameter $\varepsilon=qa/mc^2$ is very small; it is a ratio of energies which can also be written as $qe/mc\omega$, where e is the amplitude of the electric field; in which case it is a ratio of accelerations. Consequently, we can seek β_+ as a series of powers of ε. We get easily

$$\beta_+=C-i\varepsilon\cos(\omega t-kx)+\tfrac{1}{2}\varepsilon^2\cos^2(\omega t-kx)+...\ , \\ \beta_-=C+i\varepsilon\cos(\omega t-kx)+\tfrac{1}{2}\varepsilon^2\cos^2(\omega t-kx)+... \tag{3.93}$$

and

$$\beta_x = C + \frac{1}{2}\varepsilon^2 \cos^2(\omega t - kx) + \ldots \ , \ \beta_y = -\varepsilon \cos(\omega t - kx) + \ldots \ ; \quad (3.94)$$

the constant C can be determined from the condition $\overline{\beta}_x = 0$, where $\overline{\beta}_x$ is the temporal average of β_x; we get

$$\beta_x = \frac{1}{4}\varepsilon^2 \cos 2(\omega t - kx) + \ldots \ , \ \beta_y = -\varepsilon \cos(\omega t - kx) + \ldots \quad (3.95)$$

and

$$x = \frac{c}{8\omega}\varepsilon^2 \sin 2(\omega t - kx) + \ldots \ , \ y = -\frac{c}{\omega}\varepsilon \sin(\omega t - kx) + \ldots \ . \quad (3.96)$$

This is essentially a non-relativistic approximation, where x is limited by a second-order amplitude, in comparison with y which is a first-order quantity; the former corresponds to the action of the magnetic field, the latter arises from the action of the electric field; we may take approximately

$$x \simeq \frac{c}{8\omega}\varepsilon^2 \sin 2\omega t \ , \ y \simeq -\frac{c}{\omega}\varepsilon \sin \omega t \quad (3.97)$$

which describes an elongated ellipse. The (relativistic) energy $\mathcal{E} \simeq mc^2 \left[1 + \frac{1}{2}\varepsilon^2 \cos^2(\omega t - kx)\right]$ includes mainly the contribution of the electric field.

In the opposite, ultra-relativistic limit $\varepsilon \gg 1$ we may take $\beta \simeq 1 - A^2/2\varepsilon^2$; equation (3.92) becomes

$$\frac{\partial}{\partial t}e^{i\varphi} = iA\omega(1 - e^{i\varphi}) \sin(\omega t - kx) \ , \quad (3.98)$$

where A is a constant and $\beta_+ = \beta e^{i\varphi} \simeq e^{i\varphi}$. We get easily $\varphi = \pi + 2A\cos(\omega t - kx)$; the constant A can be determined from the same condition $\overline{\beta}_x = 0$. We can see that in the ultra-relativistic case the charge acted by the electromagnetic wave acquires a drift along the direction of the wave propagation. In general, the charge can emit radiation with higher-order harmonics; in the ultra-relativistic case the motion can be dominated by the radiation reaction, which points to a well-known limitation of the classical electromagnetism. High intensities of the electromagnetic field are usually attainable at the expense of high frequencies, such that the parameter $\varepsilon = qa/mc^2 = qe/mc\omega$ is still very small. The problem of high intensities of the electromagnetic field leads to the problem of high quanta, very localized both in space and time.

A pointlike charge is a quantum particle, even a relativistic quantum particle, at least because it may have a spin. In an electromagnetic field it acquires a momentum qa/c, which gives a contribution $\simeq q^2a^2/mc^2$ to energy; the Zeeman energy is of the order of $(\hbar q/mc)H \simeq (\hbar\omega)(qa)/mc^2$, so that the energy qa should be compared with energy quanta $\hbar\omega$. Usually, $qa \gg \hbar\omega$, so we may use the relativistic Schrodinger equation (Klein-Gordon equation) (or even non-relativistic Schrodinger equation if $q^2a^2 \ll mc^2$). For high field intensities and high frequencies the situation can be reversed, and the spin contribution can be high. Even so, the plane wave can allow a quasi-classical description for the spin motion, which is decoupled, to a convenient extent, from the kinematic motion of the particle; which allows again the use of the Klein-Gordon equation.

Since $\mathcal{E} = \sqrt{m^2c^4 + p^2c^2}$, $\mathcal{E} = \sqrt{m^2c^4 + (\mathbf{P} - q\mathbf{A}/c)^2c^2} + q\Phi$ and $(\mathcal{E} - q\Phi)^2 = (\mathbf{P} - q\mathbf{A}/c)^2c^2 + m^2c^4$, with $\mathcal{E} = i\hbar\partial/\partial t$ and $\mathbf{P} = -i\hbar\partial/\partial\mathbf{r}$, we have the Klein-Gordon equation

$$(i\hbar\partial/\partial t)^2\psi = (i\hbar\partial/\partial\mathbf{r} + q\mathbf{A}/c)^2c^2\psi + m^2c^4\psi \tag{3.99}$$

for the point charge in the electromagnetic plane wave; $\mathbf{p}$ is the mechanical momentum, $\mathbf{P}$ is the momentum in the electromagnetic field and Φ (in our case $= 0$) is the scalar potential of the field. The classical counter-part is the Hamilton-Jacobi equation $(\partial S/\partial t + q\Phi)^2 = (\partial S/\partial\mathbf{r} - q\mathbf{A}/c)^2c^2 + m^2c^4$ for the action S, with $\mathcal{E} = -\partial S/\partial t$ and $\mathbf{P} = \partial S/\partial\mathbf{r}$, which can be used to solve (exactly) the motion of the particle in the electromagnetic plane wave ($\Phi = 0$). With $\mathbf{A} = (0, A, 0)$ equation (3.99) becomes

$$\begin{aligned}(i\hbar\partial/\partial t)^2\psi = (i\hbar\partial/\partial x)^2c^2\psi + (i\hbar\partial/\partial z)^2c^2\psi + \\ +(i\hbar\partial/\partial y + qA/c)^2c^2\psi + m^2c^4\psi \ ;\end{aligned} \tag{3.100}$$

it is tempting to try to eliminate A by taking

$$\psi = e^{\frac{iq}{\hbar c}\int^y Ady'}\Phi = e^{\frac{iq}{\hbar c}yA}\Phi \tag{3.101}$$

(gauge invariance); unfortunately, this exponential factor introduces a y-dependence through $\partial/\partial x$, which complicates the matters. We seek solution in the form

$$\psi = e^{-i\mathcal{E}_0 t/\hbar + i\mathbf{pr}/\hbar}\Phi(x-ct) \ , \tag{3.102}$$

where $\mathcal{E}_0$ and $\mathbf{p} = (p_1, p_2, p_3)$ are constants; we may choose $\mathcal{E}_0^2 = m^2c^4 + p^2c^2$, and get

$$\Phi = const \cdot e^{-\frac{iq}{\hbar c}\int^{x-ct}\frac{p_2 A - qA^2/2c}{\mathcal{E}_0/c - p_1}d\xi} \ ; \tag{3.103}$$

the energy $(i\hbar\partial\psi/\partial t = \mathcal{E}\psi)$ and momentum $(-i\hbar\partial\psi/\partial x = P_1\psi)$ are

$$\mathcal{E} = \mathcal{E}_0 - q\frac{p_2 A - \frac{qA^2}{2c}}{\mathcal{E}_0/c - p_1} \ , \ P_1 = p_1 - \frac{q}{c}\frac{p_2 A - \frac{qA^2}{2c}}{\mathcal{E}_0/c - p_1} \ . \tag{3.104}$$

We can consider the particle initially at rest: $\mathcal{E}_0 = mc^2$, $\mathbf{p} = 0$; then,

$$\mathcal{E} = mc^2 + \frac{q^2A^2}{2mc^2} \ , \ P_1 = \frac{q^2A^2}{2mc^3} \ ; \tag{3.105}$$

averaging over time, we may view the particle as acquiring an effective mass in the field, $m^* = m(1+q^2a^2/4m^2c^4)$. It is worth noting that the solution given by equation (3.103) looks like a quasi-classical solution, with action of the order of $(q^2A^2/mc^2)t$, $(p_2qA/mc^2)x$, etc. A current $J = qP_1c/\sqrt{m^2c^2 + P_1^2}$ gives rise to electromagnetic radiation.

These results coincide partialy with Volkov states of a Dirac charge in an external electromagnetic plane wave (the spin contributes an A-change in the amplitude of the spinor).[5] The time-dependence of the energy levels obtained here applies also to the negative-energy (positron) states; therefore, the vacuum is polarized in the presence of an external electromagnetic plane wave, as it is for (quasi-) uniform fields; modified Maxwell equations can be obtained also in this case.

[5] D. Volkov, "Ueber eine Klasse von Loesungen der Diracschen Gleichung", Z. Phys. **94** 250 (1935).

4 Charges and Fields

4.1 Charge in electromagnetic field

4.1.1 Equations of motion

Let us consider the motion of a charge q with mass m in an electromagnetic field generated by the vector potential $A = A_z = A_0 \cos\frac{\omega}{c}(ct - x)$, propagating along the x-direction; ω denotes the radiation frequency and c is the speed of light in vacuum. With usual notations the equation of motions are

$$mc\frac{du^i}{ds} = \frac{q}{c}F^{ik}u_k \ , \tag{4.1}$$

where u^i is the relativistic velocity. The non-vanishing components of the field intensity (electric field $\mathbf{E} = -(1/c)\partial\mathbf{A}/\partial t$, magnetic field $\mathbf{H} = curl\mathbf{A}$) are $F^{03} = F^{13} = -F^{31} = -F^{32} = -E$, where $E = E_z = \frac{\omega}{c}A_0 \sin\frac{\omega}{c}(ct - x)$ (and $H_y = -E$). Consequently, the equations of motion are re-written as

$$\begin{gathered} mc\frac{du^0}{ds} = \frac{qE}{c}u^3 \ , \ mc\frac{du^1}{ds} = \frac{qE}{c}u^3 \ , \\ mc\frac{du^3}{ds} = \frac{qE}{c}(u^0 - u^1) \end{gathered} \tag{4.2}$$

($u^2 = 0$), where $s = k_i x^i = ct - x$ is the relativistic length and $k_i = (1, -1, 0, 0)$ is the wavevector; we get immediately $u^0 - u^1 = const = 1$, for a charge initially at rest at the origin (we recall $u^0 = \gamma$, $\mathbf{u} = \gamma\mathbf{v}/c$, where $\gamma = (1 - v^2/c^2)^{-1/2}$); it follows

$$\begin{gathered} u^1 = \frac{q^2A^2}{2m^2c^4} \ , \ u^0 = \gamma = 1 + \frac{q^2A^2}{2m^2c^4} \ , \\ u^3 = -\frac{q}{mc^2}A \end{gathered} \tag{4.3}$$

(we can check $u_i u^i = 1$). The velocities are given by

$$v_x = c\frac{q^2A^2/2m^2c^4}{1+q^2A^2/2m^2c^4} \ , \ v_z = -c\frac{qA/mc^2}{1+q^2A^2/2m^2c^4} \tag{4.4}$$

and the energy is

$$\mathcal{E} = mc^2\gamma = mc^2 + \frac{q^2A^2}{2mc^2} \ . \tag{4.5}$$

We can see that for high fields the charge reaches the ultra-relativistic limit along the direction of propagation of the radiation field ($v_x \to c$, $v_z \to 0$). From $u^i = dx^i/ds$ we get the coordinates

$$\begin{aligned} x &= \tfrac{q^2A_0^2/4m^2c^3}{1+q^2A_0^2/4m^2c^4}\left[t + \tfrac{1}{2\omega}\sin 2\tfrac{\omega}{c}(ct-x)\right] \ , \\ z &= -\tfrac{qA_0}{mc\omega}\sin\tfrac{\omega}{c}(ct-x) \ . \end{aligned} \tag{4.6}$$

4.1.2 Hamilton-Jacobi equation

The same results can be obtained from the Hamilton-Jacobi equation

$$g^{ij}\left(\frac{\partial S}{\partial x^i} + \frac{q}{c}A_i\right)\left(\frac{\partial S}{\partial x^j} + \frac{q}{c}A_j\right) = m^2c^2 \ , \tag{4.7}$$

where the metrics is $g^{ij} = (1,-1,-1,-1)$, $\mathcal{E} = -\partial S/\partial t$ is the energy, $\mathbf{P} = \mathbf{p} + \frac{q}{c}\mathbf{A} = \partial S/\partial \mathbf{R}$ is the generalized momentum and S is the mechanical action ($\mathbf{R}$ is the position vector). We seek a solution of the form $S = -f_i x^i + F(s)$, where f^i is the momentum of a free particle, $f_i f^i = m^2c^2$ (the initial momentum, the interaction is introduced adiabatically). Since $k_i k^i = 0$ and $k^i A_i = 0$ (transversality condition, with zero scalar potential), we get

$$F' = -\frac{q}{c\alpha}f_i A^i + \frac{q^2}{2\alpha c^2}A_i A^i \ , \tag{4.8}$$

where $\alpha = k_i f^i$. The solution is well known.[1] Since $\alpha = f^0 - f^1$ and $f^i = (f^0, f^1, \boldsymbol{\kappa})$, $f_i f^i = (f^0)^2 - (f^1)^2 - \kappa^2 = m^2c^2$, we get

$$f^0 = \frac{1}{2}\alpha + \frac{m^2c^2+\kappa^2}{2\alpha} \ , \ f^1 = -\frac{1}{2}\alpha + \frac{m^2c^2+\kappa^2}{2\alpha} \tag{4.9}$$

[1] L. Landau and E. Lifshitz, *Course of Theoretical Physics*, vol. 2, *The Classical Theory of Fields*, Elsevier (2004).

and

$$S = -\frac{1}{2}\alpha(ct+x) - \frac{m^2c^2+\kappa^2}{2\alpha}s + \boldsymbol{\kappa}\mathbf{r} +$$
$$+\frac{q}{c\alpha}\int^s ds'\boldsymbol{\kappa}\mathbf{A} - \frac{q^2}{2c^2\alpha}\int^s ds' A^2 \ , \qquad (4.10)$$

where $\boldsymbol{\kappa}$ is the transverse momentum and $\mathbf{r} = (y, z)$ is the transverse position vector. The derivatives of S with respect to the momentum $\boldsymbol{\kappa}$ and the parameter α are constants of motion; they may be set equal to zero. We get

$$y = \frac{\kappa_y}{\alpha}s - \frac{q}{c\alpha}\int^s ds' A_y \ ,$$
$$z = \frac{\kappa_z}{\alpha}s - \frac{q}{c\alpha}\int^s ds' A_z \ , \qquad (4.11)$$
$$x = \frac{1}{2}\left(\frac{m^2c^2+\kappa^2}{\alpha^2} - 1\right)s - \frac{q}{c\alpha^2}\int^s ds'\boldsymbol{\kappa}\mathbf{A} + \frac{q^2}{2c^2\alpha^2}\int^s ds' A^2$$

and

$$p_y = \kappa_y - \frac{q}{c}A_y \ ,$$
$$p_z = \kappa_z - \frac{q}{c}A_z \ , \qquad (4.12)$$
$$p_x = -\frac{1}{2}\alpha + \frac{m^2c^2+\kappa^2}{2\alpha} - \frac{q}{c\alpha}\boldsymbol{\kappa}\mathbf{A} + \frac{q^2}{2c^2\alpha}A^2 \ ;$$

the energy is given by

$$\mathcal{E} = c(\alpha + p_x) \ ; . \qquad (4.13)$$

we note the linear dependence on the momentum. For a charge initially at rest at $x = y = z = 0$ at the initial moment of time $t = 0$ ($\boldsymbol{\kappa} = 0$, $f^1 = 0$), and a vector potential $A = A_z = A_0\cos(\omega t - kx) = A_0\cos\frac{\omega}{c}(ct - x) = A_0\cos\frac{\omega}{c}s$ (linear polarization) we get $\alpha = mc$ ($\alpha^2 = m^2c^2$) and

$$z = -\frac{qA_0}{mc^2}\lambda\sin(\omega t - kx) \ , \ y = 0 \ ,$$
$$x = \frac{q^2A_0^2/4m^2c^4}{1+q^2A_0^2/4m^2c^4}\left[ct + \frac{\lambda}{2}\sin 2(\omega t - kx)\right] \ , \qquad (4.14)$$
$$p_x = \frac{q^2A_0^2}{2mc^3}\cos^2(\omega t - kx) \ ,$$
$$p_z = -\frac{qA_0}{c}\cos(\omega t - kx) \ , \ p_y = 0$$

and

$$\mathcal{E} = mc^2 + \frac{q^2 A_0^2}{2mc^2} \cos^2(\omega t - kx) \ , \tag{4.15}$$

where $\lambda = c/\omega$ is the radiation wavelength. For time-averaged quantities an effective mass can be defined by

$$\mathcal{E}^2/c^2 - p_x^2 = m^{*2}c^2 \ , \ m^{*2} = m^2 \left(1 + \frac{q^2 A_0^2}{2m^2 c^4}\right) \ , \tag{4.16}$$

by analogy with the free particle. Equations (4.14) and (4.15) coincide with equations (4.6), (4.3) and (4.5), respectively (we recall $p^i = mcu^i$).

We can see that, apart from oscillations, the charge exhibits a drift motion along the direction of propagation of the wave, governed by the ratio of the field energy qA_0 to the rest energy mc^2. There exist also solutions with negative energy $\mathcal{E} = c(-\alpha - p_x)$ and negative momentum p_x (corresponding to $\alpha = -mc$ in the above calculations), which move in opposite direction. The oscillations of the charge give rise to radiation (Compton effect); the Lorentz reaction force is extremely small; in the limit of high fields, where x is approximately ct, the phase is very small and the oscillations and the radiation are fadding out.

We introduce the parameter

$$\eta = \frac{qA_0}{2mc^2} \tag{4.17}$$

(or $\eta = qE_0/2m\omega c$, where $E_0 = \omega A_0/c$ is the electric field, $q > 0$); from equation (4.14) the drift velocity of the charge is given approximately by

$$v_x \simeq \frac{\eta^2}{1+\eta^2} c \ , \tag{4.18}$$

and the coordinates x and z can be written as

$$\begin{aligned} x &\simeq v_x t + \tfrac{1}{2}\tfrac{v_x \lambda}{c} \sin 2(\omega - kv_x)t = \\ &= vt + \tfrac{1}{2}\tfrac{v_x \lambda}{c} \sin 2\omega(1 - \tfrac{v_x}{c})t \ , \\ z &= -2\eta\lambda \sin \omega(1 - \tfrac{v_x}{c})t \ ; \end{aligned} \tag{4.19}$$

a current $J = qv_x$ occurs, along the direction of propagation of the radiation. The charge gets rapidly ultra-relativistic, as shown by equations (4.19).

Usually, the parameter η is small ($\eta \ll 1$). However, a high laser intensity $I = 10^{22}w/cm^2$, focalized in a pulse of dimension d, generates an electric field $E_0 = \sqrt{4\pi I/c} \simeq 10^{10}esu$ ($E_0^2 d^3 = 4\pi I d^2 \tau = 4\pi I d^3/c$, where $\tau = d/c$ is the duration of the pulse). For a linear dimension of the pulse $d = 10\mu m$, $\tau \simeq 30fs$; the intensity $10^{22}w/cm^2$ corresponds to $\simeq 10Pw$, *i.e.* a total energy per pulse $\simeq 300w$. The vector potential is $A_0 = cE_0/\omega = 10^{-5}E_0 = 10^5 esu$ for the optical frequency $\omega \simeq 3\times 10^{15}s^{-1}$ ($\nu = \omega/2\pi$, $\lambda = 0.5\mu m$); the corresponding energy for an electron is $qA_0 = 4.8 \times 10^{-5}erg \simeq 30MeV$. This energy is much higher than the rest energy of the electron $mc^2 = 0.5MeV$, such that the ratio $\eta = qA_0/2mc^2 = 30$ is much larger than unity. It follows that an electron can be accelerated, during the short duration τ of the pulse, up to velocities close to the speed of light, along the direction of propagation of the radiation field, and up to energies of the order $1GeV$.

If the radiation is propagated in gaseous plasmas,[2] then a radiation pulse is a wavepacket; when focalized, it is a three-dimensional wavepacket which distributes the electrons over its surface, such as to compensate the radiation field. Under such circumstances, the charges are accelerated by the transport motion of the wavepacket (pulse; pulsed polariton[3]). The mechanism of charge acceleration presented here is distinct from, and complementary to the mechanism of charge acceleration in plasmas.

4.1.3 Damping force

We estimate here, by a direct method, the magnitude of the damping force and show that in realistic cases the effect of the damping is small

[2] M. Apostol, "Propagation of electromagnetic pulses through the surface of dispersive bodies", Roum. J. Phys. **58** 1298 (2013).

[3] M. Apostol and M. Ganciu, "Polaritonic pulse and coherent X- and gamma rays from Compton (Thomson) backscattering", J. Appl. Phys. **109** 013307 (2011).

The damping force is given by[4]

$$g^i = \frac{2q^3}{3mc^3}\frac{\partial F^{ik}}{\partial x^l}u_k u^l -$$
$$-\frac{2q^4}{3m^2c^5}\left[F^{il}F_{kl}u^k - (F_{kl}u^l)(F^{km}u_m)u^i\right] \ ; \tag{4.20}$$

it must be much smaller than $q^2/c\gamma a^2$, where $a = q^2/mc^2$ is the classical electromagnetic radius of the charge; this condition, which indicates the limits of applicability of the Electromagnetism, does not imply always that the damping (reaction) force is smaller than the electromagnetic (Lorentz) force; in the extreme limit $v \simeq c$ the damping may be higher than the driven force. Making use of the field given by equations (4.1) we get the reaction force

$$g^0 = \frac{2q^3}{3mc^3}(u^0 - u^1)\left\{\frac{\omega^2}{c^2}Au^3 + \frac{qE^2}{mc^2}\left[1 - (u^0 - u^1)u^0\right]\right\} \ ,$$
$$g^1 = \frac{2q^3}{3mc^3}(u^0 - u^1)\left\{\frac{\omega^2}{c^2}Au^3 + \frac{qE^2}{mc^2}\left[1 - (u^0 - u^1)u^1\right]\right\} \ , \tag{4.21}$$
$$g^3 = \frac{2q^3}{3mc^3}(u^0 - u^1)^2\left(\frac{\omega^2}{c^2}A - \frac{qE^2}{mc^2}u^3\right)$$

(we can check $u_i g^i = 0$). Making use of the four-velocity given in equation (4.3) we get

$$g^0 = -\frac{2a^2\omega^2A^2}{3c^3}\left(1 + a\frac{E^2}{2m\omega^2}\right) \ ,$$
$$g^1 = -\frac{2a^2\omega^2}{3c^3}\left[A^2 - \frac{c^2E^2}{\omega^2}\left(1 - \frac{q^2A^2}{2m^2c^4}\right)\right] \ , \tag{4.22}$$
$$g^3 = \frac{2aqA\omega^2}{3c^3}\left(1 - a\frac{E^2}{m\omega^2}\right) \ ,$$

or, in the limit of high-intensity fields,

$$g^0 = g^1 \simeq -\frac{a^3A^2E^2}{3mc^3} \ , \ g^3 \simeq -\frac{2a^2qAE^2}{3mc^3} \ . \tag{4.23}$$

[4]L. Landau and E. Lifshitz, *Course of Theoretical Physics*, vol. 2, *The Classical Theory of Fields*, Elsevier (2004).

Using the numerical data given above ($E_0 = 10^{10} esu$, $A_0 = 10^5 esu$, $\omega \simeq 3 \times 10^{15} s^{-1}$ ($\lambda = 0.5 \mu m$)) for an electron, we can see that the damping force ($g^0 = g^1 \simeq 10^{-12} g/s$, $g^3 \simeq 10^{-13} g/s$) is much smaller than the Lorentz force ($qEu_0^3/c \simeq 10^{-8} g/s$, $qE_0/c \simeq 10^{-10} g/s$; $\gamma \simeq 10^4$) and the validity condition $g^i \ll q^2/c\gamma a^2$ is satisfied ($q^2/c\gamma a^2 \simeq 10^{-8} g/s$).

4.1.4 "Stopping" point

Let us assume that initially the charge moves with momentum $f^1 = -bmc$, $b > 0$, in direction opposite to the laser beam. Then, from equations (4.11)-(4.13), we get

$$\begin{gathered}
x = \tfrac{1+2\eta^2-\beta^2}{1+2\eta^2+\beta^2} ct + \tfrac{\eta^2}{1+2\eta^2+\beta^2} \lambda \sin \tfrac{2\omega}{c} s \ , \\
z = -\tfrac{2\eta}{\beta} \lambda \sin \tfrac{\omega}{c} s \ , \\
p_x = \tfrac{mc}{2\beta} \left(1 - \beta^2 + 4\eta^2 \cos^2 \tfrac{\omega}{c} s\right) \ , \\
\mathcal{E} = \tfrac{mc^2}{2\beta} \left(1 + \beta^2 + 4\eta^2 \cos^2 \tfrac{\omega}{c} s\right) \ ,
\end{gathered} \tag{4.24}$$

where $\beta = b + \sqrt{1+b^2}$. We can see that there exists a value $\beta^2 = 1 + 2\eta^2$ for which

$$\begin{gathered}
x = \tfrac{\eta^2}{2(1+2\eta^2)} \lambda \sin \tfrac{2\omega}{c} s \ , \ p_x = \tfrac{mc}{\sqrt{1+2\eta^2}} \eta^2 \cos \tfrac{2\omega}{c} s \ , \\
\mathcal{E} = \tfrac{mc^2}{\sqrt{1+2\eta^2}} \left(1 + \eta^2 + 2\eta^2 \cos^2 \tfrac{\omega}{c} s\right) \ ,
\end{gathered} \tag{4.25}$$

where the drift motion ceases. If the initial action of the launched particle is maintained, the charge oscillates indefinitely at the stopping point, and radiates (mainly with frequencies ω and 2ω); if not, this is a turning point, where the charge is reflected. The associated stochastic photon emission in the so-called radiation-dominated regime (non-linear Compton effect) was recently considered by numerical simulations.

4.2 Quantum charge

As it is well known,[5] the wavefunction of a relativistic quantum charge in an electromagnetic plane wave is

$$\psi = \left[1 + \frac{q}{2c}\frac{(\gamma k)(\gamma A)}{(pk)}\right] e^{\frac{i}{\hbar}S} u \ , \tag{4.26}$$

where

$$S = -px - \int^{\xi} d\xi^{'} \left[\frac{q}{c(pk)}(pA) - \frac{q^2}{2(pk)c^2}A^2\right] \tag{4.27}$$

is the mechanical action, p is the charge momentum vector, k is the field wavevector, γ denotes the Dirac matrices and u is a constant bispinor; the notation (pk) stands for the scalar product of the two vectors p and k (Volkov wavefunction; compare with the classical action given by equation (4.10), where p is replaced by f). The electromagnetic wave depends only on the phase $s = (kx)$. We assume that the interaction is introduced adiabatically; then u is the solution of the free Dirac equation $(\gamma p - mc)u = 0$, *i.e.* it is the plane wave constant bispinor. Therefore, the wavefunction can be written as

$$\psi_{p\sigma} = \frac{1}{\sqrt{2\varepsilon V}}\left[1 + \frac{q}{2c}\frac{(\gamma k)(\gamma A)}{(pk)}\right] e^{\frac{i}{\hbar}S} u_{p\sigma} \ , \tag{4.28}$$

where $\sigma = \pm 1$ is the spin label, V is the volume, $\varepsilon = \left(m^2c^4 + p^2c^2\right)^{1/2}$ and $u_{p\sigma}$ is normalized such as $\overline{u}_{p\sigma}u_{p\sigma'} = 2mc^2\delta_{\sigma\sigma'}$, $\overline{u}_{-p\sigma}u_{-p\sigma'} = -2mc^2\delta_{\sigma\sigma'}$.[37] We give here these constant bispinors

$$\begin{aligned} u_{p\sigma} &= \begin{pmatrix} (\varepsilon + mc^2)^{1/2} w_\sigma \\ (\varepsilon - mc^2)^{1/2} (\boldsymbol{n\sigma}) w_\sigma \end{pmatrix} , \\ u_{-p-\sigma} &= \begin{pmatrix} (\varepsilon - mc^2)^{1/2} (\boldsymbol{n\sigma}) w_\sigma^{'} \\ (\varepsilon + mc^2)^{1/2} w_\sigma^{'} \end{pmatrix} , \end{aligned} \tag{4.29}$$

[5] D. M. Volkov, "Ueber eine Klasse von Loesungen der Diracschen Gleichung", Z. Phys. **94** 250 (1935); L. Landau and E. Lifshitz, *Course of Theoretical Physics*, vol. 4, V. B. Berestetskii, E. M. Lifshitz and L. P. Pitaevskii, *Quantum Electrodynamics*, Elsevier (2002).

where $\boldsymbol{n} = \mathbf{p}/p$, $w_\sigma' = -\sigma_y w_{-\sigma}$ and w_σ can be taken as the eigenvectors of the matrix σ_z; $\boldsymbol{\sigma}$ denotes the spin matrices. We note that $u^*_{p\sigma} u_{p\sigma'} = \overline{u}_{p\sigma}\gamma^0 u_{p\sigma'} = 2\varepsilon\delta_{\sigma\sigma'}$. The wavefunctions $\psi_{p\sigma}$ are orthonormal; also, the completeness of these wavefunctions can be proved. The phase S given by equation (4.27) is the classical mechanical action; it includes the drift motion of the charge along the propagation of the radiation wave, while the pre-exponential factor in the wavefunction given by equation (4.26) includes the oscillations of the charge in the radiation field. The current $j^i = c\overline{\psi}\gamma^i\psi$ (with the probability density $\rho = \overline{\psi}\gamma^0\psi = \psi^*\psi = j^0/c$) and the momentum $q^i = \psi^*_{p\sigma}(p^i - \frac{q}{c}A^i)\psi_{p\sigma}$ can be computed straightforwardly, and an effective mass m^* can be derived, identical with the effective mass in the classical case.

Let us consider an electromagnetic wave propagating along the x-direction, $k^i = (1,1,0,0)$, $s = k_i x^i = ct - x$, with the electromagnetic potentials $A^i = (0,0,0,A)$, $A = A_0 \cos\frac{\omega}{c}s$ (linear polarization) and a charge moving initially along the y-direction, $p^i = (mc, 0, p_y, 0)$, with a small p_y; then, the pre-exponential factor of the wavefunction in equation (4.26) is

$$1 + \frac{q}{2c}\frac{(\gamma k)(\gamma A)}{(pk)} = 1 - \frac{qA_0\cos\frac{\omega}{c}s}{2mc^2}\begin{pmatrix} -i\sigma_y & \sigma_z \\ \sigma_z & -i\sigma_y \end{pmatrix} \tag{4.30}$$

(we note that the matrix $(\gamma^0 - \gamma^1)\gamma^3$ entering the pre-exponential factor is twofold degenerate); the mechanical action given by equation (4.27) becomes

$$S = -\left(mc^2 + \frac{q^2A_0^2}{4mc^2}\right)t + p_y y + \frac{q^2A_0^2}{4mc^3}x - \frac{q^2A_0^2}{8mc^2\omega}\sin\frac{2\omega}{c}s\ . \tag{4.31}$$

The charge acquires an average drift momentum

$$P_x \simeq \frac{q^2A_0^2}{4mc^3} = mc\eta^2\ , \tag{4.32}$$

an energy

$$\mathcal{E} \simeq mc^2 + \frac{q^2A_0^2}{4mc^2} = mc^2(1+\eta^2) \tag{4.33}$$

and a phase velocity

$$v_x \simeq \frac{\mathcal{E}}{P_x} = \frac{1 + e^2A_0^2/4m^2c^4}{e^2A_0^2/4m^2c^4}c = \frac{1+\eta^2}{\eta^2}c\ , \tag{4.34}$$

which is higher than the speed of light in vacuum c; the group velocity may attain values as large as c.

We can see that for high-intensity fields, $\eta = qA_0/2mc^2 \gg 1$, the charge becomes ultra-relativistic; leaving aside the pre-exponential factor given by equation (4.30), the wavefunction can be written as

$$\begin{aligned} \psi_{p\sigma} &\simeq \frac{1}{\sqrt{2V}} \begin{pmatrix} w_\sigma \\ \sigma_y w_\sigma \end{pmatrix} e^{\frac{i}{\hbar}S} , \\ \psi_{-p-\sigma} &\simeq -\frac{1}{\sqrt{2V}} \begin{pmatrix} w_{-\sigma} \\ \sigma_y w_{-\sigma} \end{pmatrix} e^{-\frac{i}{\hbar}S} , \end{aligned} \tag{4.35}$$

where

$$S \simeq -\frac{q^2A_0^2}{4mc^2}t + \frac{q^2A_0^2}{4mc^3}x = -\frac{q^2A_0^2}{4mc^3}(ct-x) = -mc\eta^2(ct-x) \; ; \tag{4.36}$$

(the two bispinors are not independent, which corresponds to the ultra-relativistic case). The extremely fast oscillations produced by the large phase S indicate that we may view the charge as being in the quasi-classical limit (ray approximation); for practical purposes the charge may be viewed as a classical charge. A similar result can be obtained by taking the limit $m \rightarrow 0$, either formally in the Volkov wavefunction given by equation (4.26), or in the initial Dirac equation (Weyl equation, with only one spinor). We can check that the current is $\overline{\psi}_{p\sigma}\gamma^i\psi_{p\sigma} = \frac{1}{V}(1,1,0,0)$, corresponding to a plane wave which describes an ultra-relativistic particle. The spinor $\psi_{-p-\sigma}$ corresponds to negative energy (and momentum). The negative-energy electrons in the Dirac Fermi sea get lower and lower (negative) energy (as if they would have a negative mass); such that the gap between the negative-energy states and positive-energy states is increased by radiation. Similar considerations are valid for pair creation in laser fields in the presence of a Coulomb potential (Bethe-Heitler process[6]). It is also worth noting that the accelerated charge "feels" not anymore the radiation for very strong fields; in the rest frame of an ultra-relativistic particle the electromagnetic fields are vanishing. From equation (4.36)

[6]H. Bethe an W. Heitler, "On the stopping of fast particles and on the creation of positive electrons", Proc. Roy. Soc. **A146** 83 (1934); W. Heitler, *The Quantum Theory of Radiation*, Dover, NY (1984).

we get the charge wavelength $\lambda \simeq \hbar/\eta^2 mc$; we can see that for large η the wavelength is much shorter than the Compton wavelength; it follows again that for many practical purposes we may use the classical approach for the accelerated charge. Within this approach small interference effects or rapidly varying spin dynamics are lost; in the limit of high-intensity fields these effects are small, so they can be disregarded. An ultra-relativistic particle is practically a "radiation" field; as such, it does not radiate, and does not "feel" the accelerating electromagnetic field.[7]

4.2.1 Standing wave

Let us consider now a classical relativistic charge in a standing electromagnetic wave with the vector potential

$$A = A_z = \frac{1}{2}A_0[\cos(\omega t - kx) + \cos(\omega t + kx)] = A_0 \cos\omega t \cos kx \quad (4.37)$$

(linear polarization); the frequency of the wave is in the optical range, $\nu = \omega/2\pi \simeq 10^{15}s^{-1}$, and the wavelength is $\lambda = 2\pi/k = c/\nu \simeq 3\times 10^{-5}cm = 0.3\mu m$. A charge in a standing wave spends there more time than the wave period $(1/\nu = 10^{-15}s)$; consequently, in the Hamilton-Jacobi equation

$$\frac{1}{c^2}(\partial S/\partial t)^2 = (gradS - \frac{q}{c}\mathbf{A})^2 + m^2c^2 \quad (4.38)$$

we may take the average of $\mathbf{A}$ and $\mathbf{A}^2$ with respect to the time; equation (4.38) becomes

$$\begin{aligned}\tfrac{1}{c^2}(\partial S/\partial t)^2 = (\partial S/\partial x)^2 + (\partial S/\partial y)^2 + (\partial S/\partial z)^2 + \\ + \tfrac{q^2A_0^2}{2c^2}\cos^2 kx + m^2c^2 \;;\end{aligned} \quad (4.39)$$

(a similar result holds for a circularly polarized wave). The solution of this equation is the mechanical action

$$S = \pm\frac{qA_0}{\sqrt{2}\omega}\cos kx + p_y y + p_z z - \mathcal{E}t \;, \quad (4.40)$$

[7] C. Miron and M. Apostol, "On the motion of an electric charge in high-intensity electromagnetic radiation", Roum. J. Phys. **62** 117 (2017).

where $p_{y,z}$ are constant transverse momenta and $\mathcal{E}$ is the energy, given by

$$\mathcal{E}^2 = m^2c^4 + \frac{1}{2}q^2A_0^2 + (p_y^2 + p_z^2)c^2 \ . \tag{4.41}$$

We note that the energy acquires the form of the energy of a free particle, at rest along the longitudinal x-direction (wave direction), with a renormalized mass. This indicates that the assumption of a relativistic classical charge in a standing electromagnetic wave is not warranted, as expected. In particular, we can see that the coordinate x is not determined.

Indeed, contradictions can appear from such an assumption. For instance, the longitudinal momentum is

$$P_x = p_x = \frac{\partial S}{\partial x} = \pm\frac{qA_0}{\sqrt{2}c}\sin kx \ , \tag{4.42}$$

whence, making use of $p_x = mv_x/(1 - v_x^2/c^2)^{1/2}$, we get a "velocity"

$$\frac{dx}{dt} = v_x = c\frac{\frac{qA_0}{\sqrt{2}mc^2}\sin kx}{\sqrt{1 + \frac{q^2A_0^2}{2m^2c^4}\sin^2 kx}} \tag{4.43}$$

and a "force"

$$\begin{aligned} \frac{dp_x}{dt} &= v_x\frac{dp_x}{dx} = \frac{1}{2}c\frac{d}{dx}\sqrt{m^2c^2 + p_x^2} = \\ &= mc^2k\frac{2\eta^2\sin kx\cos kx}{\sqrt{1+2\eta^2\sin^2 kx}} \ , \end{aligned} \tag{4.44}$$

which looks like a ponderomotive force. However, the motion under the action of such a force is meaningless. In particular, both the "velocity" and the "force" given by equations (4.43) and (4.44), respectively, vanish at $kx = n\pi$, where n is any integer.

The situation is different for a non-relativistic charge; in this case the energy reads

$$\mathcal{E} = \frac{1}{2m}(\mathbf{P} - \frac{q}{c}\mathbf{A})^2 = \frac{p_x^2}{2m} + \frac{p_y^2}{2m} + \frac{p_z^2}{2m} - \frac{q}{mc}p_zA + \frac{q^2}{2mc^2}A^2 \ , \tag{4.45}$$

or, taking the temporal average,

$$\mathcal{E} = \frac{p_x^2}{2m} + \frac{p_y^2}{2m} + \frac{p_z^2}{2m} + \frac{e^2 A_0^2}{4mc^2} \cos^2 kx \; ; \tag{4.46}$$

we can see that the charge moves in a periodic potential

$$U(x) = \frac{q^2 A_0^2}{4mc^2} \cos^2 kx \; ; \tag{4.47}$$

the standing wave acts as an optical lattice for the charge. It follows that the quantum-mechanical motion of a charge in a standing electromagnetic wave generates energy bands, as it is well known; and, in addition, a charge may suffer diffraction on a standing wave (Kapitza-Dirac effect[8]). The classical motion is confined within the potential wells centered at $kx = (2n+1)\pi/2$, according to the potential energy included in the expression of the hamiltonian given by equation (4.46), subject to a ponderomotive force $mc^2 k\eta^2 \sin 2kx$.

4.2.2 Compton effect

It is worthwhile commenting upon the Compton effect in a classical electromagnetic wave. It is instructive at this moment to have an estimation of the (mean) photon density in a laser pulse with moderate intensity $I = 10^{18} w/cm^2$; this intensity corresponds to an electric field of the order $E_0 \simeq \sqrt{4\pi I/c} = 10^8 esu$ (and a similar magnetic field). The energy density is of the order $w \simeq I/c = 10^{14} erg/cm^3$, with a density of photons $n \simeq 10^{25} cm^{-3}$ with energy, say, $\hbar\omega = 1eV$; the photon flow (flux) is $cn \simeq 10^{35}/cm^2 \cdot s$. Currently, electrons can be injected in an electromagnetic wave to give an electric current of, at most, the order $\simeq 100mA$, which corrresponds to $\simeq 10^{17}$ electrons per second (we may admit that such an electric current can be produced experimentally over a cross-sectional area $1cm^2$); it follows that we may have an electron flow $\simeq 10^{17}/cm^2 \cdot s$. We may see that the electron flow is much weaker than the photon flow. Therefore, we may conclude that the disruption of an electromagnetic wave by electron beams is unlikely and the Compton effect is not likely to

[8] P. L. Kapitza and P. A. M. Dirac, "The reflection of electrons from standing light waves", Proc. Cambridge Phil. Soc. **29** 297 (1933).

disturb appreciably either the electromagnetic wave or the electron dynamics. We may view the Compton scattering in an electromagnetic wave as a statistical-mechanical effect, where the mean free path of the electron is of the order of the mean separation distance between the photons ($\simeq 10^{-8}cm$), the Compton cross-section σ is of the order of the square of the classical electromagnetic radius of the electron ($a = e^2/mc^2 \simeq 2.8 \times 10^{-13}cm$), for the radiation wavelength $\simeq 3 \times 10^{-5}cm$ ($\hbar\omega = 1eV$).

In conclusion, we may say that the acceleration of a classical relativistic charge by a travelling high-intensity electromagnetic wave is analyzed here. Such a charge acquires quickly a fast drift motion and becomes ultra-relativistic, moving with energies in the GeV's range; under these circumstances it ceases to radiate and does not "feel" anymore the carrying wave. Also, it is shown that a relativistic quantum charge in high-intensity laser radiation (Volkov state) becomes rapidly "localized", such that a (quasi-) classical treatment is more adequate for many practical purposes. Similarly, it is shown that a charge in a standing electromagnetic wave behaves quantum-mechanically and non-relativistically. This situation leads to energy bands and charge diffraction on a standing electromagnetic wave (Kapitza-Dirac effect), as it is well known in an optical lattice.

4.3 Non-relativistic charge

4.3.1 Schroedinger equation

The motion of pointlike electric charges subject to high-intensity electromagnetic fields is currently receiving a great deal of attention, especially in connection with the development of high-power optic lasers.[9] Apart from charge acceleration, many other novel phenomena are envisaged, like vacuum polarization, vacuum breakdown, photon-photon scattering, electron-positron pair creation or non-linear quantum electrodynamics effects, as well as multiple Compton scattering, generation of higher harmonics, atomic and even nuclear effects. A quantum

[9] A. Di Piazza, C. Muller, K. Z. Hatsagortsyan and C. H. Keitel, "Extremely high-intensity laser interactions with fundamental quantum systems", Revs. Mod. Phys. **84** 1177 (2012).

relativistic charge in a plane wave of classical electromagnetic radiation in vacuum is described by the well-known Volkov wavefunction derived as early as 1930.[10] The classical relativistic Hamilton-Jacobi equation helps illuminating many interesting points of this problem.[11] The non-relativistic quantum charge in a moderately intense electromagnetic field exhibits acceleration along the direction of propagation of the radiation and oscillations, as well as its own radiation and, for bound states, ionization. Two particular features are related to these problems. One refers to the possible non-linearities brought about by an intense field of classical radiation, as a consequence of the photon high density and undetermined number of photons; this would be an important departure point from the usual treatment of electron-photon interaction in quantum electrodynamics. Another feature refers to the presence of many electrons in ionized gaseous plasmas (a usual experimental situation in laser physics), where the internal polarization field and plasma oscillations are present.

Let a non-relativistic particle with electric charge q and mass m be subject to a field of electromagnetic radiation with the vector potential $\mathbf{A}$, $div\mathbf{A} = 0$. The Schrodinger equation reads

$$i\hbar\frac{\partial\psi}{\partial t} = \frac{1}{2m}\left(\mathbf{P} - \frac{q}{c}\mathbf{A}\right)^2\psi + V\psi \ , \tag{4.48}$$

where $\mathbf{P} = -i\hbar\partial/\partial\mathbf{r}$ is the canonical (generalized) momentum and V is an external, static potential; c is the speed of light in vacuum and $\hbar$ is Planck's constant. If the particle is an electron and spin is considered, then Pauli's equation can be used instead of equation (4.48). The problem discussed here is to see whether, and in what conditions, a solution of the form

$$\psi = F\phi \tag{4.49}$$

is possible (exists), where ϕ satisfies the Schrodinger equation without

[10] D. M. Volkov, "Ueber eine Klasse von Loesungen der Diracschen Gleichung", Z. Phys. **94** 250 (1935); V. I. Ritus, "Quantum effects of the interaction of elementary particles with an intense electromagnetic field", J. Sov. Las. Res. **6** 497 (1985).

[11] L. Landau and E. Lifshitz, *Course of Theoretical Physics*, vol. 2, *The Classical Theory of Fields*, Elsevier (2004).

electromagnetic field,

$$i\hbar\frac{\partial\phi}{\partial t}=\frac{p^2}{2m}\phi+V\phi\ ,\ \mathbf{p}=-i\hbar\partial/\partial\mathbf{r}\ , \tag{4.50}$$

and the function F becomes unity ($F=1$) for $\mathbf{A}=0$.

We note immediately that if the field would derive from a gauge transformation, *i.e.* if $\mathbf{A}=grad\chi$, (*i.e.*, it would be a gauge field), there would exist a scalar potential $\varphi=-\frac{1}{c}\frac{\partial\chi}{\partial t}$, such that $\frac{1}{c}\frac{\partial\varphi}{\partial t}+div\mathbf{A}=0$; then $\mathbf{A}$ changes the momentum, $\mathbf{P}\rightarrow\mathbf{P}+\frac{q}{c}\partial\chi/\partial\mathbf{r}$ ($\partial/\partial\mathbf{r}\rightarrow\partial/\partial\mathbf{r}+\frac{iq}{c\hbar}\partial\chi/\partial\mathbf{r}$), an additional term $q\varphi=-\frac{q}{c}\frac{\partial\chi}{\partial t}$ appears in the hamiltonian and $\psi=e^{\frac{iq}{c\hbar}\chi}\phi$; the function F would be $F=e^{\frac{iq}{c\hbar}\chi}$ in this case. However, a radiation field, for instance of the form $A_x=A_y=0$, $A_z=A=A_0\cos\frac{\omega}{c}(ct-x)$, $\varphi=0$, where ω is the frequency of the radiation, is not a gauge field.

We note also that the problem formulated above is the non-relativistic counterpart of the relativistic charge in an electromagnetic wave, whose solution is the Volkov wavefunction; therefore, it may appear that a path to the solution of our problem here would be the non-relativistic limit of the Volkov wavefunction. This is not so, because in the non-relativistic limit the contribution of the anti-particles, which is included in the Volkov wavefunction, is lost. The situation is similar with the zero mass limit of the ultra-relativistic case, which cannot be taken directly on the Volkov wavefunction, because this wavefunction is not analytic in the particle mass.

Inserting equation (4.49) into equation (4.48) and making use of equation (4.50) we get

$$\begin{gathered}i\hbar\frac{\partial F}{\partial t}\phi=\frac{1}{2m}(p^2F)\phi-\frac{q}{mc}\mathbf{A}(\mathbf{p}F)\phi+\\+\frac{q^2}{2mc^2}A^2F\phi+\frac{1}{m}(\mathbf{p}\phi)(\mathbf{p}F)-\frac{q}{mc}\mathbf{A}(\mathbf{p}\phi)F\ ,\end{gathered} \tag{4.51}$$

where $\mathbf{p}=-i\hbar\partial/\partial\mathbf{r}$ and we use $div\mathbf{A}=0$. We take $A_x=A_y=0$, $A_z=A=A_0\cos\frac{\omega}{c}(ct-x)$ and equation (4.51) becomes

$$\begin{gathered}i\hbar cF^{'}\phi=-\frac{\hbar^2}{2m}F^{''}\phi+\frac{q^2}{2mc^2}A^2F\phi-\\-\frac{i\hbar}{m}(p_x\phi)F^{'}-\frac{q}{mc}A(p_z\phi)F\ ,\end{gathered} \tag{4.52}$$

where F is a function of $\xi = ct - x$ only. Introducing the (reduced) Compton wavelength $\lambda_c = \hbar/mc$ of the charge with mass m and the parameter $\eta = qA_0/2mc^2$, equation (4.52) can also be written as

$$iF'\phi = -\tfrac{1}{2}\lambda_c F''\phi + \tfrac{2\eta^2}{\lambda_c} F\phi\cos^2\tfrac{\omega}{c}\xi - i\lambda_c\left(\tfrac{p_x}{\hbar}\phi\right)F' - \\ -2\eta\left(\tfrac{p_z}{\hbar}\phi\right)F\cos\tfrac{\omega}{c}\xi \; . \tag{4.53}$$

The term $\lambda_c F''$ compared with F' shows the variation of F'over distances of the order λ_c; we may neglect such variations, as they imply relativistic corrections; in addition, the uncertainty principle implies a limitation to distances much larger than λ_c. For instance, for an electron $\lambda_c \simeq 3\times 10^{-11}cm$, which is such a short distance that the non-relativistic approximation becomes meaninglesss. Therefore we neglect the term $\lambda_c F''$ in equation (4.53). The term $\lambda_c(p_x\phi/\hbar)F'$ should be compared to $F'\phi$; it is of the order λ_c/λ_q, where λ_q is the characteristic wavelenghth in the potential V; for instance, for a bound state it is of the order of the characteristic wavelength of the bound state; for scattering, it is of the order of the scattering wavelength ($\hbar^2/m\lambda_q^2 \simeq |V|$; for a Coulomb potential it is of the order of the Bohr radius, $\hbar^2/m\lambda_q^2 \simeq q^2/\lambda_q$, $\lambda_q \simeq \hbar^2/mq^2$); for a free motion $\lambda_c/\lambda_q = v_x/c$, where v_x is the velocity along the x-direction. We can see that this term also can be neglected in equation (4.53). Therefore, we can write approximately

$$iF'\phi \simeq \frac{2\eta^2}{\lambda_c}F\phi\cos^2\frac{\omega}{c}\xi - 2\eta\left(\frac{p_z}{\hbar}\phi\right)F\cos\frac{\omega}{c}\xi \; . \tag{4.54}$$

The term $\eta(p_z\phi/\hbar)F$ may imply variations over large distances, such that it may bring relevant contributions. In the weak coupling limit $\eta \ll 1$ it brings the main contribution. However, for usual fields of interest (*i.e.* for fields which are not extremely weak) the last term in equation (4.54) may be neglected. Under these circumstances, the approximation which turns out to be relevant for determining the function F is the approximation of a free charge, where F depends only on the interaction and does not depend on the wavefunction ϕ.

In this context, we also note that the ansatz given by equation (4.49) may look too restrictive, since, in general, we expect the function F

to be a matrix. If we label the states ϕ by a suffix n, we would expect, in general, that the wavefunction ϕ_n be taken by interaction into a superposition of wavefunctions $F_{n'n}\phi_{n'}$. The assumption embedded into equation (4.49) amounts to what is known as the quasi-adiabatic hypothesis. It implies that the interaction varies slowly in comparison with the periods of transitions among the bound states of the charge (this may be not true for the ionization process).

4.3.2 Free particle

For a free particle ($V = 0$) $p_z\phi$ is a constant momentum multiplied by ϕ ($\phi \sim e^{\frac{i}{\hbar}p_z z}$); we get from equation (4.54)

$$F = e^{-i\frac{\eta^2}{\lambda_c}\xi - i\frac{\eta^2\lambda}{2\lambda_c}\sin\frac{2\omega}{c}\xi + i\frac{2\eta\lambda}{\lambda_q}\sin\frac{\omega}{c}\xi} , \tag{4.55}$$

where $\lambda = \omega/c$ is the wavelength of the radiation field.

For a free charge an interesting situation appears for moderately high fields; we may consider that the non-relativistic limit is preserved up to $\eta \simeq 1$; for an electron $\eta = 1$ corresponds to $A_0 = 10^3 esu$ and an electric field $E_0 \simeq 10^8 esu$ for an optical frequency $\omega = 10^{15}s^{-1}$ (by comparison the atomic fields are of the order $10^6 esu$); the intensity of the radiation in this case is $I = cE_0^2/4\pi \simeq 10^{18} w/cm^2$. The Compton wavelength for the electron is $\lambda_c \simeq 3 \times 10^{-11} cm$; we take the optical wavelength $\lambda = 1\mu m = 10^{-4} cm$ and the electron wavelength $\lambda_q = 1\text{Å} = 10^{-8} cm$ (corresponding to an initial energy $\simeq 3eV$ and an initial momentum $\simeq 10^{-19} g \cdot cm/s$). Under these circumstances we may neglect the last term in the phase of the function F in equation (4.55), and write

$$F \simeq e^{-i\frac{\eta^2}{\lambda_c}\xi - i\frac{\eta^2\lambda}{2\lambda_c}\sin\frac{2\omega}{c}\xi} . \tag{4.56}$$

Making use of

$$e^{iz\cos\varphi} = \sum_n i^n e^{in\varphi} J_n(z) , \tag{4.57}$$

where J_n are the Bessel functions, we can write

$$F = \sum_n e^{i\left(\frac{2n\omega}{c} - \frac{\eta^2}{\lambda_c}\right)\xi} J_{-n}(\eta^2\lambda/2\lambda_c) . \tag{4.58}$$

We can see that a free charge acquires a superposition of plane waves in the presence of the radiation field, with energies

$$\mathcal{E}_n = -2n\hbar\omega + \eta^2 mc^2 + \mathcal{E}_0 \ , \tag{4.59}$$

momenta (along the direction of propagation of the radiation)

$$p_n = -2n\hbar k + \eta^2 mc + p_{x0} \tag{4.60}$$

and very small weights $J_n(\eta^2\lambda/2\lambda_c) \simeq J_n(10^6)$ (for our numerical estimates); $\mathcal{E}_0$ is the initial energy, p_{x0} is the initial momentum along the direction of propagation of the radiation and $k = \omega/c$ is the wavevector of the radiation.

It is worth noting that the charge has a distribution of momentum (momentum density) along the direction of propagation of the radiation; it can be computed most conveniently from the wavefunction given by equation (4.56); we get

$$p(\mathbf{r}, t) = \frac{1}{V}\eta^2 mc(1 + \cos\frac{2\omega}{c}\xi) \ , \tag{4.61}$$

where V is the volume and p_{x0} is set equal to zero; the total momentum is

$$\int p(\mathbf{r}, t)d\mathbf{r} = \sum_n p_n J_n^2(\eta^2\lambda/2\lambda_c) = \eta^2 mc \ . \tag{4.62}$$

We can see that the charge is "accelerated" along the direction of propagation of the radiation, in the sense that it acquires a constant momentum $\eta^2 mc$ and a variable momentum corresponding to the oscillating term in equation (4.61); the momentum and energy of the particle depend on the time. Similarly, there exists a distribution of velocity

$$v(\mathbf{r}, t) = \frac{1}{V}\eta^2 c(1 + \cos\frac{2\omega}{c}\xi) \ . \tag{4.63}$$

We note that both the energy and the momentum scales of the charge in the radiation field are the relativistic quantities mc and mc^2, respectively, according to equations (4.59) and (4.61), although we are in the non-relativistic limit. This is due to the fact that the electromagnetic radiation (even in its classical limit) is essentially a relativistic entity; the non-relativistic limit is expressed by the inequality $\eta^2 \ll 1$. It is also worth noting that the higher harmonics appearing in equations (4.60) are not present in equation (4.61).

A charge which oscillates according to equation (4.61) radiates; however, for a charge distribution extended to the whole space this radiation is not observable (the radiating charge distribution absorbs its own radiation). In accordance with experimental situation we assume that the charge in the radiation field is confined in fact to a small spatial region, of linear dimension d, $d \gg \lambda$ (and λ_q). Inside this region we can view the charge as a wavepacket, localized at $\mathbf{r}_0$ and moving with velocity v given by equation (4.63); we neglect the motion along the other two directions. The wavepacket assumption is not necessary, though even for moderate values of η the wavelength of the wavefunction in equation (4.56) is of the order λ_c/η^2, which is much smaller than our distances of interest. This amounts to a quasi-classical approximation. Therefore, we assume a current density of the form

$$j_x(\mathbf{r},t) = \eta^2 qc\left[1+\cos\frac{2\omega}{c}(ct-x_0)\right]\delta(\mathbf{r}-\mathbf{r}_0) \ , \qquad (4.64)$$

as given by equation (4.63). We get immediately the vector potential

$$\begin{aligned} A_x &= \eta^2 q\left[1+\cos\tfrac{2\omega}{c}(ct-|\mathbf{r}-\mathbf{r}_0|-x_0)\right]\tfrac{1}{r} \simeq \\ &\simeq \eta^2 q\left[1+\cos\tfrac{2\omega}{c}(ct-r)\right]\tfrac{1}{r} \end{aligned} \qquad (4.65)$$

for $r \gg r_0$. From the gauge condition $\partial\Phi/c\partial t + div\mathbf{A} = 0$ we get the leading contribution $\Phi = (\eta^2 qx/r^2)\cos\frac{2\omega}{c}(ct-r)$. We compute now the fields according to $\mathbf{E} = -\partial\mathbf{A}/c\partial t - grad\Phi$, $\mathbf{H} = curl\mathbf{A}$. We get

$$\begin{aligned} E_x &= \tfrac{2\eta^2 q\omega}{c}\sin\tfrac{2\omega}{c}(ct-r)\tfrac{1}{r}(1-x^2/r^2) \ , \\ E_y &= -\tfrac{2\eta^2 q\omega}{c}\sin\tfrac{2\omega}{c}(ct-r)\tfrac{xy}{r^3} \ , \\ E_z &= -\tfrac{2\eta^2 q\omega}{c}\sin\tfrac{2\omega}{c}(ct-r)\tfrac{xz}{r^3} \end{aligned} \qquad (4.66)$$

and

$$\begin{aligned} H_x &= 0 \ , \ H_y = \tfrac{2\eta^2 q\omega}{c}\sin\tfrac{2\omega}{c}(ct-r)\tfrac{z}{r^2} \ , \\ H_z &= -\tfrac{2\eta^2 q\omega}{c}\sin\tfrac{2\omega}{c}(ct-r)\tfrac{y}{r^2} \ . \end{aligned} \qquad (4.67)$$

The Poyinting vector $\mathbf{S} = (c/4\pi)\mathbf{E} \times \mathbf{H}$ leads to the radiated differential intensity

$$dI = Sdf = H_y = \frac{\eta^4 q^2 \omega^2}{\pi c} \sin^2 \frac{2\omega}{c}(ct - r)(1 - x^2/r^2)do \ , \qquad (4.68)$$

where do is the element of the solid angle and $df = r^2 do$ is the element of area. We can see that the maximum radiated field is at the angle $\theta = \pi/2$, where $\cos\theta = x/r$; it is a "lateral" radiation. It is worth noting that this is a classical radiation with frequency 2ω, and an effective charge reduced by the factor η^2. If we consider the rather unphysical case of very weak fields, such that $\eta^2 \ll 1$, the function F given by equation (4.55) reduces to

$$F = e^{i\frac{2\eta\lambda}{\lambda_q} \sin \frac{\omega}{c}\xi} \ ; \qquad (4.69)$$

the charge radiates frequency ω, and it is renormalized by a factor $\sim (\lambda_c/\lambda_q)\eta$. However, the last term $i\frac{2\eta\lambda}{\lambda_q} \sin\frac{\omega}{c}\xi$ in the phase of the function F in equation (4.55) becomes important in comparison with the second term for very low fields ($\eta^2\lambda/\lambda_c \ll \eta\lambda/\lambda_q$, $E_0 \simeq 10^4 esu$): we may neglect it in our subsequent discussion.

4.3.3 Ionization

The theory described here offers the opportunity of formulating a simple model of ionization under the action of an electromagnetic wave. We assume a bound state $e^{-\frac{i}{\hbar}\mathcal{E}_b t}\varphi_b$ of a charge, where the $\mathcal{E}_b$-energy factor is shown explicitly, as, for instance, a peripheric electron in a heavy atom. Under the action of the electromagnetic radiation this state becomes $Fe^{-\frac{i}{\hbar}\mathcal{E}_b t}\phi_b$; it contains many states, among which a free particle state, which may be viewed as the ionization state; it may be taken as being of the form $\phi_i = \frac{1}{\sqrt{V}} e^{-\frac{i}{\hbar}\mathcal{E}_i t + i\mathbf{Qr}}$, where $\mathcal{E}_i$ is the energy and $\hbar\mathbf{Q}$ is the momentum of the free charge. Then, the amplitude of ionization is given by

$$a = \int d\mathbf{r}\phi_i^* Fe^{-\frac{i}{\hbar}\mathcal{E}_b t}\phi_b =$$

$$= \frac{e^{\frac{i}{\hbar}(\mathcal{E}_i - \mathcal{E}_b)t}}{\sqrt{V}} \sum_n \int d\mathbf{r} e^{-i\mathbf{Qr}}\phi_b e^{i\left(\frac{2n\omega}{c} - \frac{\eta^2}{\lambda_c}\right)\xi} J_{-n}(\eta^2\lambda/2\lambda_c) \ , \qquad (4.70)$$

where we use the function F given by equation (4.58) for moderately strong fields. We can insert in equation (4.70) the Fourier expansion of the function ϕ_b

$$\phi_b = \frac{1}{\sqrt{V}} \sum_{\mathbf{q}} c(\mathbf{q}) e^{i\mathbf{qr}} \ , \ c(\mathbf{q}) = \frac{1}{\sqrt{V}} \int d\mathbf{r} \phi_b e^{-i\mathbf{qr}} \ , \tag{4.71}$$

which leads to

$$\begin{gathered} a = e^{\frac{i}{\hbar}(\mathcal{E}_i - \mathcal{E}_b)t} \sum_n c(Q_x + 2n\omega/c - \eta^2/\lambda_c, \mathbf{Q}_t) \cdot \\ \cdot e^{i\left(\frac{2n\omega}{c} - \frac{\eta^2}{\lambda_c}\right)ct} J_{-n}(\eta^2\lambda/2\lambda_c) \ , \end{gathered} \tag{4.72}$$

where $\mathbf{Q}_t$ is the transverse momentum of the free particle in the (y, z)-plane.

First, we can check that for a very localized wavefunction ϕ_b, for which all the coefficients $c(\mathbf{q})$ may be taken equal to unity, the ionization amplitude $|a|^2 = 1$, since

$$1 = \left| e^{iz\cos\varphi} \right|^2 = \sum_{nn'} i^{n-n'} e^{i(n-n')\varphi} J_n(z) J_{n'(z)} \ ; \tag{4.73}$$

this indicates that such a very localized wavefunction (corresponding to a point particle) is not a bound state. For a reasonably bound state the coefficients $c(\mathbf{q})$ decrease with increasing q, such that we may assume that the main contribution to the summation in equation (4.72) comes from $\mathbf{Q}_t = 0$ and $Q_x + 2n_0\omega/c - \eta^2/\lambda_c \simeq 0$, where n_0 is the integral part of $(c/2\omega)(\eta^2/\lambda_c - Q_x)$; under these circumstances we may take $c(0) \simeq 1$ and the amplitude a becomes

$$\begin{gathered} a \simeq e^{\frac{i}{\hbar}(\mathcal{E}_i - \mathcal{E}_b)t} e^{-iQ_x ct} J_{-n_0}(\eta^2\lambda/2\lambda_c) \simeq \\ \simeq \sqrt{\frac{4\lambda_c}{\eta^2\lambda}} e^{\frac{i}{\hbar}(\mathcal{E}_i - \mathcal{E}_b)t} e^{-iQ_x ct} \cos\left[\eta^2\lambda/2\lambda_c - (n_0 + 1/2)\frac{\pi}{2}\right] \end{gathered} \tag{4.74}$$

for $\eta^2\lambda/2\lambda_c \gg 1$. In this context, we can note that the asymptotic formula

$$\begin{gathered} e^{iz\cos\varphi} \simeq \sqrt{\frac{2\pi}{z}} \sum_n \{ e^{iz} \delta(\varphi - \pi/4 + 2\pi n) + \\ + e^{-iz} \delta(\varphi + 3\pi/4 + 2\pi n) \} \ , \ z \gg 1 \end{gathered} \tag{4.75}$$

(Poisson formula) is not much more helpful in estimating the ionization amplitude.

It is worthwhile noting that the ionization process analyzed here in a classical electromagnetic wave is different from the ionization which proceeds by absorbing one, or more, radiation quanta (photoelectric effect), and it is also different from the ionization in static electric or magnetic fields, in quasi-static electric fields, or in the quasi-classical tunneling approximation.[12] In addition, we note that the result presented here refers especially to a periferic electron in a heavy atom.

4.3.4 Absence of quantum transitions

Let $e^{-\frac{i}{\hbar}\mathcal{E}_n t}\phi_n$ be a bound state of the charge; it is taken by interaction into $Fe^{-\frac{i}{\hbar}\mathcal{E}_n t}\phi_n$; a state $e^{-\frac{i}{\hbar}\mathcal{E}_{n'} t}\phi_{n'}$ is found in the state with interaction with a probabiliy amplitude

$$a_{nn'} = \int d\mathbf{r}\phi^*_{n'} F\phi_n e^{-i\omega_{nn'}t} = \sum_s J_s(\eta^2\lambda/2\lambda_c)\cdot \\ \cdot \int d\mathbf{r}\phi^*_{n'}\phi_n e^{i\left(\frac{2s\omega}{c}+\frac{\eta^2}{\lambda_c}\right)x} e^{i\left(\omega_{nn'}-2s\omega-\eta^2 mc^2/\hbar\right)t} , \tag{4.76}$$

where $\omega_{nn'} = (E_{n'} - E_n)/\hbar$. With the notation

$$C_{nn'}(s) = \int d\mathbf{r}\phi^*_{n'}\phi_n e^{i\left(\frac{2s\omega}{c}+\frac{\eta^2}{\lambda_c}\right)x} \tag{4.77}$$

we can write

$$a_{nn'} = \sum_s C_{nn'}(s) J_s(\eta^2\lambda/2\lambda_c) e^{i\left(\omega_{nn'}-2s\omega-\eta^2 mc^2/\hbar\right)t} . \tag{4.78}$$

Since η^2/λ_c is very large in equation (4.77) we may neglect the s-term in the phase in this equation, such that $C_{nn'}$ does not depend

[12] L. V. Keldysh, "Ionization in a field of a strong electromagnetic wave", Sov. Phys. JETP **20** 1307 (1965); A. I. Nikishov and V. I. Ritus, "Ionization of systems bound by short-range forces by the field of an electromagnetic wave", Sov. Phys. JETP **23** 168 (1966); A. M. Perelomov and V. S. Popov, "Ionization of atoms in an alternating electric filed, III.", Sov. Phys. JETP **25** 336 (1967) and References therein; M. V. Ammosov, N.B. Delone and V. P. Krainov, "Tunnel ionization of complex atoms and of atomic ions in an alternating electromagnetic field", Sov. Phys. JETP **64** 1191 (1986).

practically on s; in this case, the summation over s in equation (4.78) gives zero. Therefore, we conclude that there is no quantum transition among the bound states of a charge in a classical field of electromagnetic radiation, as expected. The ansatz given by equation (4.49) implies that the interaction is much slower than any transition among the bound states, but not necessarily so for the ionization process, especially with high energy of the ejected charge.

Also, it is worth noting that the presence of photons with energy $n\hbar\omega$ in the dressed wavefunction of the charge indicates merely the presence of virtual states, while the multiple Compton effect with absorption of n photons is cancelled out by the multiple Compton effect with emission of n photons. This process is distinct from the external Compton effect in the presence of the electromagnetic radiation.

4.3.5 Quasi-classical approximation

It is worth noting another form of the solution of equation (4.54), written as

$$\dot{F} = \frac{i}{\hbar} VF \ , \tag{4.79}$$

where we use the time derivative; V stands for the interaction generated by A. The solution can also be written as

$$F = 1 + \frac{1}{\hbar}\int^{t} dt^{'} iV + \frac{1}{\hbar^2}\int^{t} dt^{'} iV \int^{t^{'}} dt^{''} iV + ... \tag{4.80}$$

with the initial condition $F = 1$ for $t \to -\infty$. The form of the solution given by equation (4.80) is useful for the case when F and V are matrices (operators); then V includes also factors of the form $e^{i\omega_{nn'}t}$, and we can recognize easily in equation (4.80) the perturbation theory with intermediate states. As long as we require $\phi \to F\phi$ as in equation (4.49) we lose all transitions which are slower than the rate of switching on the interaction; indeed, such slow $e^{i\omega_{nn'}t}$-terms may be neglected in equation (4.80), which amounts to a quasi-classical character of the interaction V; we recover in this case the exponential form of solution which we used in the main text and which led to the absence of quantum transitions; this may not be necessarily the case for the ionization process, especially when the energy of the ejected particle is high. This circumstane may be termed a quasi-classical approximation; it may be seen formally by integrating by parts in equation (4.80) where terms like

$$\frac{1}{\hbar}\int dtVe^{i\omega_{nn'}t}=\frac{V}{\hbar\omega_{nn'}}e^{i\omega_{nn'}t}-\int dt\frac{1}{i\hbar\omega_{nn'}}\frac{\partial V}{\partial t}e^{i\omega_{nn'}t} \tag{4.81}$$

appear; they allow a direct comparison between $\hbar\omega_{nn'}$ and the rate $\partial V/\partial t$ (the first term in equation is the change in the wavefunction due to a constant interaction).

A special attention deserves the electrons in ionized gaseous plasmas subject to an electromagnetic radiation field. In normal conditions a gaseous plasma has a density of electrons $n\simeq 10^{19}cm^{-3}$; the electrons oscillate under the action of an external electric field E_{ex} and their own internal electric field E_{in}, according to the classical law

$$m\ddot{\mathbf{u}}+m\omega_0^2\mathbf{u}=q\mathbf{E}_{ex}+q\mathbf{E}_{in} \ , \tag{4.82}$$

where ω_0 is a characteristic frequency for bound electrons and $\mathbf{u}$ is their displacement from equilibrium positions; for free electrons $\omega_0 = 0$. The charge imbalance is $\delta n = -ndiv\mathbf{u}$; since the potential φ is given by $\Delta\varphi = -4\pi q\delta n = 4\pi n^2 qdiv\mathbf{u}$, we have the internal electric field $\mathbf{E}_{in} = -grad\varphi = -4\pi n^2 q\mathbf{u}$ and the equation of motion

$$\ddot{\mathbf{u}}+\omega_p^2\mathbf{u}=q\mathbf{E}_{ex} \ , \tag{4.83}$$

where $\omega_p=\sqrt{4\pi nq^2/m}$ is the plasma frequency. The displacement u is much smaller than the mean separation distance between the electrons, and, since the plasma moves as a macroscopic body (as a whole), its dynamics is entirely classical. The external radiation field superposes its oscillations with wavelength $\lambda\simeq 10^{-4}cm$ over this small, uniform displacement, according to equation (4.83). Consequently, they behave as free electrons subject to electromagnetic radiation. In a plane wave they are "accelerated", oscillate and radiate; in a focused beam they are distributed over the surface of a spherical polariton.[13]

[13] M. Apostol and M. Ganciu, "Polaritonic pulse and coherent X- and gamma rays from Compton (Thomson) backscattering", J. Appl. Phys. **109** 013307 (2011).

In conclusion, the interaction with a plane-wave field of classical electromagnetic radiation can be introduced in the dynamics of a point electric charge through an exponential factor within a quasi-classical and adiabatic approximation. This procedure allows the description of the "acceleration" and radiation of a free charge, as well as the ionization process for bound states of the charge. Since the switching on of the interaction is slower than the quantum-mechanical transitions, the latter cannot be described within this approximation.

4.4 Bound states

4.4.1 Dipole approximation

A non-relativistic charge q with mass m in an electric field $\mathbf{E}$ and potential $V(\mathbf{r})$ is governed by the dipole hamiltonian

$$H_d = \frac{1}{2m}p^2 - q\mathbf{E}\mathbf{r} + V(\mathbf{r}) \tag{4.84}$$

or the standard non-relativistic hamiltonian

$$H_s = \frac{1}{2m}\left(\mathbf{p} - \frac{q}{c}\mathbf{A}\right)^2 + V(\mathbf{r}) \ , \tag{4.85}$$

where $\mathbf{p}$ is momentum and $\mathbf{A}$ is the vector potential; for external fields we may take a zero scalar potential, $div\mathbf{A} = 0$ and $\mathbf{E} = -(1/c)\partial\mathbf{A}/\partial t$. We limit ourselves to static fields or to a radiation field of the form $\mathbf{A} = \mathbf{A}_0\cos(\omega t - \mathbf{k}\mathbf{r})$, which, for non-relativistic charges, may be approximated by $\mathbf{A} = \mathbf{A}_0\cos\omega t$, since the variations of the spatial phase $kr = \omega r/c$ are much smaller than those of the temporal phase ωt; ω is the frequency, $\mathbf{k}$ is the wavevector and c denotes the speed of light in vacuum; the non-relativistic approximation is ensured by $qA_0/mc^2 \ll 1$.

We note that $\mathbf{p} = m\mathbf{v}$ in equation (4.84) and $\mathbf{p} = m\mathbf{v} + q\mathbf{A}/c$ in equation (4.85); the former is the mechanical momentum, the latter includes the electromagnetic momentum. Consequently, in equation (4.85) the charge is immersed in the electromagnetic field, in equation (4.84) the electric field may be localized and for some time disentangled from the charge. The electric field in equation (4.84) may be unbounded in space. In general, motion in limited space regions requires special considerations.

The two hamiltonians given by equations (4.84) and (4.85) are related by the unitary (canonical) Goeppert-Mayer transform;[14] the initial moment of time is important in this equivalence. There is a further canonical transform, called the Kramers-Henneberger transform,[15] which takes the hamiltonian H_s into $p^2/2m + \widetilde{V}(\mathbf{r})$, where $\widetilde{V}(\mathbf{r}) = V(\mathbf{r} + \boldsymbol{\zeta})$ is the radiation-dressed potential; $\boldsymbol{\zeta} = q\mathbf{E}t^2/2m$ for a static field and $\boldsymbol{\zeta} = -q\mathbf{E}_0/m\omega^2$ for an oscillating field $\mathbf{E} = \mathbf{E}_0 \sin \omega t$. The former transform highlights the gauge character of the electromagnetic field, while the latter transform includes, partly, the motion of the charge in the field.

4.4.2 Static fields

Historically, the problem of bound charges in electric field appeared first for the hydrogen atom, and light atoms, in low, static electric fields. A high electric field among the static electric fields available in laboratory is typically $10^{-1}esu$ $(3kV/m)$; its mechanical work qEa upon an electron in atom is extremely small in comparison with the atomic energies; here $q = -4.8 \times 10^{-10}esu$ is the electron charge, E is the magnitude of the electric field and $a = 10^{-8}$ is the atom radius. Consequently, the field polarizes the atom, usually in the second-order of the perturbation theory (except for hydrogen, where the polarization occurs in the first order). The polarization is computed by time-independent perturbation theory, conveniently using the hamiltonian given by equation (4.84). The use of the perturbation theory assumes an adiabatic application of the field, which preserves the stationary states, causing at most linear combinations of them. And, of course,

[14] M. Goeppert-Mayer, "Ueber Elementarakte mit zwei Quantensprungen", Ann. Physik **401** 273 (1931).

[15] W. C. Henneberger, "Perturbation method for atoms in intense light beams", Phys. Rev. Lett. **21** 838 (1968); H. A. Kramers, "Non-relativistic quantum-electrodynamics and the correspondence principle", Solvay Congress 1948, in *Collected Scientific Papers*, North-Holland, Amsterdam (1956), p. 845; N. G. van Kampen, "Contribution to the quantum theory of light scattering", Kgl. Danske Videnskab. Selskab, Mat.-fys. Medd. **26** (15) (1951); W. Pauli and M. Fierz, "Zur Theorie der Emission langwelliger Lichtquanten", Nuovo Cimento **15**, 167 (1938).

the interaction strength should be small, such that the energy levels are little changed. Adiabatic introduction of the interaction is only possible for weak interaction, as required by the perturbation theory. The adiabatic application of weak static electric fields is achievable by experimental means (*e.g.*, between the plates of a capacitor slowly charging). Moreover, another problem may appear: the interaction energy $-q\mathbf{Er}$, where $\mathbf{r}$ is the charge position, exhibits a particularity: it is unbounded, which makes questionable the application of the perturbation theory; the hydrogen atom and the light atoms are "spatially extended", since the charges interact by long-range Coulomb potentials. This means, in fact, that the charge states are metastable, and the charge may go out from the atom under the action of the electric field; this is the (static) field-assisted atom ionization, treated in classical works.[16] We note that the field introduces a potential barrier (besides the "ion" potential barrier), which may be tunneled by the electron. The tunneling through the barrier proceeds by in and out propagating wavefunctions; this again assumes an adiabatic application of the field. The adiabatic hypothesis means that the rate of introducing the field is slower than time $\hbar/\Delta\mathcal{E}$, where $\Delta\mathcal{E}$ is the mean inter-spacing of the energy levels, such that the interaction does not produce transitions between the states.[17] The ionization rate is proportional to $e^{-cons/E}$, for small E; it is very small.

This is valid for the hydrogen atom, and, to some extent, for light atoms; these bound assemblies have well-separated energy levels and long-range Coulomb interaction. Heavy atoms and, in general, many-particle assemblies require a special discussion. Bound many-particle assemblies have at least two particularities. First, their single-particle levels are rather dense, almost, in some cases, they form a continuum of relevant single-particle states. It is convenient to work with single-particle states, as generated by a mean-field potential, as they govern the single-particle class of elementary excitations in these assemblies,

[16] J. R. Oppenheimer, "Three notes on the quantum theory of aperiodic effects", Phys. Rev. **13** 66 (1928); C. Lanczos, "Zur Intensitaetsschwachung der Spektrallinien in hoehen elektrischen Feldern", Z. Phys. **68** 204 (1931); H. A. Bethe and E. E. Salpeter, *Quantum Mechanics of One- and Two-Electron Atoms*, Springer, Berlin (1957).

[17] A. M. Dykhne, "Quantum transitions in the adiabatic approximation", Sov. Phys.-JETP **11** 411 (1960) (ZHETF **38** 570 (1960)).

the only relevant states for our problem. A convenient picture for heavy atoms is provided by the Thomas-Fermi model. In this circumstance, any perturbation, of a reasonably small magnitude, is rather rapidly accommodated as a valid stationary state, such that the effect of the electric field is rather rapidly incorporated in the stationary states; accordingly, the electric field produces polarization. In a heavy atom and neutral many-charge bound states the Coulomb interaction is screened to short-range forces, such that these assemblies may be viewed as "spatially finite"; consequently, the unboundedness of the interaction potential does not matter too much. This "finiteness" of these assemblies is their second particularity. Tunneling through the potential barrier of the electric field may appear, in a field-assisted ionization (decay, charge emission, dissociation, fragmentation). Some bound many-particle assemblies may exhibit spontaneous decay, like the nuclear alpha-particle decay or proton disintegration (emission). It is worth stressing the fact that the inclusion of the field in the energy levels may be done by perturbation theory.

The above considerations apply, in general, to heavy atoms and bound many-particle assemblies like molecules and atomic clusters, or atomic nuclei. Small molecules may be viewed as a particular case of heavy atoms, large molecules and atomic clusters look more close to solids, with extended electron states and (bounding, screened) short-range inter-nuclear potentials; spontaneous charge emission from these bound states, both electronic and ionic charges, is little known, ion fragmentation has an extremely small rate. Low static electric fields produce mainly polarization of these assemblies. It is worth noting that the adiabatic hypothesis holds for low fields applied to these assemblies.

4.4.3 Oscillating fields

The advent of high-power lasers brings three important new circumstances in this problem. First, the field is oscillating,, second, it may be high and, thirdly, it is applied suddenly. The experiments with laser beams proceed by placing a collection of bound charges (atoms, molecules, atomic clusters, atomic nuclei, ...) in the laser focus and firing up the laser pulse upon that collection of particles. Initially, the laser beams were not focused, they were lasting long, their succession was slow and the field intensity was low, in comparison with the field frequency. In this case, the charges may be viewed as fully immersed in the radiation field; the standard non-relativistic hamiltonian given by equation (4.85) may be used. The particles are (dynamically) polarized, may be ionized, and the radiation may affect the scattering cross-section of collisions, by absorbing and emitting multiple photons; this later process is the well-known Kroll-Watson effect. The calculations are done by time-dependent perturbation-theoretical methods, which lead to transitions among non-stationary states, or tunneling through classically unavailable states, as in the imaginary-time tunneling method; the ionization rate is of the same form $\sim e^{-const/E_0}$ as for static fields; the underlying presuposition is the adiabatic introduction of the (weak) interaction.[18]

Nowadays, the laser beams are focused; consequently, the particles may collide upon the laser focus, exchanging multiple photons; this is a new scattering process; the dipole hamiltonian given by equation (4.84) is suitable for this problem.[19] Laser beams introduce an important new feature: the interaction is applied in this case suddenly. Indeed, the time of setting up the interaction with the electromagnetic radiation upon a particle in a laser beam is of order a/c, where a is the dimension of the particle; for atoms $a/c \simeq 10^{-19}s$ $(a = 10^{-8}cm)$, for atomic nuclei $a/c \simeq 10^{-24}s$ $(a = 10^{-13}cm)$. These are very short times in comparison with any relevant time. They produce perturbation energies of the order $\hbar c/a \simeq 10keV$ for atoms and $\hbar c/a \simeq 10Mev$ for atomic nuclei; these energy uncertainties are much larger than mean inter-separation between energy levels, at least for the relevant single-particle states; the single-particle states are mixed up and, in

[18] L. V. Keldysh, "Ionization in the field of a strong electromagnetic wave", Sov. Phys.-JETP **20** 1307 (1965) (ZhETF **47** 1945 (1965)); A. M. Perelomov, V. S. Popov and M. V. Terentiev, "Ionization of atoms in an alternating electric field", Sov. Phys.-JETP **23** 924 (1966) (ZhETF **50** 1393 (1966)); A. I. Nikishov and V. I. Ritus, "Ionization of systems bound by short-range forces by the field of an electromagnetic wave", Sov. Phys.-JETP **23** 168 (1966) (ZhETF **50** 255 (1966)); M. V. Ammosov, N. B. Delone and V. P. Krainov, "Tunnel ionization of complex atoms and of atomic ions in an alternating electromagnetic field", Sov. Phys.-JETP **64** 1191 (1986) (ZhETF **91** 2008 (1986)); H. R. Reiss, "Complete Keldysh theory and its limiting cases", Phys. Rev. **A42** 1476 (1990).

[19] M. Apostol, "Scattering of non-relativistic charged particles by electromagnetic radiation", Z. Naturforschung **A72** 1173 (2017).

fact, undefined, a circumstance which precludes any application of the perturbation theory at this stage.

In the subsequent duration of time, after the sudden application of the interaction, the assembly of charges may follow two distinct courses. If the field strength is small, the charges oscillate, emit higher-order harmonics of electromagnetic radiation, according to the variation of the dressed potential $\widetilde{V}(\mathbf{r}) = V(\mathbf{r} + \boldsymbol{\zeta})$ and may tunnel;[20] if there exists a spontaneous ionization (decay, charge emission), the decay rate is affected;[21] after each process of this type, the charges are re-arranged (reconfigured), the mean-field potential is modified, and the radiation-dressing process is resumed. These processes imply a gradual accommodation of the charges to the interaction, which becomes incorporated in the assembly's dynamics. The accommodation proceeds in a long time (since the perturbation is small), but more rapidly in many-particle assemblies, where the perturbation has the chance to be close to a quasi-stationary state. The calculations in this case proceed by slowly time-depending quasi-stationary states, linear combinations of these states, slow transition rates responsible of radiation emission, tunneling, etc, generally, perturbation-theoretical type of calculations. The adiabatic hypothesis holds in this regime.

If the field strength is large, the charges bound state is rapidly dissociated (ionized, fragmented);[22] the criterion of separation between the two regimes, of weak and high field intensity, is provided by comparing ζ with the dimension of the bound state; for a reasonably large ζ, the ejected charge is taken and removed by the electric fields of extraction, employed in this kind of experiments.[23]

[20] M. Lewenstein, P. Balcou, M. Y. Ivanov, A. L'Huillier and P. B. Corkum, "Theory of high-harmonics generation by low-frequency laser fields", Phys. Rev. **A49** 2117 (1994); B. A. Zon, "Many-electron tunneling in atoms", J. Exp. Theor. Phys. **89** 219 (1999); "Tunneling ionization of atoms with excitation of the core", J. Exp. Theor. Phys. **91** 899 (2000).

[21] M. Apostol, "Electromagnetic radiation effect on alpha decay", J. Math. Theor. Phys. **1** 155 (2018).

[22] M. Apostol, "Fast atom ionization in strong electromagnetic radiation", Z. Naturforschung **A73** 461 (2018).

[23] M. Apostol, *Structure of Matter*, Nova, NY (2019).

4.5 Electric and magnetic pulses

4.5.1 Pulses of electric charge

Sharp pulses of electric charge can be constructed in a thermoelectric sample, either metallic or semiconducting, by means of the thermoelectric (Seebeck) effect.[24] In a metallic sample such pulses are much higher than in a semiconductor sample (though the voltage is much lower), but a connected series of p-n-semiconducting samples can give comparatively high pulses. The scheme of operating such a pulsed thermoelement is shown in Figs. 4.1-4.3.

According to the diffusion theory such a pulse has a width (linear spatial extension)

$$l' = \sqrt{v\Lambda\tau_{off}} \tag{4.86}$$

where v is the velocity of the charge carriers, Λ is their mean-free path and τ_{off} is the time duration needed to construct the pulse; during this time the external circuit is open (off-switched). Typically, $v = 10^7 cm/s$ for metals and $v = 10^6 cm/s$ for semiconductors (at room temperature); $\Lambda = 10^3 \text{Å}$ for metals and $\Lambda = 10^2 \text{Å}$ (both at room temperature). In order to construct a sharp pulse it is preferrable to keep τ_{off} as short as possible. This way, the gradients are very high and the sample can absorb a high power from the heat source. Typically, for $\tau_{off} = 10^{-6}s$ we get a pulse extension $l' = 10^{-2}cm$ (100 microns) for metals and $l' = 3 \cdot 10^{-3}cm$ (30 microns) for semiconductors. The off-time is limited by a δ-like extension $l' = \Lambda$ which corresponds to $\tau_{off} = \Lambda/v$ (lower bound) and by an extension l' as long as the length l of the sample (upper bound). In the latter case we have a stationary (equilibrium) transport, which dissipates heat (as in the traditional way of operating the thermoelectric samples), in contrast to a pulsed (non-equilibrium) transport whereby the dissipation is circumvented to a large extent.

According to the theoretical estimates the density of electric charge

[24] M. Apostol and M. Nedelcu, "Pulsed thermoelectricity", J. Appl. Phys. **108** 023702 (2010).

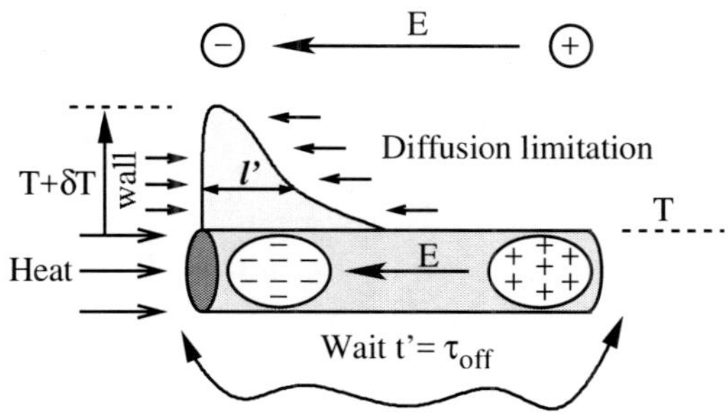

Figure 4.1: An electric pulse is constructed at the hot end of a thermoelement.

in a pulse created by a difference of temperature δT is given by[25]

$$\varrho = -e(\pi^2/4)\frac{nT\delta T}{\mu^2} \tag{4.87}$$

for metals and

$$\varrho = -e(3/2 - \mu/T)\frac{n\delta T}{T} \tag{4.88}$$

for semiconductors, where $-e$ is the electron charge ($e = 4.8 \cdot 10^{-10} esu = 1.6 \cdot 10^{-19} C$), n is the density of the charge carriers, T is the operating temperature (temperature of the cold end) and μ is the chemical potential. Typically, for metals $n = 10^{22} cm^{-3}$ and $\mu \sim 1eV$, while for semiconductors $n = 10^{17} cm^{-3}$ and $\mu/T \sim -5.7$ at room temperature ($T = 300K$; it is derived from the density of the charge carriers at equilibrium (slight modifications may appear in comparison with equation (4.88), according to specific circumstances)). For a typical $\delta T = 100K$ we get $\varrho \sim 1C/cm^3$ for metals and $\varrho \sim 10^{-2} C/cm^3$ for semiconductors. These are very high charge concentrations.

A word of caution must be inserted here. First, the above formulae are valid for $\delta T \ll T$, so their application to $\delta T/T = 1/3$ as we did above is questionable and may lead to important overestimates. Second, the equations used above are valid for thermal equilibrium.

[25] M. Apostol, *Physical Kinetics*, Cambridge Scholars Publ., Newcastle upon Tyne (2020).

and their application to a gaussian pulse as the one given in equation (4.86) is again questionable. For a gaussian pulse one may reasonably consider a random-walk extension of thermal equilibrium, as given by an effective width $\lambda = \sqrt{\Lambda l'}$. It is this effective width which is used below in estimating the total electric charge in such a pulse. Typically, both for metals and semiconductors, with the illustrative numbers given above, the effective width of the pulse is of the order $\lambda \sim 10^{-4}cm$ (1 micron).

Such pulses of electric charge can be created at the hot end of the sample by applying a heat flow. As it was said above one needs a duration τ_{off} to create such a pulse. Thereafter, by switching on the electric circuit the pulses are able to fly through the external circuit with velocity v, until they reach the cold end of the sample. The flight time is therefore $\tau_{on} = l_e/v$, where l_e is the length of the external circuit. The operation can be repeated periodically, with frequency $f = 1/(\tau_{off} + \tau_{on})$. The time needed to deflate the pulse of its heat at the cold end is equal to the time τ_{off} needed for creating another, identical pulse at the hot end, so the device can be operated cyclically. The concept of a pulsed thermomachine has been developed recently, as based on this mechanism of non-equilibrium transport, which is superior to the traditional thermoelectric devices in that it may produce more energy (Joule-Lenz heat of the pulses flying through the external circuit) and may exhibit a higher efficiency quotient. Of course, the energy produced is taken, in a superior fashion, from the heat pump, by circumventing the dissipation inherently associated with the stationary, equilibrium transport. For a typical pulse-operating regime, an increase factor of approximately 5 is obtained for metallic samples and 15 for semiconductors. For the typical values given above one may obtain pulse-current densities as high as $10^4 A/cm^2$ for metals and $10 A/cm^2$ for semiconductors. The pulsed thermomachine can also be operated reversibly such as to produce a significant imbalance of temperature along the sample (Peltier effect).

It was shown that the maximal mean-power such a pulsed thermomachine can deliver into the external circuit is obtained for $\tau_{on} = \tau_{off} = l_e/v$ (frequency $f = v/2l_e$) and for a pulse extension $l' = \sqrt{\Lambda l_e}$. It is given by $P = P_{eq}(l/2l')$ where P_{eq} is the power delivered in the equilibrium regime (l is the length of the sample). One can see that it

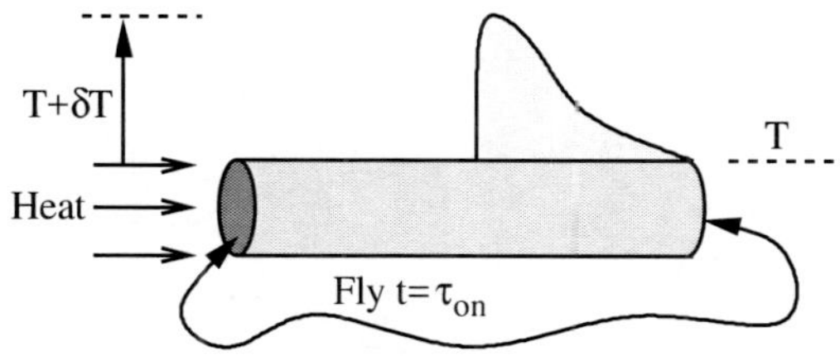

Figure 4.2: The pulse flyes through the external circuit at the cold end of the thermoelement.

is advantageous to have sharp pulses ($l'/l \ll 1$) in order to get a high power. However, since $l' = \sqrt{\Lambda l_e}$, the width of the pulses is limited by the length l_e of the external circuit. Therefore, the pulsed thermomachine requires short external circuits and adequate procedures of taking out the delivered power, *e.g.* by magnetic storage. During its working cycle the heat contacts can be maintained continuously.

4.5.2 Remote pulses

Let us envisage a rather long external circuit, as for a remote operation (say, of the order of meters), while keeping the pulses as sharp as possible. The time τ_{off} may then be neglected in comparison with time $\tau_{on} = l_e/v$, and the operating frequency is given approximately by $f = 1/\tau_{on} = v/l_e$. The power delivered into the external circuit is not maximal in this case[26] but we still can have high local pulses of electric and magnetic fields, as shown below. For an excessive length of the external circuit one should also consider the problem of switching periodically the heat contacts too, which may raise important technical problems. In addition, we remark upon the energy loss associated with a long external circuit, in the sense that the pulse energy can be dissipated (through the Joule-Lenz heat) and it can also be radiated in the whole space (though at a very low rate), while the usable energy is still confined to a comparatively small region of space.

The total electric charge in a pulse is given by $q = \varrho\lambda A$, where A is the cross-section of the sample. In order to have a high electric charge

[26]The frequency dependence of the power is $P = P_{eq}(l/\sqrt{v\Lambda})\sqrt{f(1-\tau_{on}f)}$.

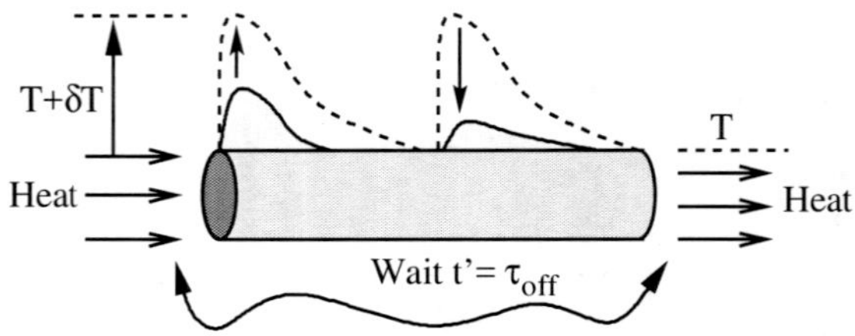

Figure 4.3: The pulse is deflated of heat at the cold end and another pulse is constructed at the hot end.

the electric contacts and the external circuit must have a cross-section as high as possible (which may also raise technical problems). It may seem from the above formula that it would be preferable to have an extended pulse too (large λ) in order to have a high electric charge. This is however misleading because for extended pulses the dissipation diminishes considerably the effective power and the local effect of high electric and magnetic fields is greatly reduced. What we envisage here is highly concentrated electric and magnetic effects, as arising from sharp electric pulses. We use here for an illustrative purpose a charge density $\varrho \sim 1C/cm^3$ (as for metals), a cross-section $A = 1cm^2$ and an effective width $\lambda = 10^{-4}cm$ (1 micron). We get a total charge $q = 10^{-4}C$. For semiconductors it may be smaller by a factor 10^{-2}.

Therefore, we consider electric pulses as sharp as to approximate them by δ-like distributions (particles), and write the density of electric charge as $\varrho = q\delta(\mathbf{r} - \mathbf{r}_0)$, where $\mathbf{r}_0$ is the location of the pulse. It is worth noting that we cannot have a rapid succession of pulses along the external circuit, since the operating frequency is limited by the length of the external circuit ($f = v/l_e$). The device cannot work cyclically for time durations smaller than the time needed by the pulse to arrive at the other (cold) end of the thermoelectric sample. In the following we shall consider therefore only one pulse, the results being repeated by frequency $f = v/l_e$ at most.

4.5.3 Wave equations

With usual notations the first pair of Maxwell equations

$$curl\mathbf{E} = -\frac{1}{c}\frac{\partial \mathbf{H}}{\partial t} \ , \ div\mathbf{H} = 0 \qquad (4.89)$$

are satisfied by representing the fields with potentials

$$\mathbf{E} = -grad\Phi - \frac{1}{c}\frac{\partial \mathbf{A}}{\partial t} \ , \ \mathbf{H} = curl\mathbf{A} \ . \qquad (4.90)$$

The second pair of Maxwell equations

$$curl\mathbf{H} = \frac{1}{c}\frac{\partial \mathbf{E}}{\partial t} + \frac{4\pi}{c}\mathbf{j} \ , \ div\mathbf{E} = 4\pi\varrho \qquad (4.91)$$

leads to the wave equations

$$\Delta\Phi - \frac{1}{c^2}\frac{\partial^2\Phi}{\partial t^2} = -4\pi\varrho \ , \ \Delta\mathbf{A} - \frac{1}{c^2}\frac{\partial^2\mathbf{A}}{\partial t^2} = -\frac{4\pi}{c}\mathbf{j} \qquad (4.92)$$

providing the Lorenz gauge

$$div\mathbf{A} + \frac{1}{c}\frac{\partial \Phi}{\partial t} = 0 \qquad (4.93)$$

is satisfied.

4.5.4 Retarded potentials

Consider a localized charge $\varrho = e(t)\delta(\mathbf{r})$. Outside this charge the first of the wave equations (4.92) has spherical symmetry and reads therefore $(1/r^2)\partial(r^2\partial\Phi/\partial r)/\partial r - (1/c^2)\partial^2\Phi/\partial t^2 = 0$ or $\chi'' - \ddot{\chi}/c^2 = 0$ with $\Phi = \chi/r$. The (outgoing) solution is $\chi = f(t - r/c)$ and $\Phi = f(t - r/c)/r$, where f is determined from the boundary condition at $r = 0$. For $r \to 0$ the time derivative in the wave equation can be neglected and the wave equation become the potential equation. We get therefore $\Phi = e(t - r/c)/r$. Similarly, $\mathbf{A} = \mathbf{j}(t - r/c)/cr$. In general,

$$\begin{aligned} \Phi(\mathbf{r}, t) &= \int d\mathbf{r}' \cdot \tfrac{\varrho(\mathbf{r}', t - R/c)}{R} \ , \\ \mathbf{A}(\mathbf{r}, t) &= \tfrac{1}{c}\int d\mathbf{r}' \cdot \tfrac{\mathbf{j}(\mathbf{r}', t - R/c)}{R} \ , \end{aligned} \qquad (4.94)$$

where $\mathbf{R} = \mathbf{r} - \mathbf{r}'$. These are the retarded (Kirchhoff) electromagnetic potentials.

Let $\mathbf{r}_0(t)$ be the trajectory of a pointlike electric charge q moving with velocity $\mathbf{v} = \dot{\mathbf{r}}_0(t)$. The charge density in equations (4.94) is $\varrho = q\delta(\mathbf{r}' - \mathbf{r}_0(t'))$ where $t' = t - R(t')/c$ and $\mathbf{R} = \mathbf{r} - \mathbf{r}'$. The gradient of the argument of the delta function is $1 - \mathbf{vR}/cR$, so the integral in (4.94) gives

$$\Phi(\mathbf{r}, t) = \frac{q}{R - \mathbf{vR}/c} \tag{4.95}$$

and similarly

$$\mathbf{A}(\mathbf{r}, t) = \frac{\mathbf{v}q}{c(R - \mathbf{vR}/c)} \tag{4.96}$$

where $\mathbf{R} = \mathbf{r} - \mathbf{r}_0$. All the quantities in the *rhs* of (4.95) and (4.96) are taken at the retarded time $t' = t - R(t')/c$. These are the Lienard-Wiechert potentials.

From these potentials we derive the fields. The electric field produced by such a moving charge is given by

$$\mathbf{E} = \mathbf{E}_v + \mathbf{E}_r \tag{4.97}$$

where

$$\mathbf{E}_\mathbf{v} = q\frac{1 - v^2/c^2}{(R - \mathbf{vR}/c)^3}(\mathbf{R} - \mathbf{v}R/c) \tag{4.98}$$

is the velocity field and

$$\mathbf{E}_r = \frac{q}{c^2(R - \mathbf{vR}/c)^3}\mathbf{R} \times [(\mathbf{R} - \mathbf{v}R/c) \times \dot{\mathbf{v}}] \tag{4.99}$$

is the radiation (acceleration) field; as it is said above, $\mathbf{R} = \mathbf{r} - \mathbf{r}_0$ and all the functions and derivatives in the *rhs* of (4.98) and (4.99) are taken at the retarded time t' given by $t' + R(t')/c = t$. Similarly, the magnetic field is given by

$$\mathbf{H} = \frac{1}{R}\mathbf{R} \times \mathbf{E}\ . \tag{4.100}$$

The velocity of the pulses is much lower than the velocity of light (typically $v = 10^7 cm/s$ for metals or $v = 10^6 cm/s$ for semiconductors

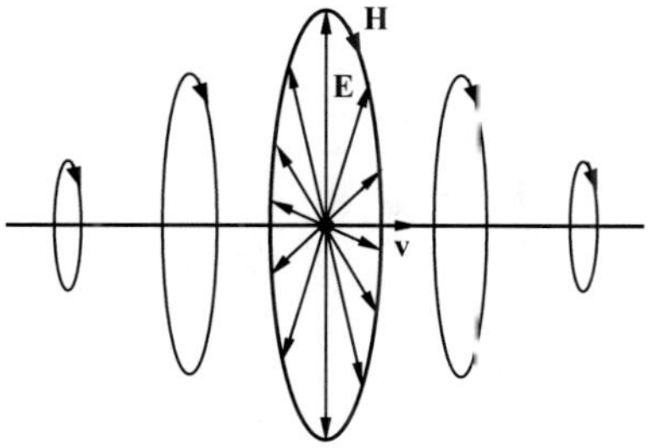

Figure 4.4: Electric and magnetic fields around a pulse.

as compared with the light velocity $c = 3 \cdot 10^{10} cm/s$). We are interested in distances R of the same order of magnitude as those covered by the pulses in their flight through the external circuit. It follows that we can safely neglect the retardation and put $t' = t$ in the above equations.

4.5.5 Uniform motion of the pulses

We choose cylindrical coordinates $\mathbf{r} = (x, y, z)$, $\rho^2 = x^2 + y^2$ and $\mathbf{r}_0 = (0, 0, z_0)$, $\rho_0 = 0$, $z_0 = vt$ with uniform velocity $v = const$. The radiation field given by (4.99) is vanishing. The electric field reduces to the static limit

$$\mathbf{E} = \frac{q}{R^3}\mathbf{R} = \frac{q}{[\rho^2 + (z - vt)^2]^{3/2}}(x, y, z - vt) \tag{4.101}$$

while the magnetic field is given by

$$\mathbf{H} = \frac{q}{cR^3}\mathbf{v} \times \mathbf{R} = \frac{vq}{c\,[\rho^2 + (z - vt)^2]^{3/2}}(-y, x, 0) \ . \tag{4.102}$$

In cylindrical coordinates we have a radial electric field

$$E_\rho = \frac{q\rho}{[\rho^2 + (z - vt)^2]^{3/2}} \tag{4.103}$$

and a longitudinal electric field

$$E_z = \frac{q(z - vt)}{[\rho^2 + (z - vt)^2]^{3/2}} \ . \tag{4.104}$$

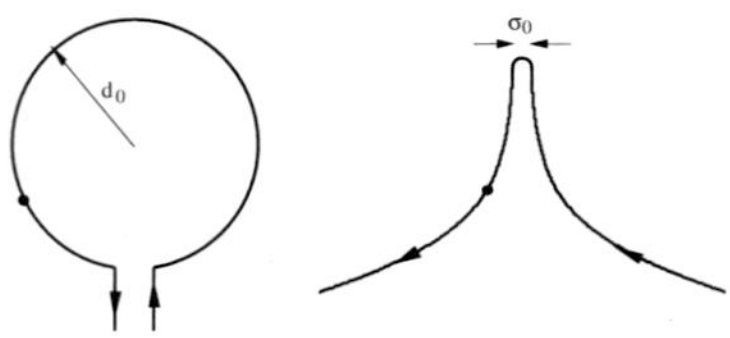

Figure 4.5: A loop and a needle for flying pulses.

Similarly, we have an azimuthal magnetic field

$$H_\varphi = \frac{vq\rho}{c\left[\rho^2 + (z - vt)^2\right]^{3/2}} \ . \tag{4.105}$$

These fields are shown in Fig. 4.4.

We can represent the fields as $E = q/d^2$ and $H = vE/c$, where d is the relevant distance from the pulse. As it was said above, in $q = \varrho\lambda A$ we take, as an illustrative example, $\varrho = 1C/cm^3$, $\lambda = 10^{-4}cm$ and $A = 1cm$ (as for metals). We get $q = 10^{-4}C$ and an electric field $E = 9 \cdot 10^5/d^2 \cdot V/m$. This is an extremely high electric field.[27] For comparison, the mean electric field at $1m$ from a $100w$ light bulb is $\sim 50V/m$. For $d = 1m$ the electric field is $E \sim 10^6 V/m$. Its energy is $\mathcal{E} \sim 10^4 J$[28] and its maximal power $P \sim 10^{13}w$ ($P = \mathcal{E}/(l'/v)$ for $v = 10^5 m/s$). The average power is $\sim 10^9 w$ (for length scale $d = 1m$). The magnetic field is by a factor $v/c = 10^{-3}$ smaller. It is given by $H = 300/d^2 \cdot Gs$ (with d in cm). The magnetic energy is negligible. For an external circuit length $l_e = 10m$ such pulses may succeed at a rate of $l_e/v = 10^{-4}s$ at most ($10kHz$).

The dissipated energy per volume and unit time (Joule-Lenz heat) is j^2/σ, where j is the electric flow (charge per unit area of the cross-section and per unit time) and σ is the electrical conductivity. The electric flow is given by $j = \varrho v = qv/Al'$. Making use of $q = 10^{-4}C$, $A = 1cm^2$ and $l' = 10^{-2}cm$(100 microns) we get $j = 10^5 A/cm^2$. This is a very high electric flow. In Gauss units we get $j = 3 \cdot 10^{14}$. The electrical conductivity is given by $\sigma = e^2 nv\Lambda/\mu$ (for metals, in Gauss

[27] With SI units a factor $1/4\pi\varepsilon_0 = 9 \times 10^9$ appears in Coulomb law with respect to the Gauss equation $E = q/d^2$.

[28] $\mathcal{E} = q^2/a$, where $a^3 = l'A$.

units). Making use of $e = 4.8 \cdot 10^{-10} esu$, $n = 10^{22} cm^{-3}$, $v = 10^7 cm/s$, $\Lambda = 10^{-5} cm$ (1000Å) and $\mu = 1eV = 1.6 \cdot 10^{-12} erg$, we get $\sigma \sim 10^{17} s^{-1}$. The dissipated energy is therefore $j^2/\sigma \sim 10^{12} erg/cm^3 \cdot s = 10^5 w/cm^3$. For one flight of a pulse through the external circuit we must multiply this result by Al_e times l_e/v. For $l_e = 1m$ we get the dissipated energy $\mathcal{E}_{dis} \sim 10^2 J$. We can see that it is much smaller than the electric energy of the pulse estimated above (q^2/a), so we may conclude that the pulse is still preserved during its flight to the cold end of the sample, thus enabling a repeated operation of the device. However, if the external circuit is excessively long the repetition can be significantly delayed.

4.5.6 A loop and a needle

We consider the pulses flying along a circular loop of radius d_0 with angular velocity $\Omega = v/d_0$, as shown in Fig. 4.5. The velocity fields (both electric and magnetic) remain of the same order of magnitude as the static fields given above, on the average. In addition, there appear radiation fields, of average magnitude

$$E_r \sim H_r \sim \frac{q d_0^2 \Omega^2}{c^2 d^2} \tag{4.106}$$

where d is the distance from the centre of the loop (they go like $1/R$ at large distance, in contrast to the static fields which go like $1/R^2$). For a loop $\Omega = v/d_0$ and the radiation fields go like $E_r \sim H_r \sim q d_0^2 v^2/c^2 d^2$. The relativistic factor v^2/c^2 makes them very weak. However, for a needle of length d_0 and small radius σ_0 the angular velocity is $\Omega = v/\sigma_0$ and the relativistic factor v^2/c^2 can be attenuated to some extent by the factor σ_0^2/d_0^2. Such a needle is shown in Fig. 4.5. It is worth emphasizing that such fields are oscillating with frequency Ω (*e.g.* typically $\Omega \sim 100kHz$), which may have specific effects on localized targets.

In conclusion we may say that high electric and magnetic fields can be obtained locally by pulse-operating a thermoelectric device. The pulses can succeed periodically along the external circuit with a definite frequency, which is upper bounded. Such pulses may be efficiently used in a remote scheme, for various purposes.

4.6 Electric pulse on a metallic wire

It is well known that electromagnetic waves may be guided along metallic wires and focused in metallic structures with a restricted geometry, beyond the diffraction limit. These phenomena have been associated with Sommerfeld surface waves[29] and surface plasmon-polaritons.[30] Recently, there is an increasing interest in propagating electric pulses along metallic wires, especially in the terahertz frequency range. Their propagation is almost dispersionless and their velocity is close to the speed of light in vacuum.

We present herein an idealized model of plasmon-polaritons excited in a long metallic wire by electric pulses, which may propagate over a long distance, practically without energy loss. The effects of an electric pulse propagating in a metallic wire are different from those associated with the propagation of a surface wave. The propagation of the pulse and its associated effects are not restricted to the surface; they occur inside the wire.

We consider a metallic wire, with a shape close to a straight line (*i.e.*, without many tight bends). The wire has a circular cross-section with radius r_0, much smaller than its length. We denote by x the coordinate along the wire. An electric pulse is applied at $x = 0$, uniformly distributed over the cross-section. The charge density of the pulse may be represented as

$$\rho_0 = Q\tau\delta(t)\delta(x)\theta(r_0 - r) \ , \tag{4.107}$$

where Q is the surface charge density, τ is the short duration of the pulse, t denotes the time and r is the radius in the plane of the cross-section. Similarly, the current density of the pulse may be represented as

$$j_{0x} = Qa\delta(t)\delta(x)\theta(r_0 - r) \ , \tag{4.108}$$

where a is the small spatial extension of the pulse along the x-direction. The δ-functions in equations above should be viewed as equal to $1/\tau$,

[29]A. Sommerfeld, "Ueber die Fortpflanzung elektrodynamischer Wellen laengs eines Drahtes", Ann. Phys. Chem. **67** 233 (1899); *Lectures on Theoretical Physics*, vol. 3, *Electrodynamics*, Academic Press, London (1964).

[30]M. Apostol, "Surface plasmon-polaritons. Superficial modes", Roum. J. Phys. **65** 604 (2020).

$1/a$ over a duration τ or a distance a, respectively. Outside the pulse the charge and the current are zero, while inside the pulse the charge density is $\rho_0 = Q/a$ and the current density is $j_{0x} = Q/\tau = v\rho_0$, where $v = a/\tau$ is the velocity of charge injection (pulse building). We can see that the charge conservation is verified. The charge and the current densities given above determine the scalar potential Φ_0 and the x-component $A_{0x} = \frac{a}{c\tau}\Phi_0$ of the vector potential, where c is the speed of light in vacuum. The scalar potential satisfies the wave equation

$$\frac{1}{c^2}\ddot{\Phi}_0 - \Delta\Phi_0 = 4\pi\rho_0 = 4\pi Q\tau\delta(t)\delta(x)\theta(r_0 - r) \; ; \tag{4.109}$$

by means of the Fourier transformations we get the solution

$$\Phi_0 = 2\pi Q c\tau r_0 \int d\kappa dk \frac{J_1(kr_0)J_0(kr)}{K} \sin cKt \cdot e^{i\kappa x} \quad , \tag{4.110}$$

where κ is the x-component of the wavevector $\boldsymbol{K} = (\kappa, \boldsymbol{k})$, $\boldsymbol{k}$ is the transverse wavevector and $J_{0,1}$ are the Bessel functions of zeroth and first order, respectively. It is easy to see that the potential Φ_0 has a slow variation with r, except for $r \simeq r_0$ where the surface effects appear. Inside the wire ($r < r_0$), and leaving aside the surface effects, we may approximate the potential given by equation (4.110) by

$$\begin{aligned} \Phi_0 &\simeq 2\pi Q c\tau \textstyle\int d\kappa dz J_1(z) \frac{\sin c\kappa t}{\kappa} e^{i\kappa x} = \\ &= 2\pi Q c\tau\theta(t - \mid x \mid /c) \; . \end{aligned} \tag{4.111}$$

Indeed, under the conditions given above equation (4.109) becomes the one-dimensional wave equation

$$\frac{1}{c^2}\ddot{\Phi}_0 - \frac{\partial^2\Phi_0}{\partial x^2} = 4\pi Q\tau\delta(t)\delta(x) \; , \tag{4.112}$$

whose solution is the one-dimensional Green function $4\pi Q\tau \times \frac{c}{2}\theta(t - \mid x \mid /c))$.

It is easy to check that the potentials Φ_0 (given by equation (4.111)) and $A_{0x} = \frac{a}{c\tau}\Phi_0$ satisfy the Lorenz gauge; for instance, outside the pulse the derivatives are zero, while inside the pulse $\partial\Phi_0/c\partial t$ and

$\partial A_{0x}/\partial x$ may be replaced by $2\pi Qc\tau/a$ and $-2\pi Qc\tau/a$. The solution given by equation (4.111) is valid inside the wire, close to the axis of the wire. Near the surface the function θ in equation (4.111) spreads out over a distance of the order r_0 (in all directions), such that, for thin wires, we may still view the solution as close to a θ-pulse. According to equation (4.110), the potentials are continuous with continuous derivatives at the surface of the wire. In the outside region, far away from the (thin) wire, we may view the pulse as being localized at a point in space ($x = 0$, $r = 0$), in a very short time τ, such that, at large distances, it generates the well-known spherical waves. We can estimate the energy radiated in space by such a pulse, and find out that the total energy radiated in time τ is extremely small. The assumptions implied by equations (4.107) and (4.108) and the approximation given by equation (4.111) define the idealized model of pulse propagation presented herein.

Making use of equation (4.111), the electric field along the wire ($E_{0x} = -\partial A_{0x}/c\partial t - \partial\Phi_0/\partial x$) is given by

$$\begin{aligned} E_{0x} &= 2\pi Q\tau \left[sgn(x) - v/c\right]\delta(t- \mid x \mid /c) \simeq \\ &\simeq 2\pi Q\tau \cdot sgn(x)\delta(t- \mid x \mid /c) \end{aligned} \tag{4.113}$$

(we assume $v \ll c$); the other two components of the electric field and the magnetic field are zero. This is a pulse of electric field which propagates (dispersionless) along the wire with the speed of light in vacuum. It may viewed as an external electric field. We can see that the pulse of the electric field is more narrow in time (a/c) than the source pulse (τ); indeed, two charges separated by distance a emit radiation with a time delay a/c, and $v = a/\tau \ll c$.[31]

We note that the above results are obtained by using the charge and the current densities given by equations (4.107) and (4.108), which may be viewed as corresponding to charges injected into the wire from the outside. If a pulse-like charge and current imbalance is created (uniformly) by an external field (without charge transfer), the densities may be taken as $\rho_0 = Qa\tau\delta(t)\delta'(x)$ and $j_{0x} = -Qa\tau\delta'(t)\delta(x)$ (for $r < r_0$). The results obtained for such pulses are similar with those given by a δ-representation for ρ_0 and j_{0x}.

[31] M. Apostol, "Electric pulse on a metallic wire", Optik **204** 164095 (2020).

The electric field E_{0x} generates an external force acting on the mobile charges in the wire (electrons). In general, a displacement $\boldsymbol{u}$ of the charges generates an imbalance of charge density $-nqdiv\boldsymbol{u}$, where n is the charge concentration in the wire and q is the electron charge ($\simeq 4.8 \times 10^{-10} esu$). At the same time, an internal electric field $\boldsymbol{E}_i$ appears in the wire, given by $div\boldsymbol{E}_i = -4\pi nqdiv\boldsymbol{u}$ (polarization $\boldsymbol{P} = nq\boldsymbol{u}$); the internal field generates a force $-4\pi nq^2\boldsymbol{u}$. The equation of motion for the displacement of a charge along the wire is

$$m\ddot{u} + m\omega_p^2 u + m\gamma\dot{u} = qE_0\tau \cdot sgn(x)\delta(t- \mid x \mid /c) \ , \qquad (4.114)$$

where $\omega_p = (4\pi nq^2/m)^{1/2}$ is the plasma frequency, γ is a damping coefficient ($\gamma \ll \omega_p$) and $E_0 = 2\pi Q$ is a notation introduced from equation (4.113). By a temporal Fourier transformation of equation (4.114) we get

$$u(\omega, x) = -\frac{qE_0\tau}{m} \cdot \frac{sgn(x)}{\omega^2 - \omega_p^2 + i\gamma\omega} e^{i\frac{\omega}{c}|x|} \qquad (4.115)$$

and the displacement along the wire

$$\begin{aligned} u(t,x) &= \tfrac{1}{2\pi} \int d\omega u(\omega) e^{-i\omega t} = \\ &= \tfrac{qE_0\tau}{m\omega_p} sgn(x) \sin\omega_p(t- \mid x \mid /c) e^{-\frac{\gamma}{2}(t-|x|/c)} \ , \ t > \mid x \mid /c \end{aligned} \qquad (4.116)$$

and $u(t,x) = 0$ for $t < \mid x \mid /c$ (both poles in equation (4.115) are placed in the lower half-plane).

The displacement in equation (4.116) represents a wave of charge density (and internal electric field, polarization), which propagates inside the wire with the speed of light in vacuum; it may be called a polariton. The frequency of this polariton is the plasma frequency, such that we may call this wave a plasmon-polariton. Its wavelength is $\lambda = c/\omega_p$. Typical plasma frequency ω_p in metals is a few eV, *e.g.* $\omega_p \simeq 5 \times 10^{15} s^{-1}$ for $n = 10^{22} cm^{-3}$ (electron mass $\simeq 10^{-27} g$), such that λ is of the order $10^{-5} cm$ ($0.1\mu m$); it is comparable with the mean freepath of electrons in typical metals at room temperature, $\Lambda \simeq 0.1 - 1\mu m$. We note that this wave affects all the charges in the cross-section of the wire. We can see that the electric pulse propagating along the wire excites a tail of plasmon-polaritons, which are propagating along the wire with the speed of light in vacuum and wavefronts placed at $t =\mid x \mid /c$. The fact that the wavelength λ is much smaller than the diameter of the wire is in accordance with our assumption that the propagation does not depend appreciably on the transverse dimensions of the wire (and surface effects may be neglected).

Let us introduce the notation $u(t,x) = u_0 sgn(x)\sin\omega_p(t- \mid x \mid /c)e^{-\frac{\gamma}{2}(t-|x|/c)}$, where $u_0 = qE_0\tau/m\omega_p$ is the amplitude of the displacement (equation (4.116)). In usual conditions this is a very small displacement; for instance, for a field $E_0 = 10^{-2}esu$ $(300V/m)$ and $\tau = 1ns$ $(10^{-9}s)$ we get $u_0 \simeq 0.1\mathring{A}$. The maximal velocity of the displacement $u_0\omega_p$ is of the order $10^6 cm/s$ (much smaller than the Fermi velocity in metals).

The plasmon-polariton solution given above is valid outside the pulse $(t- \mid x \mid /c > 0)$. Let us now estimate the motion of a charge inside the pulse. The period of oscillation of a charge ω_p^{-1} $(\simeq 2\times 10^{-16}s)$ is much shorter than the time a/c spent by the pulse upon a charge; for instance, for a pulse width $a = 3\times 10^{-2}cm$ $(300\mu m)$ we get $a/c = 10^{-12}s \gg \omega_p^{-1}$. Consequently, we may estimate the displacement of a charge inside the pulse by $w = qE_0\tau c/ma\omega_p^2$ (from equation (4.114); we note that the effective field inside the pulse is increased by the factor $\tau c/a$). Making use of the numerical data given here, we get $w \simeq 2\times 10^{-13}cm$. We may see that the restoring force $(m\omega_p^2 w)$ is very strong and the energy spent by the field to set the charge in motion is extremely small; practically, the field does not lose energy in this process.

However, we must include the damping force in the energy balance. The energy conservation derived from equation (4.114) reads

$$\begin{aligned}&\frac{\partial}{\partial t}\left(\frac{1}{2}m\dot{u}^2 + \frac{1}{2}m\omega_p^2u^2\right) + m\gamma\dot{u}^2 = \\ &= qE_0\tau\cdot sgn(x)\delta(t- \mid x \mid /c)\dot{u}\ .\end{aligned} \tag{4.117}$$

If we integrate equation (4.117) over a small interval about the point $t- \mid x/c$ (across the pulse), we can see that the first term on the left side (time derivative of the energy) may be neglected and the velocity of the displacement inside the pulse is of the form $\dot{u} = \dot{s}\tau\cdot sgn(x)\delta(t- \mid x \mid /c)$; it follows that the energy loss of the pulse (per charge and per unit time) is $m\gamma\dot{s}^2(\tau c/a)^2$, equal with the work done by the field upon a charge $qE_0\dot{s}(\tau c/a)^2$ per unit time; hence, we get $\dot{s}=qE_0/m\gamma$ and the energy loss $(q^2E_0^2/m\gamma)(\tau c/a)^2$ (per charge and unit time). Since the duration of the pulse on a charge is a/c, we may estimate this displacement as $s=qE_0a/mc\gamma$. In order to compute this displacement we need the damping coefficient γ. In cuasi-static conditions, the damping coefficient is given by $\gamma=\omega_p^2/4\pi\sigma$, where σ is the static conductivity; in typical metals $\gamma\simeq 10^{13}-10^{14}s^{-1}$ (at room temperature). Using $\gamma=10^{13}s^{-1}$ and the numerical data given here, we get $s\simeq 5\times 10^{-10}cm$. We can see that $s\ll\Lambda$ ($\Lambda\simeq 0.1-1\mu m$), such that the damping is ineffective. Therefore, we may say that the pulse is propagated practically without loss, over large distances.

Outside the pulse, equation (4.117) is the energy conservation for a damped oscillator, with the rate of energy loss γ. In general, the damping coefficient is diminished in oscillations, in comparison with the static conditions. We may view the damping coefficient as being given by $\gamma=1/t_{lf}$, where t_{lf} is the lifetime of the electron excitations; it defines the mean freepath by $\Lambda=v_Ft_{lf}$, where v_F is the Fermi velocity. The electron-phonon interaction brings, usually, a comparable contribution (besides the effects of impurities, etc). The electron lifetime can be estimated by $\gamma=1/t_{lf}\simeq T^2/\hbar\mu$, where T is the temperature and μ is the Fermi level. For typical metals at room temperature we get $\gamma\simeq 10^{12}s^{-1}$ ($\mu=1eV$) and $\Lambda\simeq 10^{-4}cm$ ($v_F=10^8cm/s$). Since the displacement amplitude u_0 is much smaller than the mean freepath Λ, the plasmon-polariton is a non-thermal (adiabatic) process, where the damping coefficient is, practically, not effective. The energy transferred by the electric pulse to the plasmon-polaritons may be transported by the latter over large distances.

It is worth analyzing the energy radiated by plasmon-polaritons. The electromagnetic potentials are given by the Kirchhoff formula for the charge density $\rho=-nqdiv\boldsymbol{u}$ and the current density $\boldsymbol{j}=nq\dot{\boldsymbol{u}}$. The potentials of the field generated by the plasmon-polaritons are given by

$$\Phi=A_x=nqu_0\tfrac{\omega_p}{c}S\cdot$$
$$\cdot\int_0^{ct}dx'\frac{\cos\omega_p\left(t-\sqrt{r^2+(x-x')^2}/c-x'/c\right)}{\sqrt{r^2+(x-x')^2}}\ , \tag{4.118}$$

where r is the distance from the wire, $(x, \boldsymbol{r})$ is the position of the observation point and S denotes the area of the cross-section of the wire (for a semi-infinite wire). We introduce the new variables $z = \omega_p(x - x')/c$, $y = \omega_p r/c$ and note that the limit $\omega_p x/c$ may be taken to infinity, since the wavelength $\lambda = c/\omega_p$ is much smaller than any relevant distance. Then, the integral in equation (4.118) becomes a function of $t - x/c$, which leads to vanishing fields (and a vanishing radiated energy).

In conclusion, an electric pulse localized both in space and time on a thin metallic wire may propagate along the wire with the speed of light in vacuum, practically without energy loss and without dispersion; the pulse excites behind a tail of plasmon-polaritons which, also, propagate along the wire with the speed of light in vacuum and plasma frequency without energy loss. The reason for such a lossless propagation is the high plasma frequency, which leads to charge displacements much smaller than the mean freepath. The results presented here can be extended to dielectric wires.

The model discussed here is an idealized model. The idealization resides in assuming δ-functions for the pulse, which depend only on the coordinate x along the wire (and time t). Also, the building of the pulse is assumed to be achieved by a uniform injection of charges, or a uniform imbalance of charges over the cross-section of the wire, and the surface effects are neglected. These are important elements of idealization. In experiments, the pulse has a finite extension in all three directions and in time; usually, in terahertz experiments the pulses are radially polarized. In these conditions, the propagation should be considered in all three space dimensions and surface effects should be included, as in cylindrical wires. Then, we may have dispersion, energy loss and contributions from surface waves and surface plasmon-polaritons. The main effect of such more realistic conditions is an attenuation of the pulse with distance.

5 Relativity and Quantum Mechanics

5.1 Maxwell equations

Let us assume that pointlike electric charges q are distributed in space (vacuum) with a uniform concentration n; also, we assume that a rigid neutralizing background of charges $-q$ is present. A local displacement $\boldsymbol{u}$ produces a net charge density $\rho = -nq div\boldsymbol{u}$. Let us call electric field the quantity $\boldsymbol{E} = -4\pi nq\boldsymbol{u}$. We get the Gauss law

$$div\boldsymbol{E} = 4\pi\rho \ . \tag{5.1}$$

The factor 4π is chosen for convenience. Indeed, for a pointlike charge density $\rho = Q\delta(\boldsymbol{r})$ the solution of the Poisson equation (5.1) is obtained by $4\pi r^2 E = 4\pi Q$, $E = Q/r^2$, where Q is the pointlike charge (Coulomb law). It is worth noting that r is not allowed to go to zero for such pointlike charges. For a small displacement $\boldsymbol{u}$ we can define a density of dipole moments $nq\boldsymbol{u}$, *i.e.* a polarization $\boldsymbol{P} = nq\boldsymbol{u}$; consequently, the procedure described above works also for polarizable matter. We may imagine that we have a charge density ρ which generates an electric field $\boldsymbol{E}$ given by $div\boldsymbol{E} = 4\pi\rho$. In polarizable matter there appears an additional charge density $-nq div\boldsymbol{u} = -div\boldsymbol{P}$, such that the total field is given by $div\boldsymbol{E} = 4\pi\rho - 4\pi div\boldsymbol{P}$, or $div(\boldsymbol{E} + 4\pi\boldsymbol{P}) = 4\pi\rho$; $\boldsymbol{D} = \boldsymbol{E} + 4\pi\boldsymbol{P}$ is called the electric displacement.

The velocity $\dot{\boldsymbol{u}}$ gives a density of electric current $\boldsymbol{j} = nq\dot{\boldsymbol{u}}$, such that $\dot{\rho} = -nq div\dot{\boldsymbol{u}}$, *i.e.* the continuity equation of the electric charge

$$\dot{\rho} + div\boldsymbol{j} = 0 \ . \tag{5.2}$$

we note also that $\frac{\partial \boldsymbol{E}}{\partial t} = -4\pi\boldsymbol{j}$. A particular case is $\boldsymbol{u} \neq 0$, $\dot{\boldsymbol{u}} \neq 0$, $\boldsymbol{j} \neq 0$, $\partial\boldsymbol{E}/\partial t \neq 0$, but $div\boldsymbol{u} = 0$, $\rho = 0$, $div\boldsymbol{E} = 0$ and $div\boldsymbol{j} = 0$.

Let us assume that the elementary charges q perform small rotations. Specifically, a charge q placed at $\boldsymbol{r}_a$ rotates about the center $\boldsymbol{r}$ with a velocity $\boldsymbol{v}_a$. The quantity $(\boldsymbol{r}_a - \boldsymbol{r}) \times q\boldsymbol{v}_a$ gives rise to a dipole moment, providing we divide it by a velocity c. We call the quantity $\boldsymbol{m} = (\boldsymbol{r}_a - \boldsymbol{r}) \times q\boldsymbol{v}_a/2c$ magnetic moment. The velocity c is a universal velocity. For small distances $\mid \boldsymbol{r}_a - \boldsymbol{r} \mid$ we may define a density $\boldsymbol{M} = n\boldsymbol{m} = (\boldsymbol{r}_a - \boldsymbol{r}) \times nq\boldsymbol{v}_a/2c$ of magnetic moments, which we call magnetization. It has two remarkable properties: first, $div\boldsymbol{M} = 0$ and, second, $curl\boldsymbol{M} = nq\boldsymbol{v}_a/c$, which is a current density $\boldsymbol{j}_m = nq\boldsymbol{v}_a$ divided by c, with the property $div\boldsymbol{j}_m = 0$. It is worth noting that the radius $\boldsymbol{r}_a - \boldsymbol{r}$ of the orbits of these rotating charges cannot be made zero, *i.e.* the magnetic moments have a small, but finite radius. Let us call the quantity $\boldsymbol{B} = 4\pi\boldsymbol{M}$ magnetic induction. Then,

$$div\boldsymbol{B} = 0 \ , \tag{5.3}$$

which is Gauss law for magnetic induction; and

$$curl\boldsymbol{B} = \frac{4\pi}{c}\boldsymbol{j}_m = 4\pi curl\boldsymbol{M} \ . \tag{5.4}$$

We note that the magnetic induction is given from this equation by $2\pi r B = \frac{4\pi}{c}\oint d\boldsymbol{s}\boldsymbol{j}_m$, where the integral is the flux of current through a surface bounded by the circle with radius $\boldsymbol{r}$, *i.e.* $B = \frac{2}{cr}\oint d\boldsymbol{s}\boldsymbol{j}_m$, or $\boldsymbol{B} = \frac{2}{cr^2}\oint d\boldsymbol{s}\boldsymbol{j}_m \times \boldsymbol{r}$ with an oriented surface. Therefore, we see again that we cannot take a zero lateral dimension of the current. We may add the quantity $\frac{\partial \boldsymbol{E}}{\partial t} + 4\pi\boldsymbol{j} = 0$ on the right of equation (5.4), and we may also view $\boldsymbol{E}$, $\boldsymbol{j}$ and $\boldsymbol{M}$ as independent quantities; $\partial\boldsymbol{E}/\partial t$ and $c \cdot curl\boldsymbol{M}$ are current densities. We get the equation

$$curl\boldsymbol{B} = \frac{1}{c}\frac{\partial \boldsymbol{E}}{\partial t} + \frac{4\pi}{c}\boldsymbol{j} + 4\pi curl\boldsymbol{M} \tag{5.5}$$

for a magnetizable sample of matter, whence

$$curl(\boldsymbol{B} - 4\pi\boldsymbol{M}) = \frac{1}{c}\frac{\partial \boldsymbol{E}}{\partial t} + \frac{4\pi}{c}\boldsymbol{j} \ , \tag{5.6}$$

or

$$curl\boldsymbol{H} = \frac{1}{c}\frac{\partial \boldsymbol{E}}{\partial t} + \frac{4\pi}{c}\boldsymbol{j} \ , \tag{5.7}$$

where $\boldsymbol{H} = \boldsymbol{B} - 4\pi\boldsymbol{M}$ is the magnetic field. If matter is not only magnetizable but also polarizable, an additional term $\frac{4\pi}{c}\frac{\partial \boldsymbol{P}}{\partial t}$ appears on the right in equation (5.7), which becomes

$$curl\boldsymbol{H} = \frac{1}{c}\frac{\partial \boldsymbol{D}}{\partial t} + \frac{4\pi}{c}\boldsymbol{j} \ . \tag{5.8}$$

Equation (5.7) is the Maxwell-Ampere equation. In vacuum $\boldsymbol{M} = 0$, $\boldsymbol{P} = 0$ and $\boldsymbol{B} = \boldsymbol{H}$, $\boldsymbol{E} = \boldsymbol{D}$. The two equations above become

$$div\boldsymbol{H} = 0 \ , \ curl\boldsymbol{H} = \frac{1}{c}\frac{\partial \boldsymbol{E}}{\partial t} + \frac{4\pi}{c}\boldsymbol{j} \ . \tag{5.9}$$

The displacement field $\boldsymbol{u}$ has a more general motion, not only that given by $div\boldsymbol{u}$, which leads to the electric charge. Indeed, apart from translations, the local variation of u_i is

$$du_i = \partial_j u_i dx_j = \frac{1}{2}\left(\partial_j u_i + \partial_i u_j\right) dx_j + \frac{1}{2}\left(\partial_j u_i - \partial_i u_j\right) dx_j \ , \tag{5.10}$$

where $i, j, ... = 1, 2, 3$ are cartesian labels. $u_{ij} = \frac{1}{2}\left(\partial_j u_i + \partial_i u_j\right)$ is the strain tensor. It is related to the deformation of the length element $dl = (dx_i dx_i)^{1/2}$. Indeed,

$$\begin{aligned} dl'^2 &= dx'_i dx'_i = (dx_i + \partial_j u_i dx_j)(dx_i + \partial_k u_i dx_k) = \\ &= dl^2 + 2\partial_j u_i dx_j dx_i + \partial_j u_i \partial_k u_i dx_j dx_k \ , \end{aligned} \tag{5.11}$$

or

$$dl'^2 = dl^2 + (\partial_i u_j + \partial_j u_i + \partial_i u_k \partial_j u_k) dx_i dx_j \ , \tag{5.12}$$

which, for small variations, reduces to

$$dl'^2 = dl^2 + 2u_{ij} dx_i dx_j \ . \tag{5.13}$$

The second term in equation (5.10) can be written as

$$du_i^{rot} = \frac{1}{2}\left(\partial_j u_i - \partial_i u_j\right) dx_j = \frac{1}{2}\left(curl\boldsymbol{u} \times d\boldsymbol{r}\right)_i \ , \tag{5.14}$$

i.e. $d\boldsymbol{u}^{rot}$ is a rotation of angle $\boldsymbol{\varphi} = \frac{1}{2}curl\boldsymbol{u}$. The decomposition of the variations of the local displacement in translations, deformations and rotations given by equation (5.10) is the Helmholtz theorem.

From equation (5.14) we can write

$$curl\boldsymbol{u} = \tfrac{2}{dr^2}d\boldsymbol{r} \times d\boldsymbol{u}^{rot} = \tfrac{2dt}{dr^2}d(\boldsymbol{r}_a - \boldsymbol{r}) \times \boldsymbol{v}_a = \\ = \tfrac{4cdt}{nqdr^2}d(\boldsymbol{r}_a - \boldsymbol{r}) \times \boldsymbol{j}_m/2c = \tfrac{cdt}{\pi nqdr^2}d\boldsymbol{B} \ ; \tag{5.15}$$

the variation dr is $2cdt$, such that we get

$$curl\boldsymbol{u} = \frac{1}{4\pi nqcdt}d\boldsymbol{B} \ , \tag{5.16}$$

or

$$curl\boldsymbol{E} = -\frac{1}{c}\frac{\partial \boldsymbol{B}}{\partial t} \ . \tag{5.17}$$

In vacuum $\boldsymbol{B} = \boldsymbol{H}$ and we have

$$curl\boldsymbol{E} = -\frac{1}{c}\frac{\partial \boldsymbol{H}}{\partial t} \ . \tag{5.18}$$

This is the Faraday equation.
The fields in vacuum can be expressed as $\boldsymbol{E} = -\frac{1}{c}\frac{\partial \boldsymbol{A}}{\partial t} - grad\Phi$, $\boldsymbol{H} = curl\boldsymbol{A}$, where the potentials satisfy the wave equations

$$\tfrac{1}{c^2}\tfrac{\partial^2 \Phi}{\partial t^2} - \Delta\Phi = 4\pi\rho \ , \\ \tfrac{1}{c^2}\tfrac{\partial^2 \boldsymbol{A}}{\partial t^2} - \Delta\boldsymbol{A} = \tfrac{4\pi}{c}\boldsymbol{j} \ , \tag{5.19}$$

providing the Lorenz gauge

$$\frac{1}{c}\frac{\partial \Phi}{\partial t} + div\boldsymbol{A} = 0 \tag{5.20}$$

is satisfied. The electromagnetic field has an energy density, whose time variation is given by the mechanical work density $\boldsymbol{jE}$ per unit time. This work leads to the field-charge interaction density

$$V = \rho\Phi - \frac{1}{c}\boldsymbol{j}\boldsymbol{A} \ . \tag{5.21}$$

The Lorentz force density $\boldsymbol{f} = \rho\boldsymbol{E} + \frac{1}{c}\boldsymbol{j} \times \boldsymbol{H}$ exists between charges and field, which equals a field stress force and the time variation of the field momentum density. The self-interaction should be avoided. A pointlike charge carries with it its longitudinal field, the self-interaction cancels out the field energy, such that the longitudinal field is irrelevant for self-interaction, and for interaction in general. The self-interaction with radiation requires a finite radius of the charge $a = q^2/mc^2$, where m is the charge m and q is the electric charge, such that the self-interaction is vanishing.

5.2 Hamilton-Jacobi equation

The lagrangian of a pointlike charge q with mass m, moving with velocity $\boldsymbol{v}$, in an electromagnetic field is

$$L = -mc^2\sqrt{1 - v^2/c^2} - q\Phi + \frac{q}{c}\boldsymbol{vA} \ ; \tag{5.22}$$

if we add the field lagrangian $\frac{1}{8\pi}\int d\boldsymbol{r}(E^2 - H^2)$ and take the variation of the action $S = \int Ldt$ with respect to the field, we get the Maxwell equations. Similarly, the variation of the action with respect to the particle coordinates leads to the equation of motion of the charge. The equation of motion is

$$\frac{d}{dt}\left(\frac{\partial L}{\partial \boldsymbol{v}}\right) - gradL = 0 \ , \tag{5.23}$$

or

$$\frac{d}{dt}\frac{m\boldsymbol{v}}{\sqrt{1 - v^2/c^2}} = q\boldsymbol{E} + \frac{1}{c}\boldsymbol{v} \times \boldsymbol{H} \ . \tag{5.24}$$

The momentum of the particle is

$$\boldsymbol{P} = \frac{\partial L}{\partial \boldsymbol{v}} = \frac{m\boldsymbol{v}}{\sqrt{1 - v^2/c^2}} + \frac{q}{c}\boldsymbol{A} \ , \tag{5.25}$$

where $\boldsymbol{p} = m\boldsymbol{v}/\sqrt{1 - v^2/c^2}$ is the mechanical momentum. The energy (hamiltonian) of the particle is

$$\mathcal{E} = \boldsymbol{vP} - L = \frac{mc^2}{\sqrt{1 - v^2/c^2}} + q\Phi \ . \tag{5.26}$$

By using the equation of motion we get

$$\mathcal{E} = -\frac{\partial S}{\partial t} ; \tag{5.27}$$

it is conserved along the trajectory. Similarly, from

$$\begin{aligned} \delta S = \delta \int L dt = \int [\tfrac{\partial L}{\partial \boldsymbol{r}} \delta \boldsymbol{r} dt + \tfrac{\partial L}{\partial \boldsymbol{v}} \delta (\tfrac{d\boldsymbol{r}}{dt}) dt] = \\ = \tfrac{\partial L}{\partial \boldsymbol{v}} \delta \boldsymbol{r} + \int [\tfrac{\partial L}{\partial \boldsymbol{r}} \delta \boldsymbol{r} dt - \tfrac{d}{dt} \tfrac{\partial L}{\partial \boldsymbol{v}} \delta \boldsymbol{r} dt] = \tfrac{\partial L}{\partial \boldsymbol{v}} \delta \boldsymbol{r} \end{aligned} \tag{5.28}$$

we get the momentum

$$\boldsymbol{P} = \frac{\partial S}{\partial \boldsymbol{r}} . \tag{5.29}$$

S is a function of t and $\boldsymbol{r}$ at every point in time and space, its extremum leading to the equation of motion and the trajectory.
From equations (5.25) and (5.26) we get

$$(\mathcal{E} - q\Phi)^2 - c^2 \left(\boldsymbol{P} - \frac{q}{c} \boldsymbol{A} \right)^2 - m^2 c^4 = 0 , \tag{5.30}$$

which is the relativistic energy-momentum relation. In the absence of the field

$$\mathcal{E}^2 = c^2 p^2 + m^2 c^4 . \tag{5.31}$$

Inserting the action in equation (5.30) we get the Hamilton-Jacobi equation

$$\left(\frac{\partial S}{\partial t} + q\Phi \right)^2 - c^2 \left(gradS - \frac{q}{c} \boldsymbol{A} \right)^2 - m^2 c^4 = 0 , \tag{5.32}$$

or, in the absence of the field,

$$\left(\frac{\partial S}{\partial t} \right)^2 - c^2 (gradS)^2 - m^2 c^4 = 0 . \tag{5.33}$$

In the non-relativistic limit $S = S^{'} - mc^2 t$ and the Hamilton-Jacobi equation becomes

$$\begin{aligned} \tfrac{\partial S^{'}}{\partial t} + q\Phi + \tfrac{1}{2m} \left(gradS^{'} - \tfrac{q}{c} \boldsymbol{A} \right)^2 - \tfrac{1}{2mc^2} \left(\tfrac{\partial S^{'}}{\partial t} + q\Phi \right)^2 \simeq \\ \simeq \tfrac{\partial S^{'}}{\partial t} + q\Phi + \tfrac{1}{2m} \left(gradS^{'} - \tfrac{q}{c} \boldsymbol{A} \right)^2 = 0 , \end{aligned} \tag{5.34}$$

is the non-relativistic energy

$$\mathcal{E}' = \frac{1}{2m}\left(\boldsymbol{P} - \frac{q}{c}\boldsymbol{A}\right)^2 + q\Phi \ , \tag{5.35}$$

where

$$\boldsymbol{P} \simeq m\boldsymbol{v} + \frac{q}{c}\boldsymbol{A} \tag{5.36}$$

and corrections of the order $1/c^2$ are omitted.

5.3 Quantum Mechanics

The solution $S(t, \boldsymbol{r})$ of the Hamilton-Jacobi equation (equations (5.32), (5.34)) depends on integration constants; its derivatives with respect to these constants should be equal to zero (or constants which can be set equal to zero by adequate initial values of the time and position). These equations give the trajectory, thus solving the motion problem. From the definitions $\boldsymbol{p} = \partial S/d\boldsymbol{r}$ and $\mathcal{E} = -\partial S/\partial t$, we get $d\delta S = d\boldsymbol{r}\delta\boldsymbol{p} = -dt\delta\mathcal{E}$, *i.e.* $\frac{d\boldsymbol{p}}{dt} = -\frac{\partial\mathcal{E}}{\partial\boldsymbol{r}}$, which is the Hamilton equation of motion (which determines the trajectory). On the other hand the variation of the action

$$\Delta S = \frac{\partial S}{\partial \boldsymbol{r}}\Delta\boldsymbol{r} + \frac{\partial S}{\partial t}\Delta t = \boldsymbol{p}\Delta\boldsymbol{r} - \mathcal{E}\Delta t \tag{5.37}$$

leads to $\delta\boldsymbol{p}\Delta\boldsymbol{r} - \delta\mathcal{E}\Delta t = 0$, which gives the velocity $\boldsymbol{v} = \Delta\boldsymbol{r}/\Delta t = \frac{\partial\mathcal{E}}{\partial\boldsymbol{p}}$ along the trajectory; this is the other Hamilton equation, where $\mathcal{E}(\boldsymbol{r}, \boldsymbol{p}, t) = H$ is the hamiltonian. This is the description of the classical motion (as it is made by Classical Mechanics). The main assumption of the classical motion is large variations of the mechanical action, and large values of the mechanical action, which is a continuous and differentiable function with large values of its derivatives; this is a reasonable assumption for bodies at the macroscopic scale.

At the atomic and subatomic scale, the mechanical action varies in quanta $\hbar$ of action, in an undetermined way, without a determined trajectory. This is the quantum-mechanical motion. It follows that the classical motion occurs for $\Delta S \gg \hbar$; this limit is also called the quasi-classical motion. The lack of trajectory suggests inevitably a wave, of the form $\Psi = Ae^{\frac{i}{\hbar}S}$. In the quasi-classical motion $\Delta S \gg \hbar$ and S may be taken as the classical action; the expansion $\Delta S = \boldsymbol{p}\Delta\boldsymbol{r} - \mathcal{E}\Delta t$ is possible, and it shows that there exists a velocity, *i.e.* a trajectory, which is the ray of the wave. Therefore the wavefunction Ψ recovers the classical motion in the limit $\Delta S \gg \hbar$. Making use of this Ψ for the classical motion is the quasi-classical description of the motion. Moreover,

$$
\begin{aligned}
-i\hbar\tfrac{\partial}{\partial \boldsymbol{r}}\Psi &= -i\hbar\tfrac{\partial A}{\partial \boldsymbol{r}} + \boldsymbol{p}\Psi \ , \\
i\hbar\tfrac{\partial \Psi}{\partial t} &= i\hbar\tfrac{\partial A}{\partial t}e^{\frac{i}{\hbar}S} + \mathcal{E}\Psi \ ,
\end{aligned}
\tag{5.38}
$$

where $\boldsymbol{p} = \partial S/\partial\boldsymbol{r}$ and $\mathcal{E} = -\partial S/\partial t$. Since $\boldsymbol{p}$ and $\mathcal{E}$ are large and $\hbar$ may be neglected in comparison with S, it follows that A has a slow variation, such that we recover the definition of the momentum and the energy by using $-i\hbar\frac{\partial}{\partial \boldsymbol{r}}$ for $\boldsymbol{p}$ and $i\hbar\frac{\partial}{\partial t}$ for $\mathcal{E}$ with the wavefunction Ψ. It is worth noting that in this limit $\hbar \longrightarrow 0$ the derivatives $\frac{\partial}{\partial \boldsymbol{r}}$ and $\frac{\partial}{\partial t}$ are large. Therefore, the dynamics, in the non-relativistic case, is given by the equation

$$
i\hbar\frac{\partial \Psi}{\partial t} = H(\boldsymbol{p} = -i\hbar\partial/\partial\boldsymbol{r}, \boldsymbol{r}, t)\Psi \ , \tag{5.39}
$$

which is the Schroedinger equation; it is transcribed by means of the wavefunction from $\mathcal{E} = \frac{1}{2m}(\boldsymbol{P} - q\boldsymbol{A}/c)^2 + q\Phi$, which is the Hamilton-Jacobi equation for a charge q with mass m, in electromagnetic field, written with energy and momentum. Obviously, the generalization to ensembles of many particles, or other (mechanically describable) situations, is immediate (though for condensed matter a field description is preferable, in order to account for the motion of the particle density, and the density of other physical quantities).

Schroedinger equation leads immediately to

$$
\begin{aligned}
&\tfrac{\partial S}{\partial t} + \tfrac{1}{2m}(gradS)^2 - \tfrac{\hbar^2}{2mA}\Delta A = 0 \ , \\
&\tfrac{\partial}{\partial t}(A^2) + \tfrac{1}{m}div(A^2 gradS) = 0 \ ,
\end{aligned}
\tag{5.40}
$$

by using the quasi-classical wavefunction $\Psi = Ae^{\frac{i}{\hbar}S}$. The first equation (5.40) in the limit $\hbar \rightarrow 0$ (and slow variations of A) is the

Hamilton-Jacobi equation, as expected. The second equation (5.40) is the conservation of the quantity A^2 which moves with velocity $\boldsymbol{v} = \boldsymbol{p}/m$; in the limit of slow variations of A, we get $div\boldsymbol{v} = 0$, as expected. This shows that $A^2 = \mid \Psi \mid^2$ is the probability of localization of the particle. Therefore, the quantum-mechanical quantities are operators, we measure their mean values on the wavefunction and statistical deviations, the wavefunction indicates uncertainties of the form $\Delta p_x \Delta x \geqq \hbar/2$, $\Delta\mathcal{E}\Delta t \geqq \hbar/2$ (half the unit radius of the semi-circle with a phase variation π). It is worth noting that the non-relativistic Schroedinger equation has a quasi-classical limit which provides its meaning. For small values of the action, comparable to $\hbar$, the motion is quantum-mechanical, as described by Schroedinger equation. For large variations of the action, as compared to $\hbar$, the Schroedinger equation passes over in the Hamilton-Jacobi equation of a classical motion. A quasi-classical description in terms of a wavefunction persists for correlated quantum-mechanical motions in coherent states. The Schroedinger equation is obtained formally by replacing the momentum by $-i\hbar\frac{\partial}{\partial \boldsymbol{r}}$ and the energy by $i\hbar\frac{\partial}{\partial t}$ in the Hamilton-Jacobi equation (interaction with fields included).

Let us do the same replacement in the relativistic Hamilton-Jacobi equation $\mathcal{E}^2 = c^2p^2 + m^2c^4$. We get the Klein-Gordon equation[1]

$$\hbar^2\frac{\partial^2\Psi}{\partial t^2} - c^2\hbar^2\Delta\Psi + m^2c^4\Psi = 0 \ . \tag{5.41}$$

Inserting $\Psi = Ae^{\frac{i}{\hbar}S}$ in this equation leads to

$$\begin{gathered}\left(\frac{\partial S}{\partial t}\right)^2 - c^2(gradS)^2 - m^2c^4 - \frac{\hbar^2}{A}\left(\frac{\partial^2 A}{\partial t^2} - c^2\Delta A\right) = 0 \ , \\ \frac{\partial}{\partial t}\left(A^2\frac{\partial S}{\partial t}\right) - c^2div\left(A^2gradS\right) = 0 \ .\end{gathered} \tag{5.42}$$

In the limit $\hbar \rightarrow 0$ and a slowly varying A the first equation (5.42) is the Hamilton-Jacobi equation, but the second equation (5.42) becomes

$$\frac{\partial^2 S}{\partial t^2} - c^2\Delta S = 0 \ , \tag{5.43}$$

[1] O. Klein, "Quantentheorie und fuenfdimensionale Relativitaetstheorie", Z. Phys. **37** 895 (1926); W. Gordon, "Der Comptoneffekt nach der Schroedingerschen Theorie", Z. Phys. **40** 117 (1926).

which shows that S is a function of $t \pm r/c$ (plane wave), such that the Hamilton-Jacobi equation reduces to $m^2c^4 = 0$. We conclude that the Klein-Gordon equation has no quasi-classical limit, which means that the quantization of the relativistic motion is impossible. This is an expected result, because a relativistic particle has a large energy and a large momentum, which make its motion a classical motion. Only the small variations of this motion might be quantized. Moreover, if we admit a quantum-mechanical motion for the relativistic particle, we cannot give a probability-conservation meaning of the second equation (5.42), since $\partial S/\partial t$ may have either sign.

5.4 Dirac equation

Let us introduce the relativistic notations p^μ for $(\mathcal{E}/c, \mathbf{p})$, $\mu = 0, 1, 2, 3$, and x^μ for $(ct, \mathbf{r})$, as well as $\partial^\mu = \partial/\partial x_\mu$ for $(\frac{\partial}{c\partial t}, -grad)$ and $\partial_\mu = \partial/\partial x^\mu$ for $(\frac{\partial}{c\partial t}, grad)$, with the metric $g_{\mu\nu} = g^{\mu\nu} = (+, -, -, -)$; such that $p^\mu = i\hbar\partial^\mu$ and $p_\mu = i\hbar\partial_\mu$. If we put

$$\gamma^\mu p_\mu \psi = mc\psi \ , \tag{5.44}$$

where γ^μ are matrices which satisfy the condition

$$\{\gamma^\mu, \gamma^\nu\} = 2g^{\mu\nu} \ , \tag{5.45}$$

then we recover the Klein-Gordon equation

$$\begin{aligned} \gamma^\mu p_\mu \gamma^\nu p_\nu \psi = \gamma^\mu \gamma^\nu p_\mu p_\nu \psi = \tfrac{1}{2}\{\gamma^\mu, \gamma^\nu\} p_\mu p_\nu \psi = \\ = g^{\mu\nu} p_\mu p_\nu \psi = p_\mu p^\mu \psi = m^2c^2\psi \ ; \end{aligned} \tag{5.46}$$

by applying twice the equation (5.44); in these equations ψ is a quantity with four components (a bispinor). Equation (5.44) is the Dirac equation.[2] It is relativistically invariant. In the presence of an electromagnetic field with potentials A_μ, if we apply twice the equation

$$\gamma^\mu \left(p_\mu - \frac{q}{c} A_\mu\right) \psi = mc\psi \tag{5.47}$$

[2]P. A. M. Dirac, "The quantum theory of the electron", Proc. Roy. Soc. **A117** 610 (1928); Part II., **A118** 351 (1928).

for a charge q, we get

$$\left(p_\mu - \frac{q}{c}A_\mu\right)\left(p^\mu - \frac{q}{c}A^\mu\right) - \frac{iq\hbar}{2c}\sigma^{\mu\nu}F_{\mu\nu} = m^2c^2 \tag{5.48}$$

(leaving aside ψ), where $\sigma^{\mu\nu} = \frac{1}{2}[\gamma^\mu, \gamma^\nu]$ and $F_{\mu\nu} = \partial_\mu A_\nu - \partial_\nu A_\mu$ is the electromagnetic field.

The matrices γ^μ are a representation of the Lorentz group; they can be chosen as

$$\gamma^0 = \begin{pmatrix} 1 & 0 & 0 & 0 \\ 0 & 1 & 0 & 0 \\ 0 & 0 & -1 & 0 \\ 0 & 0 & 0 & -1 \end{pmatrix}, \ \gamma^1 = \begin{pmatrix} 0 & 0 & 0 & 1 \\ 0 & 0 & 1 & 0 \\ 0 & -1 & 0 & 0 \\ -1 & 0 & 0 & 0 \end{pmatrix},$$
$$\gamma^2 = \begin{pmatrix} 0 & 0 & 0 & -i \\ 0 & 0 & i & 0 \\ 0 & i & 0 & 0 \\ -i & 0 & 0 & 0 \end{pmatrix}, \ \gamma^3 = \begin{pmatrix} 0 & 0 & 1 & 0 \\ 0 & 0 & 0 & -1 \\ -1 & 0 & 0 & 0 \\ 0 & 1 & 0 & 0 \end{pmatrix}, \tag{5.49}$$

or, with a shorthand notation,

$$\gamma^0 = \begin{pmatrix} 1 & 0 \\ 0 & -1 \end{pmatrix}, \ \boldsymbol{\gamma} = \begin{pmatrix} 0 & \boldsymbol{\sigma} \\ -\boldsymbol{\sigma} & 0 \end{pmatrix}, \tag{5.50}$$

where $\boldsymbol{\sigma}$ are Pauli's matrices

$$\sigma^1 = \begin{pmatrix} 0 & 1 \\ 1 & 0 \end{pmatrix}, \ , \sigma^2 = \begin{pmatrix} 0 & -i \\ i & 0 \end{pmatrix}, \ \sigma^3 = \begin{pmatrix} 1 & 0 \\ 0 & -1 \end{pmatrix}. \tag{5.51}$$

The electromagnetic field can be written as

$$F_{\mu\nu} = \begin{pmatrix} 0 & E_x & E_y & E_z \\ -E_x & 0 & -H_z & H_y \\ -E_y & H_z & 0 & -H_x \\ -E_z & -H_y & H_x & 0 \end{pmatrix} = (\mathbf{E}, \mathbf{H}), \tag{5.52}$$

and, with the same notation, $\sigma^{\mu\nu} = (\boldsymbol{\alpha}, i\boldsymbol{\Sigma})$, where

$$\boldsymbol{\alpha} = \begin{pmatrix} 0 & \boldsymbol{\sigma} \\ \boldsymbol{\sigma} & 0 \end{pmatrix}, \ \boldsymbol{\Sigma} = \begin{pmatrix} \boldsymbol{\sigma} & 0 \\ 0 & \boldsymbol{\sigma} \end{pmatrix}. \tag{5.53}$$

We can see that two labels of ψ belong to the spin (1/2), while the other two belong to the $\pm\mathcal{E}$; for $-\mathcal{E}$ the momentum should be $-\mathbf{p}$ too, such that the charge is $-q$; these are the antiparticles (positrons, for electrons); the Dirac equation predicts the existence of the spin 1/2 and the antiparticles.

If we multiply the Dirac equation (5.44) by γ^0 on the left, we get

$$p_0\psi = (\boldsymbol{\alpha}\mathbf{p} + mc\beta)\,\psi \ , \tag{5.54}$$

or

$$i\hbar\frac{\partial\psi}{\partial t} = \left(-i\hbar c\boldsymbol{\alpha} grad + mc^2\beta\right)\psi \ , \tag{5.55}$$

where $\boldsymbol{\alpha} = \gamma^0\boldsymbol{\gamma}$ is given by equation (5.53) and β is another notation for γ^0. This equation may be viewed as a relativistic Schroedinger equation. In the presence of the electromagnetic field Dirac's equation (5.47) becomes

$$p_0\psi = (\boldsymbol{\alpha}\mathbf{p} + mc\beta)\,\psi + \frac{q}{c}\gamma^0\gamma^\mu A_\mu\psi \ ; \tag{5.56}$$

the last term in equation (5.56) is the energy of interaction of the charge with the radiation. The interaction hamiltonian is

$$V = \frac{q}{c}\int d\mathbf{r}\psi^*\gamma^0\gamma^\mu\psi A_\mu \ . \tag{5.57}$$

We define the current $j^\mu = \psi^*\gamma^0\gamma^\mu\psi$ and introduce the notation $\overline{\psi} = \psi^*\gamma^0$; then

$$V = \frac{q}{c}\int d\mathbf{r}j^\mu A_\mu \ . \tag{5.58}$$

If we multiply equation (5.56) by ψ^* on the left, and the equation for ψ^* by ψ on the right, we get the conservation of the current

$$\frac{\partial}{\partial t}(\psi^*\psi) + c\cdot grad(\psi^*\boldsymbol{\alpha}\psi) = 0 \tag{5.59}$$

(or $\partial_\mu j^\mu = 0$). This equation may suggest that we can view ψ as a wavefunction, with a conserved probability $\psi^*\psi$. However, this current moves with velocity $c\boldsymbol{\alpha}$, which is an unphysical, undetermined motion.

5.5 Zitterbewegung

The Dirac equation

$$i\hbar\frac{\partial\psi}{\partial t} = \left(-i\hbar c\boldsymbol{\alpha} grad + mc^2\beta\right)\psi \tag{5.60}$$

becomes in the quasi-classical limit

$$\mathcal{E} = c\boldsymbol{\alpha}\,\boldsymbol{p} + mc^2\beta \ ; \tag{5.61}$$

this is not a valid equation. The result is preserved in the presence of the electromagnetic field with potentials Φ, $\boldsymbol{A}$ (equation (5.48) for $\hbar \to 0$), where the Dirac equation becomes

$$\mathcal{E} = c\boldsymbol{\alpha}\left(\boldsymbol{p} - \frac{q}{c}\boldsymbol{A}\right) + mc^2\beta + q\Phi \ . \tag{5.62}$$

It follows that the Dirac equation is a purely quantum-mechanical equation, which describes a motion with action comparable to $\hbar$. On the other hand, the motion described by the Dirac equation has energies of the order mc^2 and momenta of the order mc, such that, in fact, the motion described by the Dirac equation requires times of the order $\hbar/mc^2$ and distances of the order $\hbar/mc$. The distance $\lambda_c = \hbar/mc$ is the Compton wavelength of the particle. For an electron $\lambda_c = 3.8 \times 10^{-11}cm$. This distance is covered by the particle in time $\hbar/mc^2$, such that the particle would have a velocity of the order c. Indeed, the same result is obtained from the conservation law of probability given by equation (5.59).[3] This is an unphysical result.

If we seek solutions of the form $\psi \sim e^{-\frac{i}{\hbar}Et+\frac{i}{\hbar}\boldsymbol{pr}}$, we get from the Dirac equation (5.60) four eigenvectors with double eigenvalues $\mathcal{E} = \pm\sqrt{c^2p^2+m^2c^4}$. The negative-energy solutions correspond to antiparticles, the double degeneracy corresponds to the two states of the

[3]G. Breit, "An interpretation of Dirac's theory of the electron", Proc. Nat. Acad. Sci. **14** 553 (1928).

1/2-spin. Equation (5.60) reads

$$\begin{pmatrix} mc^2 & 0 & cp_3 & c(p_1 - ip_2) \\ 0 & mc^2 & c(p_1 + ip_2) & -cp_3 \\ cp_3 & c(p_1 - ip_2) & -mc^2 & 0 \\ c(p_1 + ip_2) & -cp_3 & 0 & -mc^2 \end{pmatrix} \psi \tag{5.63}$$
$$= \mathcal{E}\psi \; .$$

The matrix on the left in this equation has a vanishing mean value $\overline{\mathcal{E}} = 0$ and a deviation $\sqrt{\overline{\mathcal{E}^2}} = \sqrt{c^2p^2 + m^2c^4}$. It follows that in the motion described by the Dirac equation we have an uncertainty of the order mc^2 in energy and mc in momenta. This motion mixes up the momentum components and the energy, as well as the particle-antiparticle and the spin states. But this uncertainty is of the same order of magnitude as the values determined by the Dirac equation. It follows that this undefined, mixed-up motion described by the Dirac equation is undetermined. It is the unphysical Zitterbewegung (or trembling motion).

The Dirac equation predicts the existence of the antiparticles, the existence of the spin, but the quantum-mechanical motion it describes is undetermined and unphysical over distances of the order λ_c. The application of the Dirac equation to various physical problems (like the hydrogen atom) amounts to using in fact the Schroedinger (or Pauli) equation with relativistic corrections. With the limitations induced by the cutoff length λ_c the Dirac equation can be used to describe quantum-mechanical motion. However, it includes both an undefined motion and a cutoff.

5.6 Quantum field

Leaving aside the Zitterbewegung, a quantum-mechanical motion which proceeds with an action comparable to $\hbar$ should take place over distances longer than $\lambda_c = \hbar/mc$ and during times longer than $\hbar/mc^2$. It follows that we need a momentum cutoff mc and an energy cutoff mc^2.[4] This is a motion which smooths out the Zitterbewegung,

[4] L. Landau and R. Peierls, "Erweiterung des Unbestimmtheitsprinzips fur die relativistische Quantentheorie", Z. Phys. **69** 56 (1931); L. Landau and E.

or a quantum-mechanical motion super-imposed over the Zitterbewegung, which averages it. It can be achieved by the following technical procedure.

We give a variation δx^μ to the coordinates x^μ, and denote it by u^μ, according to the scheme $x^\mu \longrightarrow x^\mu + \delta x^\mu$, $\delta x^\mu = u^\mu$, $x^\mu \longrightarrow x^\mu + u^\mu$. We will take the first-order variations with respect to u^μ of the Dirac equation

$$\gamma^\mu \partial_\mu \psi = \frac{mc}{i\hbar}\psi \tag{5.64}$$

(equation (5.44)). According to the Dirac equation the variations of the coordinates should be viewed as matrices (like γ^μ). We write $x^\mu = s^\mu \cdot 1$, where $s^0 = ct$, $\mathbf{s} = \mathbf{r}$ and 1 denotes the unit matrix; we have $x_\mu x^\mu = s_\mu s^\mu \cdot 1 = s^2 \cdot 1$, where $s^2 = c^2t^2 - \mathbf{r}^2$. For δx^μ we need $\delta x_\mu \delta x^\mu = u_\mu u^\mu = u^2 = ds^2$; the (non-trivial) solution of this equation is

$$\delta x^\mu = u^\mu = \frac{1}{2}u\gamma^\mu \tag{5.65}$$

(since $\gamma_\mu \gamma^\mu = 4$). The first-order expansion of the Dirac equation is

$$\gamma^\mu \left(\partial_\mu \psi + u^\nu \partial_\nu \partial_\mu \psi\right) = \frac{mc}{i\hbar}\left(\psi + u^\nu \partial_\nu \psi\right) \ , \tag{5.66}$$

or

$$\partial^\mu \partial_\mu \psi = -\frac{m^2c^2}{\hbar^2}\psi \ , \tag{5.67}$$

which is the Klein-Gordon equation $p_\mu p^\mu \psi = m^2c^2\psi$ (equation (5.46)). This equation describes the motion smoothed over the Zitterbewegung; we note that u is absorbed into ψ, hence the bosonic character of the latter. The ψ in equation (5.67) is u multiplied by a bispinor. It is a boson field with four components. In the presence of the electromagnetic field, we need to use the covariant derivative $D_\mu = \partial_\mu - \frac{q}{ic\hbar}A_\mu$, according to $\psi \longrightarrow \psi + u^\mu D_\mu \psi$; we get the Klein-Gordon equation (5.48) with electromagnetic field. The boson occupancy number is the number of electrons; the cutoff length λ_c ensures the exclusion principle.

Lifshitz, *Course of Theoretical Physics*, vol. 4, *Quantum Electrodynamics* (V Berestetskii, E. Lifshitz, L. Pitaevskii), Butterworth-Heinemann (1971).

The quantum-mechanical motion of a relativistic particle is effected by bosonic quanta of a field with four components.[5] For a field the localization probability is meaningless. The conservation laws are the energy-momentum conservation, the charge conservation, etc. The avoidance of the self-interaction is achieved by the cutoff λ_c.

[5] M. Apostol, *A lame Duck: Quantum Electrodynamics*, apoma, Magurele (2020).

6 Electron-Photon Interaction

6.1 Maxwell equations

The electric charge density ρ and the electric current density $\boldsymbol{j}$ are related by the charge conservation (continuity) equation

$$\frac{\partial\rho}{\partial t} + div\boldsymbol{j} = 0 \ . \tag{6.1}$$

The Maxwell equations in vacuum are

$$\begin{gathered} div\mathbf{E} = 4\pi\rho \ , \ div\mathbf{H} = 0 \ , \\ curl\mathbf{E} = -\frac{1}{c}\frac{\partial\mathbf{H}}{\partial t} \ , \ curl\mathbf{H} = \frac{1}{c}\frac{\partial\mathbf{E}}{\partial t} + \frac{4\pi}{c}\mathbf{j} \ , \end{gathered} \tag{6.2}$$

where $\boldsymbol{E}$ is the electric field, $\boldsymbol{H}$ is the magnetic field and c is the speed of light. All the quantities in these equations are real. We may view these equations as providing the electromagnetic fields $\boldsymbol{E}$ and $\boldsymbol{H}$ produced by the sources ρ and $\boldsymbol{j}$. Apart from the particular solution generated by sources (the self-field), the equations have also the free-field solution. Sources are localized, while the fields extend to the whole space. Therefore, we may imagine a space region with fields and without charges and currents. These are external fields.

The homogeneous Maxwell equations are solved by

$$\boldsymbol{E} = -\frac{1}{c}\frac{\partial\boldsymbol{A}}{\partial t} - grad\Phi \ , \ \boldsymbol{H} = curl\boldsymbol{A} \ , \tag{6.3}$$

where the vector potential $\boldsymbol{A}$ and the scalar potential Φ are given by the wave equations

$$\frac{1}{c^2}\frac{\partial^2\mathbf{A}}{\partial t^2} - \Delta\mathbf{A} = \frac{4\pi}{c}\mathbf{j} \ , \ \frac{1}{c^2}\frac{\partial^2\Phi}{\partial t^2} - \Delta\Phi = 4\pi\rho \ , \tag{6.4}$$

providing the Lorenz gauge

$$\frac{1}{c}\frac{\partial\Phi}{\partial t}+div\boldsymbol{A}=0\ , \tag{6.5}$$

is satisfied.

Electric charges q are pointlike particles with mass m which move with velocity $\boldsymbol{v}$; the charge density of a pointlike particle placed at $\boldsymbol{r}_0$ is $\rho = q\delta(\boldsymbol{r}-\boldsymbol{r}_0)$ and its current density is $\boldsymbol{j} = q\boldsymbol{v}\delta(\boldsymbol{r}-\boldsymbol{r}_0)$, where $\boldsymbol{v}=\dot{\boldsymbol{r}}_0$. The δ-functions generate fields which are singular at the position of the charge, at the retarded moment when the field is produced. The δ-functions and the field singularities imply that the pointlike charges have an infinitesimal dimension, such that the point at which the charge is placed is avoided.

The Maxwel equations lead to the charge-field energy conservation, according to the equation

$$\frac{1}{8\pi}\frac{\partial}{\partial t}(E^2+H^2)+\frac{c}{4\pi}div(\mathbf{E}\times\mathbf{H})+\mathbf{jE}=0\ . \tag{6.6}$$

In this equation $u=(E^2+H^2)/8\pi$ is the density of electromagnetic energy,

$$\mathbf{S}=\frac{c}{4\pi}\mathbf{E}\times\mathbf{H} \tag{6.7}$$

is the density of electromagnetic energy flux and $\boldsymbol{jE}$ is the rate of density of the mechanical work of the source. The fields in this equation include both the self-field and an external field. Similarly, the Maxwell equations lead to the momentum conservation

$$\mathbf{f}-\mathbf{t}+\frac{\partial\mathbf{g}}{\partial t}=0\ , \tag{6.8}$$

where

$$\mathbf{f}=\rho\mathbf{E}+\frac{1}{c}\mathbf{j}\times\mathbf{H} \tag{6.9}$$

is the density of the Lorentz force, $t_i=\partial_j\sigma_{ij}$, the tensor σ_{ij} given by

$$\sigma_{ij}=\frac{1}{4\pi}\left[E_iE_j+H_iH_j-\frac{1}{2}\delta_{ij}(E^2+H^2)\right]\ , \tag{6.10}$$

is the electromagnetic stress tensor and

$$\mathbf{g} = \frac{1}{4\pi c}\mathbf{E} \times \mathbf{H} = \frac{1}{c^2}\mathbf{S} \tag{6.11}$$

is the density of the electromagnetic momentum. A similar conservation law holds for the angular momentum.

The density of mechanical work $\boldsymbol{jE}$ in equation (6.6) leads to a charge-field interaction energy

$$V = \int d\boldsymbol{r}\left(\rho\Phi - \frac{1}{c}\boldsymbol{jA}\right) \ . \tag{6.12}$$

These equations may include external fields (solutions of the homogneous Maxwell equations) besides the (self-) field generated by the charges and the currents.

6.2 Self-interaction

Let us derive the electromagnetic interaction V (equation (6.12)) from the rate of density of the mechanical work $\boldsymbol{jE}$ by using $\boldsymbol{E} = -\frac{1}{c}\frac{\partial \boldsymbol{A}}{\partial t} - grad\Phi$. We have

$$\begin{aligned} V &= \int dtd\boldsymbol{r}\boldsymbol{j}\left(-\tfrac{1}{c}\tfrac{\partial \boldsymbol{A}}{\partial t} - grad\Phi\right) = \\ &= -\tfrac{1}{c}\int d\boldsymbol{r}\boldsymbol{j}\boldsymbol{A} - \int dtd\boldsymbol{r}\boldsymbol{j}grad\Phi \ . \end{aligned} \tag{6.13}$$

For a pointlike charge q with density $\rho = q\delta(\boldsymbol{r}-\boldsymbol{r}_0)$ the current density is

$$\boldsymbol{j} = q\frac{d\boldsymbol{r}_0}{dt}\delta(\boldsymbol{r}-\boldsymbol{r}_0) = -q\frac{d\boldsymbol{r}}{dt}\delta(\boldsymbol{r}-\boldsymbol{r}_0) = -\rho\frac{d\boldsymbol{r}}{dt} \ , \tag{6.14}$$

such that the second term on the right in equation (6.13) becomes

$$\begin{aligned} -\int dtd\boldsymbol{r}\boldsymbol{j}grad\Phi &= \int dtd\boldsymbol{r}\rho\tfrac{d\boldsymbol{r}}{dt}grad\Phi = \\ &= \int dtd\boldsymbol{r}\rho\tfrac{d\Phi}{dt} = \int d\boldsymbol{r}\rho\Phi \ . \end{aligned} \tag{6.15}$$

It is worth noting that the use of the velocity $d\boldsymbol{r}/dt$ instead of $d\boldsymbol{r}_0/dt$ means that we avoid the self-interaction of the charge. Thus, we get the electromagnetic interaction V given by equation (6.12).

A pointlike charge q produces in its immediate vicinity the Coulomb potential $\Phi = \frac{q}{|\boldsymbol{r}-\boldsymbol{r}_0|}$ and the electric field $\boldsymbol{E} = -grad\Phi$, and nothing else. The potential can be written as

$$\Phi = \int d\boldsymbol{r}' \frac{\rho(\boldsymbol{r}')}{\mid \boldsymbol{r}-\boldsymbol{r}' \mid} = \int d\boldsymbol{r}' \frac{q\delta(\boldsymbol{r}'-\boldsymbol{r}_0)}{\mid \boldsymbol{r}-\boldsymbol{r}' \mid} = \frac{q}{\mid \boldsymbol{r}-\boldsymbol{r}_0 \mid} \ . \tag{6.16}$$

The self-interaction of the charge is

$$\begin{aligned} V_s &= \int dt d\boldsymbol{r}\boldsymbol{j}\,(-grad\Phi) = \\ &= \int dt d\boldsymbol{r} q \tfrac{d\boldsymbol{r}_0}{dt}\delta(\boldsymbol{r}-\boldsymbol{r}_0)\,(-grad\Phi) = \\ &= -\int dt d\boldsymbol{r}\rho\tfrac{d\Phi}{dt} = -\int dt d\boldsymbol{r} d\boldsymbol{r}'\rho(r)\tfrac{d\rho(r')}{dt}\tfrac{1}{|\boldsymbol{r}-\boldsymbol{r}'|} = \\ &= -\tfrac{1}{2}\int d\boldsymbol{r} d\boldsymbol{r}' \tfrac{\rho(\boldsymbol{r})\rho(\boldsymbol{r}')}{|\boldsymbol{r}-\boldsymbol{r}'|} = -\tfrac{q^2}{2}\int d\boldsymbol{r}\tfrac{\delta(\boldsymbol{r}-\boldsymbol{r}_0)}{|\boldsymbol{r}-\boldsymbol{r}'|} \ . \end{aligned} \tag{6.17}$$

The integration in this singular integral can be performed over a spherical region with a small radius a. The result is of the order

$$V_s = -\frac{q^2}{2a} \ . \tag{6.18}$$

The energy of the field over the region occupied by the charge is

$$\begin{aligned} U_s &= \tfrac{1}{8\pi}\int d\boldsymbol{r} E^2 = -\tfrac{1}{8\pi}\int d\boldsymbol{r}\boldsymbol{E} grad\Phi = \\ &= \tfrac{1}{8\pi}\int d\boldsymbol{r} div\boldsymbol{E}\cdot\Phi = \tfrac{1}{2}\int d\boldsymbol{r}\rho\Phi = \\ &= \tfrac{q^2}{2}\int d\boldsymbol{r}\tfrac{\delta(\boldsymbol{r}-\boldsymbol{r}_0)}{|\boldsymbol{r}-\boldsymbol{r}'|} = \tfrac{q^2}{2a} \ . \end{aligned} \tag{6.19}$$

We can see that the self-interaction and the self-field energy cancel each other out.

In the vicinity of the charge the fields are static; therefore the time derivative of the field momentum density $\boldsymbol{g}$ is zero. By using $\boldsymbol{E} = \frac{q\boldsymbol{r}}{r^3}$, we get the stress tensor

$$\sigma_{ij} = \frac{q^2}{4\pi}\left(\frac{x_i x_j}{r^6} - \frac{1}{2r^4}\delta_{ij}\right) \tag{6.20}$$

(where x_i are the coordinates of the position $\boldsymbol{r}$). The density of the stress forces $t_i = \partial_j \sigma_{ij}$ is vanishing. Similarly, the integration of the Lorentz force density $\boldsymbol{f} = \rho \boldsymbol{E} = q\delta(\boldsymbol{r})\boldsymbol{r}/r^3$ over a small spherical region gives zero. Therefore, in its immediate vicinity a pointlike charge carries with it the Coulomb field, with a vanishing total energy (self-energy+self-interaction), which acts with a vanishing force upon the charge.

In conclusion, we should avoid the self-interaction and the self-energy by using a small but finite radius a for "pointlike" charges.

A charge in motion generates a propagating electromagnetic field. The charge moves in this field, so it interacts with its own field, as a consequence of the retardation. This is a self-interaction. Since we are interested in the field near the charge, we may leave aside the scalar potential, whose Coulomb field has no effect, as shown above. Also, according to the Lorenz gauge, the longitudinal field is left aside. The transverse vector potential is generated according to the equation

$$\frac{1}{c^2}\frac{\partial^2 \boldsymbol{A}}{\partial t^2} - \Delta \boldsymbol{A} = \frac{4\pi}{c}\boldsymbol{j} \ , \tag{6.21}$$

where $\boldsymbol{j}$ is the transverse current density. Let us assume a pointlike charge q with trajectory $\boldsymbol{r}_0(t)$, governed by an external force; its charge density is $\rho = q\delta(\boldsymbol{r} - \boldsymbol{r}_0)$ and its current density is $\boldsymbol{j} = q\boldsymbol{v}\delta(\boldsymbol{r} - \boldsymbol{r}_0)$, where $\boldsymbol{v} = \dot{\boldsymbol{r}}_0$. Equation (6.21) becomes

$$\frac{1}{c^2}\frac{\partial^2 \boldsymbol{A}}{\partial t^2} - \Delta \boldsymbol{A} = \frac{4\pi q}{c}\boldsymbol{v}\delta(\boldsymbol{r} - \boldsymbol{r}_0) \ . \tag{6.22}$$

We can see that $\boldsymbol{A}$ is parallel to $\boldsymbol{v}$. The field $\boldsymbol{A}$ acts upon the charge. Consequently, the charge with mass m acquires an additional velocity $\boldsymbol{v}'$, whose equation of motion is

$$m\frac{d\boldsymbol{v}'}{dt} = q\boldsymbol{E}_0 + \frac{1}{c}(\boldsymbol{v} + \boldsymbol{v}') \times \boldsymbol{H}_0 \ , \tag{6.23}$$

where $\boldsymbol{E}_0 = -\frac{1}{c}\frac{\partial \boldsymbol{A}_0}{\partial t}$ is the electric field and $\boldsymbol{H}_0 = curl \boldsymbol{A}_0$ is the magnetic field. The suffix 0 indicates that the fields are taken at position $\boldsymbol{r}_0$. This is the Lorentz force. Since we are interested in the region close to the charge, we may neglect the second term on the right in equation (6.23), and get

$$\frac{d\boldsymbol{v}'}{dt} = -\frac{q}{mc}\frac{d\boldsymbol{A}_0}{dt} \tag{6.24}$$

and $\boldsymbol{v}' = -\frac{q}{mc}\boldsymbol{A}_0$. It follows that the self-interaction produces a reaction current density

$$\boldsymbol{j}' = q\boldsymbol{v}'\delta(\boldsymbol{r}-\boldsymbol{r}_0) = -\frac{q^2}{mc}\boldsymbol{A}\delta(\boldsymbol{r}-\boldsymbol{r}_0) \ , \tag{6.25}$$

which, in its turn, determines a field given by

$$\frac{1}{c^2}\frac{\partial^2\boldsymbol{A}'}{\partial t^2} - \Delta\boldsymbol{A}' = -\frac{4\pi q^2}{mc^2}\boldsymbol{A}\delta(\boldsymbol{r}-\boldsymbol{r}_0) \ . \tag{6.26}$$

For $\boldsymbol{r}$ close to $\boldsymbol{r}_0$ we may limit ourselves to the static part of this field, which is $\boldsymbol{A}' = -\frac{q^2}{mc^2}\frac{\boldsymbol{A}_0}{|\boldsymbol{r}-\boldsymbol{r}_0|}$. The total field at the position of the charge is $\boldsymbol{A}_0 - \frac{q^2\boldsymbol{A}_0}{mc^2}\lim|_{\boldsymbol{r}\to\boldsymbol{r}_0}\frac{1}{|\boldsymbol{r}-\boldsymbol{r}_0|}$. This shows that the charge must be viewed as having an undetermined position inside a region with dimension of order

$$a = \frac{q^2}{mc^2} \ ; \tag{6.27}$$

the field at the position of the charge is zero and the force acting on the charge is vanishing; this way, the self-interaction is avoided. Therefore, we must limit ourselves to a cutoff length of the order $a = q^2/mc^2$, a cutoff wavevector of the order $1/a$ and cutoff frequency of the order c/a. In addition, if we see the charge as being pointlike and moving indefinitely inside a spherical region with radius a, its self-energy is of the order $q^2/a = mc^2$.

Also, the fields are limited by a cutoff $E_s = H_s = q/a^2 = m^2c^4/q^3$ (or $qE_sa = mc^2$, $m\omega^2 = q\omega H_s/c$, with $\omega = c/a$). $a = q^2/mc^2$ is the classical radius of the charge. It sets the limits of the classical Electromagnetism. For electrons $q = -e = -4.8\times10^{-10}esu$, $a \simeq 2.8\times10^{-13}cm^{-1}$ ($c/a \simeq 10^{23}s^{-1}$, $m \simeq 10^{-27}g$, $mc^2 \simeq 0.5MeV$, $1eV = 1.6\times10^{-12}erg$), $E_s = H_s \simeq 6\times10^{15}esu$ ($1esu = 3\times10^4V/m$).

We shall see that for a quantum-mechanical motion the cutoff length is the Compton wavelength $\lambda_c = \hbar/mc$, where $\hbar \simeq 10^{-27}erg\cdot s$ is Planck's constant. The cutoff momentum is mc, the cutoff frequency is $mc^2/\hbar$ and the self-energy (the cutoff energy) is mc^2. The cutoff quantum fields given above are $E_s = H_s = m^2c^3/\hbar q$; such a cutoff field is called the Schwinger limit. For electrons $\lambda_c \simeq 3.8 \times 10^{-11} cm$ and $E_s = H_s \simeq 4 \times 10^{13} esu$.

The ratio a/λ_c is $\alpha = a/\lambda_c = q^2/\hbar c$. For electrons $\alpha = e^2/\hbar c \simeq \frac{1}{137}$ is the fine-structure constant.

6.3 Longitudinal field

Let us represent the scalar potential in the volume V as the Fourier expansion

$$\Phi(t, \boldsymbol{r}) = \frac{1}{\sqrt{V}} \sum_{\boldsymbol{k}} \Phi_{\boldsymbol{k}}(t) e^{i\boldsymbol{kr}} \quad , \tag{6.28}$$

where $\Phi^*_{-\boldsymbol{k}} = \Phi_{\boldsymbol{k}}$. Similarly, the vector potential is represented as

$$\boldsymbol{A}(t, \boldsymbol{r}) = \frac{1}{\sqrt{V}} \sum_{\boldsymbol{k}} \boldsymbol{A}_{\boldsymbol{k}}(t) e^{i\boldsymbol{kr}} + \frac{1}{\sqrt{V}} \sum_{\boldsymbol{k}} \frac{i\boldsymbol{k}}{ck^2} \dot{\Phi}_{\boldsymbol{k}}(t) e^{i\boldsymbol{kr}} \quad , \tag{6.29}$$

where $\boldsymbol{k}\boldsymbol{A}_{\boldsymbol{k}} = 0$. The first term in equation (6.29) is the transverse part of the vector potential, while the second term is its longitudinal part. The Lorenz gauge is satisfied by this representation. The electric field is given by

$$\begin{gathered} \boldsymbol{E} = \frac{1}{\sqrt{V}} \sum_{\boldsymbol{k}} \left(-\frac{1}{c} \dot{\boldsymbol{A}}_{\boldsymbol{k}} \right) e^{i\boldsymbol{kr}} + \\ + \frac{1}{\sqrt{V}} \sum_{\boldsymbol{k}} \frac{i\boldsymbol{k}}{k^2} \left(-\frac{1}{c^2} \ddot{\Phi}_{\boldsymbol{k}} - k^2 \Phi_{\boldsymbol{k}} \right) e^{i\boldsymbol{kr}} = \\ = \frac{1}{\sqrt{V}} \sum_{\boldsymbol{k}} \left(-\frac{1}{c} \dot{\boldsymbol{A}}_{\boldsymbol{k}} \right) e^{i\boldsymbol{kr}} - \frac{1}{\sqrt{V}} \sum_{\boldsymbol{k}} \frac{4\pi i\boldsymbol{k}}{k^2} \rho_{\boldsymbol{k}} e^{i\boldsymbol{kr}} \quad , \end{gathered} \tag{6.30}$$

where $\rho_{\boldsymbol{k}}$ is the Fourier component of the charge density; it has both a transverse part and a longitudinal part. The magnetic field is

$$\boldsymbol{H} = \frac{1}{\sqrt{V}} \sum_{\boldsymbol{k}} i\boldsymbol{k} \times \boldsymbol{A}_{\boldsymbol{k}} e^{i\boldsymbol{kr}} \quad ; \tag{6.31}$$

it is a transverse field. The energy of the electromagnetic field can now be written as

$$U = \frac{1}{8\pi}\int d\boldsymbol{r}\left(E^2 + H^2\right) =$$
$$= \frac{1}{8\pi}\sum_{\boldsymbol{k}}\left(\frac{1}{c^2}\dot{\boldsymbol{A}}_{\boldsymbol{k}}\dot{\boldsymbol{A}}_{-\boldsymbol{k}} + k^2\boldsymbol{A}_{\boldsymbol{k}}\boldsymbol{A}_{-\boldsymbol{k}}\right) + \tag{6.32}$$
$$+\sum_{\boldsymbol{k}}\frac{2\pi}{k^2}\rho_{\boldsymbol{k}}\rho_{-\boldsymbol{k}} \; .$$

We recognize in the last row of equation (6.32) the Coulomb energy (with $4\pi/\sqrt{V}k^2$ the Fourier transform of the Coulomb potential). We have replaced $\frac{1}{c^2}\ddot{\Phi}_{\boldsymbol{k}} + k^2\Phi_{\boldsymbol{k}}$ in equation (6.30) by $4\pi\rho_{\boldsymbol{k}}$, as if the potential Φ is generated by a charge distribution. Consequently, the Coulomb energy in equation (6.32) arises from a self-interaction.

Let us now compute the interaction energy $V = \int dt d\boldsymbol{r}\boldsymbol{j}\boldsymbol{E}$, where $\boldsymbol{j}$ is the current density, by using the electric field given by equation (6.30). We get

$$V = \sum_{\boldsymbol{k}}\int dt\left(-\frac{1}{c}\boldsymbol{j}_{\boldsymbol{k}}\dot{\boldsymbol{A}}_{-\boldsymbol{k}}\right) + \sum_{\boldsymbol{k}}\int dt\frac{4\pi i\boldsymbol{k}\boldsymbol{j}_{\boldsymbol{k}}}{k^2}\rho_{-\boldsymbol{k}} \; . \tag{6.33}$$

Since $\sum_{\boldsymbol{k}}\left(-\frac{1}{c}\boldsymbol{j}_{\boldsymbol{k}}\dot{\boldsymbol{A}}_{-\boldsymbol{k}}\right)\Delta t = \sum_{\boldsymbol{k}}\left(-\frac{1}{c}\boldsymbol{j}_{\boldsymbol{k}}\Delta\boldsymbol{A}_{-\boldsymbol{k}}\right)$ is a mechanical work done in time Δt, it follows that $\sum_{\boldsymbol{k}}\left(-\frac{1}{c}\boldsymbol{j}_{\boldsymbol{k}}\boldsymbol{A}_{-\boldsymbol{k}}\right)$ is an interaction energy; therefore,

$$\begin{aligned} V &= \sum_{\boldsymbol{k}}\left(-\frac{1}{c}\boldsymbol{j}_{\boldsymbol{k}}\boldsymbol{A}_{-\boldsymbol{k}}\right) + \sum_{\boldsymbol{k}}\int dt\frac{4\pi i\boldsymbol{k}\boldsymbol{j}_{\boldsymbol{k}}}{k^2}\rho_{-\boldsymbol{k}} = \\ &= \sum_{\boldsymbol{k}}\left(-\frac{1}{c}\boldsymbol{j}_{\boldsymbol{k}}\boldsymbol{A}_{-\boldsymbol{k}}\right) - \sum_{\boldsymbol{k}}\int dt\frac{4\pi}{k^2}\dot{\rho}_{\boldsymbol{k}}\rho_{-\boldsymbol{k}} \; , \end{aligned} \tag{6.34}$$

where the continuity equation is used. In the second summation in equation (6.34) we change $\boldsymbol{k}$ into $-\boldsymbol{k}$ and write $\frac{1}{2}\left(\dot{\rho}_{\boldsymbol{k}}\rho_{-\boldsymbol{k}} + \dot{\rho}_{-\boldsymbol{k}}\rho_{\boldsymbol{k}}\right) = \frac{1}{2}\frac{d}{dt}\left(\dot{\rho}_{\boldsymbol{k}}\rho_{-\boldsymbol{k}}\right)$. Finally, we get the interaction

$$V = \sum_{\boldsymbol{k}}\left(-\frac{1}{c}\boldsymbol{j}_{\boldsymbol{k}}\boldsymbol{A}_{-\boldsymbol{k}}\right) - \sum_{\boldsymbol{k}}\frac{2\pi}{k^2}\rho_{\boldsymbol{k}}\rho_{-\boldsymbol{k}} \; . \tag{6.35}$$

The Coulomb interaction in equation (6.35) is a self-interaction. We can see that this interaction cancels the Coulomb energy out in equation (6.32). The longitudinal degrees of freedom do not participate in the dynamics of the charges, the currents and the electromagnetic field. This amounts to view the field as a free radiation field, with a vanishing scalar potential, interacting with currents, although the longitudinal part of the field remains in the electric field (equation (6.30)). We are left with the transverse field energy

$$U = \frac{1}{8\pi}\sum_{\boldsymbol{k}}\left(\frac{1}{c^2}\dot{\boldsymbol{A}}_{\boldsymbol{k}}\dot{\boldsymbol{A}}_{-\boldsymbol{k}} + k^2\boldsymbol{A}_{\boldsymbol{k}}\boldsymbol{A}_{-\boldsymbol{k}}\right) \tag{6.36}$$

and the transverse part of the interaction

$$V = \sum_{\boldsymbol{k}}\left(-\frac{1}{c}\boldsymbol{j}_{\boldsymbol{k}}\boldsymbol{A}_{-\boldsymbol{k}}\right) \quad . \tag{6.37}$$

It is worth noting that this interaction is not a self-interaction. We can see that the field is a set of harmonic oscillators with eigenfrequencies $\pm\omega_k$, $\omega_k = ck$, with momenta $\boldsymbol{P}_{\boldsymbol{k}} = \frac{1}{4\pi c^2}\dot{\boldsymbol{A}}_{\boldsymbol{k}}$ and coordinates $\boldsymbol{A}_{\boldsymbol{k}}$, whose equation of motion is the Maxwell equation

$$\frac{1}{c^2}\ddot{\boldsymbol{A}}_{\boldsymbol{k}} + k^2\boldsymbol{A}_{\boldsymbol{k}} = \frac{4\pi}{c}\boldsymbol{j}_{\boldsymbol{k}} \quad . \tag{6.38}$$

Noteworthy, in this equation $\boldsymbol{j}_{\boldsymbol{k}}$ is the transverse part of the current. Since $\boldsymbol{A}^*_{-\boldsymbol{k}} = \boldsymbol{A}_{\boldsymbol{k}}$, we may write $\boldsymbol{A}_{\boldsymbol{k}} = \alpha_k \boldsymbol{e}_{\boldsymbol{k}} a_{\boldsymbol{k}}(t) + \alpha_k \boldsymbol{e}_{-\boldsymbol{k}} a^*_{-\boldsymbol{k}}(t)$, where α_k and $a_{\boldsymbol{k}}$ are undetermined coefficients and $\boldsymbol{e}_{\boldsymbol{k}}$ is the polarization vector ($\boldsymbol{e}_{\boldsymbol{k}}\boldsymbol{e}_{\boldsymbol{k}} = 1$, $\boldsymbol{e}_{\boldsymbol{k}}\boldsymbol{k} = 0$). For free fields $a_{\boldsymbol{k}}(t) = a_{\boldsymbol{k}}e^{-i\omega_k t}$, $\omega_k = ck$, because $\boldsymbol{A}_{\boldsymbol{k}}$ should satisfy the free wave equation $\frac{1}{c^2}\ddot{\boldsymbol{A}}_{\boldsymbol{k}} + k^2\boldsymbol{A}_{\boldsymbol{k}} = 0$. The free field energy becomes

$$U = \sum_{\boldsymbol{k}}\frac{\omega_k^2\alpha_k^2}{4\pi c^2}\left(a^*_{\boldsymbol{k}}a_{\boldsymbol{k}} + a_{\boldsymbol{k}}a^*_{\boldsymbol{k}}\right) \quad . \tag{6.39}$$

Similarly, the total momentum of the free field (equation (6.11)) is

$$\begin{gathered} \boldsymbol{G} = \tfrac{1}{4\pi c}\int d\boldsymbol{r}\boldsymbol{E}\times\boldsymbol{H} = \tfrac{1}{c^2}\int d\boldsymbol{r}\boldsymbol{S} = \\ = \sum_{\boldsymbol{k}}\tfrac{\omega_k\alpha_k^2}{4\pi c^2}\boldsymbol{k}\left(a^*_{\boldsymbol{k}}a_{\boldsymbol{k}} + a_{\boldsymbol{k}}a^*_{\boldsymbol{k}}\right) \;; \end{gathered} \tag{6.40}$$

the total angular momentum is vanishing.

6.4 Photons

We choose the coefficients

$$\alpha_k = c\sqrt{\frac{2\pi\hbar}{\omega_k}} \ , \tag{6.41}$$

where $\hbar$ is Planck's constant; the transverse vector potential reads

$$\begin{aligned} \boldsymbol{A} &= \tfrac{1}{\sqrt{V}} \sum_{\boldsymbol{k}} \alpha_k \left(\boldsymbol{e_k} a_{\boldsymbol{k}} + \alpha_k \boldsymbol{e}_{-\boldsymbol{k}} a^*_{-\boldsymbol{k}} \right) e^{i\boldsymbol{kr}} = \\ &= \tfrac{1}{\sqrt{V}} \sum_{\boldsymbol{k}} \alpha_k \boldsymbol{e_k} \left(a_{\boldsymbol{k}} e^{i\boldsymbol{kr}} + a^*_{\boldsymbol{k}} e^{-i\boldsymbol{kr}} \right) \ . \end{aligned} \tag{6.42}$$

We can view the coefficients $a_{\mathbf{k}}$ and $a^*_{\mathbf{k}}$ as coordinates of a motion with energy U and momentum $\mathbf{G}$. We quantize this motion by imposing commutation relations

$$[a_{\mathbf{k}}, a^*_{\mathbf{k}'}] = \delta_{\mathbf{kk}'} \ , \ \ [a_{\mathbf{k}}, a_{\mathbf{k}'}] = 0 \ ; \tag{6.43}$$

the energy and the momentum (equations (6.39) and (6.40)) become

$$U = \sum_{\boldsymbol{k}} \frac{1}{2} \hbar\omega_k \left(a^*_{\boldsymbol{k}} a_{\boldsymbol{k}} + a_{\boldsymbol{k}} a^*_{\boldsymbol{k}} \right) = \sum_{\boldsymbol{k}} \hbar\omega_k \left(a^*_{\boldsymbol{k}} a_{\boldsymbol{k}} + 1/2 \right) \tag{6.44}$$

and

$$\boldsymbol{G} = \sum_{\boldsymbol{k}} \frac{1}{2} \hbar\boldsymbol{k} \left(a^*_{\boldsymbol{k}} a_{\boldsymbol{k}} + a_{\boldsymbol{k}} a^*_{\boldsymbol{k}} \right) = \sum_{\boldsymbol{k}} \hbar\boldsymbol{k} \left(a^*_{\boldsymbol{k}} a_{\boldsymbol{k}} + 1/2 \right) \ . \tag{6.45}$$

The time dependence of the coefficients $a_{\mathbf{k}}$ and $a^*_{\mathbf{k}}$ is obtained by using the free hamiltonian of the electromagnetic field

$$H_{em} = U = \sum_{\mathbf{k}} \hbar\omega_k \left(a^*_{\mathbf{k}} a_{\mathbf{k}} + \frac{1}{2} \right) \tag{6.46}$$

in the quantum-mechanical equations of motion

$$i\hbar \frac{da_{\mathbf{k}}}{dt} = [a_{\mathbf{k}}, H_{em}] \ . \tag{6.47}$$

We can see that the motion of the electromagnetic field is realized by quantum-mechanical harmonic oscillators. The corresponding quanta of the electromagnetic field are called photons. The eigenstates $\mid n_{\mathbf{k}} >$ of H_{em} are defined by non-negative integers $n_{\mathbf{k}} \geq 0$, such that the action of the occupation number $n_{\mathbf{k}} = a^*_{\mathbf{k}} a_{\mathbf{k}}$ is

$$n_{\mathbf{k}} \mid n_{\mathbf{k}} >= a^*_{\mathbf{k}} a_{\mathbf{k}} \mid n_{\mathbf{k}} >= n_{\mathbf{k}\alpha} \mid n_{\mathbf{k}\alpha} > ; \tag{6.48}$$

it follows immediately

$$\begin{aligned} a_{\mathbf{k}} \mid n_{\mathbf{k}} >&= \sqrt{n_{\mathbf{k}}} \mid n_{\mathbf{k}} - 1 > , \\ a^*_{\mathbf{k}} \mid n_{\mathbf{k}} >&= \sqrt{n_{\mathbf{k}} + 1} \mid n_{\mathbf{k}\alpha} + 1 > \end{aligned} \tag{6.49}$$

and the existence of a vacuum state $\mid 0_{\mathbf{k}} >$, such that $a_{\mathbf{k}} \mid 0_{\mathbf{k}} >= 0$. $a_{\mathbf{k}}$ is a photon destruction (annihilation) operator and $a^*_{\mathbf{k}}$ is a photon creation operator.

6.5 Non-relativistic Quantum Mechanics

The quantum-mechanical electromagnetic field interacts through photons with quantum-mechanical pointlike charges. The non-relativistic classical motion of a pointlike particle with mass m and charge e in an electromagnetic field with potentials Φ and $\boldsymbol{A}$ is given by the Hamilton-Jacobi equation

$$-\mathcal{E} + e\Phi + \mathcal{H}(\boldsymbol{p} - \frac{e}{c}\boldsymbol{A}) = 0 , \tag{6.50}$$

where $\mathcal{H}(\boldsymbol{p}) = \frac{1}{2m}p^2$ is the non-relativistic hamiltonian (without Φ), $\mathcal{E} = -\frac{\partial S}{\partial t}$ is the energy, $\boldsymbol{p} = \frac{\partial S}{\partial \boldsymbol{r}}$ is the momentum and S is the action. The solution S of this equation gives the trajectory of the particle. From these definitions we get $\frac{\partial \mathcal{E}}{\partial \boldsymbol{r}} = -\dot{\boldsymbol{p}}$, which is one of the Hamilton equation. On the other hand, the minimum of the variation of the action $\Delta S = \boldsymbol{p}\Delta \boldsymbol{r} - \mathcal{E}\Delta t$ along the trajectory leads to the other Hamilton equation $\frac{\partial \mathcal{E}}{\partial \boldsymbol{p}} = \frac{d\boldsymbol{r}}{dt} = \boldsymbol{v}$, where $\boldsymbol{v}$ is the velocity.

The classical motion proceeds by large variations of the action, with determined derivatives $\boldsymbol{p}$ and $\mathcal{E}$. For small variations of S, comparable to $\hbar$, the classical motion is undetermined, in the sense that there is no trajectory. This is a quantum-mechanical motion.

The lack of trajectory suggests a wave, of the form $\Psi = Ae^{\frac{i}{\hbar}S}$. In the (quasi-) classical motion $\Delta S \gg \hbar$ and S may be taken as the classical action; the expansion $\Delta S = \boldsymbol{p}\Delta\boldsymbol{r} - \mathcal{E}\Delta t$ is possible, and it shows that there exists a velocity, *i.e.* a trajectory, which is the ray of the wave. Therefore the wavefunction Ψ recovers the classical motion in the limit $\Delta S \gg \hbar$. By making use of this Ψ, with S the classical action, we get the quasi-classical description of the motion. Moreover,

$$\begin{aligned} -i\hbar\tfrac{\partial}{\partial \boldsymbol{r}}\Psi &= -i\hbar\tfrac{\partial A}{\partial \boldsymbol{r}} + \boldsymbol{p}\Psi \ , \\ i\hbar\tfrac{\partial \Psi}{\partial t} &= i\hbar\tfrac{\partial A}{\partial t}e^{\frac{i}{\hbar}S} + \mathcal{E}\Psi \ , \end{aligned} \tag{6.51}$$

where $\boldsymbol{p} = \partial S/\partial \boldsymbol{r}$ and $\mathcal{E} = -\partial S/\partial t$. Since $\boldsymbol{p}$ and $\mathcal{E}$ are large and $\hbar$ may be neglected in comparison with S, it follows that A has a slow variation, such that we recover the definition of the momentum and the energy by using $-i\hbar\frac{\partial}{\partial \boldsymbol{r}}$ for $\boldsymbol{p}$ and $i\hbar\frac{\partial}{\partial t}$ for $\mathcal{E}$ with the wavefunction Ψ. It is worth noting that in this $\hbar \longrightarrow 0$ limit the derivatives $\frac{\partial}{\partial \boldsymbol{r}}$ and $\frac{\partial}{\partial t}$ are large. Therefore, the dynamics, in the non-relativistic case, is given by the equation

$$i\hbar\frac{\partial \Psi}{\partial t} = \mathcal{H}(\boldsymbol{p} = -i\hbar\partial/\partial\boldsymbol{r})\Psi \ , \tag{6.52}$$

which is the Schroedinger equation (without field); it is transcribed by means of the wavefunction from the Hamilton-Jacobi equation (6.50). The field can be introduced straightforwardly from equation (6.50). Schroedinger equation leads immediately to

$$\begin{aligned} &\tfrac{\partial S}{\partial t} + \tfrac{1}{2m}(gradS)^2 - \tfrac{\hbar^2}{2mA}\Delta A = 0 \ , \\ &\tfrac{\partial}{\partial t}(A^2) + \tfrac{1}{m}div(A^2 gradS) = 0 \ , \end{aligned} \tag{6.53}$$

by using the wavefunction $\Psi = Ae^{\frac{i}{\hbar}S}$. The first equation (6.53) in the limit $\hbar \rightarrow 0$ (and slow variations of A) is the Hamilton-Jacobi equation, as expected. The second equation (6.53) is the conservation of the quantity A^2 which moves with velocity $\boldsymbol{v} = \boldsymbol{p}/m$; in the limit of slow variations of A, we get $div\boldsymbol{v} = 0$, as expected. This shows that $A^2 = \mid \Psi \mid^2$ is the probability of localization of the particle.

Therefore, the quantum-mechanical quantities are operators (like $\mathcal{E} \rightarrow i\hbar\frac{\partial}{\partial t}$, $p = i\partial\frac{\partial}{\partial r}$), and we measure their mean values on the wavefunction and statistical deviations. Moreover, the uncertainties in the action are of the form $\Delta p_x \Delta x$, $\Delta\mathcal{E}\Delta t$, and the wavefunction indicates the uncertainty relations $\Delta p_x \Delta x \geqq \hbar/2$, $\Delta\mathcal{E}\Delta t \geqq \hbar/2$ (half the unit radius of the semi-circle with a phase variation π). It is worth noting that the non-relativistic Schroedinger equation has a quasi-classical limit which provides its meaning.

6.6 Relativistic particle

By its definition, a relativistic particle has a large action. Consequently, a quantum-mechanical motion of a relativistic particle is meaningless. The variations of the action of a relativistic particle may be small; such variations may admit a quantum-mechanical description. But this is not a quantum-mechanical description of the relativistic motion: it is a quantum-mechanical description of the variations of the relativistic motion, *i.e.* a quantum-mechanical description of the interaction of a relativistic particle.

A relativistic particle with mass m has a minimum momentum mc and a minimum energy mc^2. Consequently, according to the quantum-mechanical uncertainties, the quantum-mechanical motion of a relativistic particle implies a minimum length of the order of the Compton wavelength $\lambda_c = \hbar/mc$ and a minimum time of the order $\hbar/mc^2$.[1] These limitations apply also to the variations of the relativistic motion.

The Hamilton-Jacobi equation for a relativistic particle with mass m is

$$\mathcal{E}^2 - c^2p^2 - m^2c^4 = 0 \ . \tag{6.54}$$

If, according to the quantum-mechanical rules, we replace $\mathcal{E}$ by $i\hbar\frac{\partial}{\partial t}$

[1] L. Landau and R. Peierls, "Erweiterung des Unbestimmtheitsprinzips fur die relativistische Quantentheorie", Z. Phys. **69** 56 (1931); L. Landau and E. Lifshitz, *Course of Theoretical Physics*, vol. 4, *Quantum Electrodynamics* (V. Berestetskii, E. Lifshitz, L. Pitaevskii), Butterworth-Heinemann (1971).

and $\boldsymbol{p}$ by $-i\hbar\frac{\partial}{\partial \boldsymbol{r}}$ in this equation, we get the Klein-Gordon equation[2]

$$\hbar^2\frac{\partial^2\Psi}{\partial t^2} - c^2\hbar^2\Delta\Psi + m^2c^4\Psi = 0 \tag{6.55}$$

for a wavefunction Ψ. For a general form $\Psi = Ae^{\frac{i}{\hbar}S}$ of this wavefunction we get from this equation

$$\begin{gathered}\left(\frac{\partial S}{\partial t}\right)^2 - c^2(gradS)^2 - m^2c^4 - \frac{\hbar^2}{A}\left(\frac{\partial^2 A}{\partial t^2} - c^2\Delta A\right) = 0 \ , \\ \frac{\partial}{\partial t}\left(A^2\frac{\partial S}{\partial t}\right) - c^2 div\left(A^2 gradS\right) = 0 \ .\end{gathered} \tag{6.56}$$

In the quasi-classical limit $\hbar \rightarrow 0$ and for a slowly-varying A the first equation (6.56) is the classical Hamilton -Jacobi equation, with S the classical action. The second equation is a continuity (conservation) equation. But $A^2\frac{\partial S}{\partial t}$ is not a positive quantity, and its motion proceeds with velocity c. We cannot give a meaning to this conservation equation. In the quasi-classical limit, when A varies slowly, this continuity equation leads to the wave equation $\frac{\partial^2 S}{\partial t^2} - c^2\Delta S = 0$ for the classical action, which is improper. This shows that it is impossible to establish a quantum-mechanical description for a relativistic particle.

As it is well known, an attempt of getting out of this situation is the Dirac equation.[3]

With the relativistic notations p^μ for $(\mathcal{E}/c, \mathbf{p})$, $\mu = 0, 1, 2, 3$, and x^μ for $(ct, \mathbf{r})$, as well as $\partial^\mu = \partial/\partial x_\mu$ for $(\frac{\partial}{c\partial t}, -grad)$ and $\partial_\mu = \partial/\partial x^\mu$ for $(\frac{\partial}{c\partial t}, grad)$, and the metric $g_{\mu\nu} = g^{\mu\nu} = (+, -, -, -)$, we have $p^\mu = i\hbar\partial^\mu$ and $p_\mu = i\hbar\partial_\mu$. The Dirac equation reads

$$\gamma^\mu p_\mu \psi = mc\psi \ , \tag{6.57}$$

where γ^μ are matrices which satisfy the condition

$$\{\gamma^\mu, \gamma^\nu\} = 2g^{\mu\nu} \ . \tag{6.58}$$

---

[2] O. Klein, "Quantentheorie und fuenfdimensionale Relativitaetstheorie", Z. Phys. **37** 895 (1926); W. Gordon, "Der Comptoneffekt nach der Schroedingerschen Theorie", Z. Phys. **40** 117 (1926).

[3] P. A. M. Dirac, "The quantum theory of the electron", Proc. Roy. Soc. **A117** 610 (1928); Part II., **A118** 351 (1928).

If we apply twice the operator $\gamma^\mu p_\mu$, we recover the Klein-Gordon equation:

$$\begin{aligned}\gamma^\mu p_\mu \gamma^\nu p_\nu \psi = \gamma^\mu \gamma^\nu p_\mu p_\nu \psi = \tfrac{1}{2}\{\gamma^\mu, \gamma^\nu\} p_\mu p_\nu \psi = \\ = g^{\mu\nu} p_\mu p_\nu \psi = p_\mu p^\mu \psi = m^2 c^2 \psi \ ;\end{aligned} \tag{6.59}$$

in these equations ψ is a quantity with four components (a bispinor). The solutions of the Dirac equation are four states with energy $\pm\sqrt{c^2p^2 + m^2c^4}$ and 1/2-spin. The spin is visible for a charge e in the presence of an electromagnetic field with potentials $A_\mu = (\Phi, -\boldsymbol{A})$, when the Dirac equation

$$\gamma^\mu \left(p_\mu - \frac{e}{c} A_\mu \right) \psi = mc\psi \tag{6.60}$$

applied twice leads to a Klein-Gordon equation

$$\left(p_\mu - \frac{e}{c} A_\mu \right) \left(p^\mu - \frac{e}{c} A^\mu \right) \Psi - \frac{ie\hbar}{2c} \sigma^{\mu\nu} F_{\mu\nu} \Psi = m^2 c^2 \Psi \ , \tag{6.61}$$

where $\sigma^{\mu\nu} = \frac{1}{2}[\gamma^\mu, \gamma^\nu]$ is the spin matrix and $F_{\mu\nu} = \partial_\mu A_\nu - \partial_\nu A_\mu$ is the electromagnetic field. The states with negative energies correspond to antiparticles. The energy of the Dirac particles requires anticommutation relations for the particle creation and destruction operators, such that the Dirac particles are fermions.

The Dirac equation (6.44) leads to the continuity equation

$$\frac{\partial}{\partial t}(\psi^* \psi) + c \cdot grad(\psi^* \gamma^0 \boldsymbol{\gamma} \psi) = 0 \ . \tag{6.62}$$

The quantity $\psi^*\psi$ is positive, so it may be viewed as a localization probability; but it propagates with the velocity $c\boldsymbol{\alpha}$, where $\boldsymbol{\alpha} = \gamma^0 \boldsymbol{\gamma}$, which shows not only that the velocity of the Dirac particle is not determined,[4] but, in fact, it is meaningless. Moreover, in the quasi-classical limit the Dirac equation becomes $\gamma^0 \mathcal{E} - c\boldsymbol{\gamma p} = mc^2$, which is not a valid equation. The Dirac equation describes an undetermined and unphysical quantum-mechanical motion.

[4]G. Breit, "An interpretation of Dirac's theory of the electron", Proc. Nat. Acad. Sci. **14** 553 (1928); W. H. Furry and J. R. Oppenheimer, "On the theory of the electron and positive", Phys. Rev. **45** 245 (1934).

The Dirac equation is a purely quantum-mechanical equation, which describes a motion with action comparable to $\hbar$. On the other hand, the motion described by the Dirac equation has energies of the order mc^2 and momenta of the order mc, such that, in fact, the motion described by the Dirac equation requires times of the order $\hbar/mc^2$ and distances of the order of the Compton wavelength $\lambda_c = \hbar/mc$. For electrons $\lambda_c = 3.8 \times 10^{-11} cm$. This distance is covered by the particle in time $\hbar/mc^2$, such that the particle would have a velocity of the order c. A velocity of the order c implies much higher momenta than mc and much higher energies than mc^2. Consequently, the motion described by the Dirac equation is unphysical. The various applications of the Dirac equation to the relativistic motion in external fields (like, *e.g.*, the Coulomb field) are in fact relativistic corrections to the Schroedinger (or Pauli) equation. With these limitations the Dirac equation may be used for relativistic motion, *e.g.* for relativistic scattering, since it is relativistically invariant and, in such cases, we describe the classical relativistic motion. However, this equation includes an indefinite motion.

By using the Dirac matrices

$$\gamma^0 = \begin{pmatrix} 1 & 0 & 0 & 0 \\ 0 & 1 & 0 & 0 \\ 0 & 0 & -1 & 0 \\ 0 & 0 & 0 & -1 \end{pmatrix} , \ \gamma^1 = \begin{pmatrix} 0 & 0 & 0 & 1 \\ 0 & 0 & 1 & 0 \\ 0 & -1 & 0 & 0 \\ -1 & 0 & 0 & 0 \end{pmatrix} ,$$

$$\gamma^2 = \begin{pmatrix} 0 & 0 & 0 & -i \\ 0 & 0 & i & 0 \\ 0 & i & 0 & 0 \\ -i & 0 & 0 & 0 \end{pmatrix} , \ \gamma^3 = \begin{pmatrix} 0 & 0 & 1 & 0 \\ 0 & 0 & 0 & -1 \\ -1 & 0 & 0 & 0 \\ 0 & 1 & 0 & 0 \end{pmatrix} , \tag{6.63}$$

the Dirac equation reads

$$\begin{pmatrix} mc^2 & 0 & cp_3 & c(p_1 - ip_2) \\ 0 & mc^2 & c(p_1 + ip_2) & -cp_3 \\ cp_3 & c(p_1 - ip_2) & -mc^2 & 0 \\ c(p_1 + ip_2) & -cp_3 & 0 & -mc^2 \end{pmatrix} \psi \tag{6.64}$$

$$= \mathcal{E}\psi \ .$$

The matrix on the left in this equation has a vanishing mean value $\overline{\mathcal{E}} = 0$ and a deviation $\sqrt{\overline{\mathcal{E}^2}} = \sqrt{c^2p^2 + m^2c^4}$. It follows that in the motion described by the Dirac equation we have an uncertainty of the order mc^2 in energy and mc in momenta. This motion mixes up the momentum components and the energy, as well as the particle-antiparticle and the spin states. This uncertainty is of the same order of magnitude as the values determined by the Dirac equation. This undefined, mixed-up motion described by the Dirac equation is the unphysical Zitterbewegung (or trembling motion).[5] It takes place on a region with dimension of the order of the Compton wavelength $\lambda_c = \hbar/mc$. We note that the Zitterbewegung is specific to fermions, where matrices occur, while the relativistic uncertainty governed by λ_c is more general; it is associated with the two, positive and negative, values of the relativistic energy (the existence of the antiparticles). The quantum-mechanical motion of pointlike relativistic particles should be described by using the cutoff length λ_c.[6]

6.7 Boson field

The bispinor ψ can be written as a superposition of elementary solutions of the Dirac equation (6.57),

$$\psi = \frac{1}{\sqrt{V}} \sum_{\mathbf{k}\sigma} \frac{1}{\sqrt{2\varepsilon_k}} \left(u_{\mathbf{k}} c_{\mathbf{k}\sigma} e^{-i\varepsilon_k t + i\mathbf{kr}} + v_{\mathbf{k}} b^*_{\mathbf{k}\sigma} e^{i\varepsilon_k t - i\mathbf{kr}} \right) , \qquad (6.65)$$

where σ is a 1/2-spin label,

$$\varepsilon_k = c\sqrt{k^2 + m^2c^2/\hbar^2} \ , \ \mathbf{p} = \hbar\mathbf{k} \ , \qquad (6.66)$$

[5] E. Schroedinger, "Ueber die kraeftefreie Bewegung in der relativistischen Quantenmechanik", Sitzungsberichte der Preussischen Akademie der Wissenschaften, Berlin, 418 (1930); "Zur Quantendynamik des Elektrons", Sitzungsberichte der Preussischen Akademie der Wissenschaften, Berlin, 63 (1931).

[6] V. F. Weisskopf, "On the self-energy and the electromagnetic field of the electron", Phys. Rev. **56** 72 (1939); T. A. Welton, "Some observable effects of the quantum-mechanical fluctuations of the electromagnetic field", Phys. Rev. **74** 1157 (1948).

$$u_{\mathbf{k}} = \begin{pmatrix} \sqrt{\varepsilon_k + mc^2/\hbar}w \\ \sqrt{\varepsilon_k - mc^2/\hbar}(\mathbf{n}\boldsymbol{\sigma})w \end{pmatrix} ,$$
$$v_{\mathbf{k}} = \begin{pmatrix} \sqrt{\varepsilon_k - mc^2/\hbar}(\mathbf{n}\boldsymbol{\sigma})w' \\ \sqrt{\varepsilon_k + mc^2/\hbar}w' \end{pmatrix} , \tag{6.67}$$

$\mathbf{n} = \mathbf{k}/k$, w and w' are orthogonal unit spinors ($w^*w = 1$, $w'^*w' = 1$, $w^*w' = 0$) and $c_{\mathbf{k}}$ and $b_{\mathbf{k}}$ are undetermined coefficients; ψ is, in general, complex. The spinors w and w' (and, consequently, $u_{\mathbf{k}}$ and $v_{\mathbf{k}}$) carry a spin label; in addition, it is convenient to take $w'^{(\sigma)} = \sigma_y w^{(-\sigma)}$.[7]

The left-hand side of the Dirac equation (6.57) looks like a first-order expansion for a coordinate variation $\delta x^\mu \sim \gamma^\mu$. Since $\overline{\gamma^\mu} = 0$ and $\overline{(\gamma^\mu)^2} = \pm 1$, the Dirac equation implies

$$\overline{\Delta\psi} = \frac{mc}{\hbar}\Delta x\overline{\psi} \ll \overline{\psi} , \tag{6.68}$$

i.e. $\Delta x \ll \hbar/mc$, where Δx is a length over which ψ varies. The Dirac equation describes an undetermined motion (the Zitterbewegung) inside a region with dimension of the order of the Compton wavelength $\lambda_c = \hbar/mc$. The next-order term in the expansion gives the variation of this motion. Therefore, the quantum-mechanical motion of a relativistic particle proceeds over distances longer than $\lambda_c = \hbar/mc$, during times longer than $\hbar/mc^2$, with momenta smaller than mc and energies smaller than mc^2. It is a quantum-mechanical change of the relativistic motion, as generated by an interaction. It requires a momentum cutoff mc and an energy cutoff mc^2. This is a motion which smooths out the Zitterbewegung, or a quantum-mechanical motion super-imposed over the Zitterbewegung, which averages it. It can be achieved by the technical procedure given below.

We give a variation δx^μ to the coordinates x^μ, and denote it by u^μ, according to the scheme $x^\mu \longrightarrow x^\mu + \delta x^\mu$, $\delta x^\mu = u^\mu$, $x^\mu \longrightarrow x^\mu + u^\mu$. We will take the first-order variations with respect to u^μ of the Dirac

[7] L. Landau and E. Lifshitz, *Course of Theoretical Physics*, vol. 4, *Quantum Electrodynamics* (V. Berestetskii, E. Lifshitz, L. Pitaevskii), Butterworth-Heinemann (1971).

equation

$$\gamma^\mu \partial_\mu \psi = \frac{mc}{i\hbar}\psi \tag{6.69}$$

(equation (6.57)). According to the Dirac equation the variations of the coordinates should be viewed as matrices (like γ^μ). We write $x^\mu = s^\mu \cdot 1$, where $s^0 = ct$, $\mathbf{s} = \mathbf{r}$ and 1 denotes the unit matrix; we have $x_\mu x^\mu = s_\mu s^\mu \cdot 1 = s^2 \cdot 1$, where $s^2 = c^2t^2 - \mathbf{r}^2$. For δx^μ we need $\delta x_\mu \delta x^\mu = u_\mu u^\mu = u^2 = ds^2$; the (non-trivial) solution of this equation is

$$\delta x^\mu = u^\mu = \frac{1}{2}u\gamma^\mu \tag{6.70}$$

(since $\gamma_\mu\gamma^\mu = 4$). The first-order expansion of the Dirac equation is

$$\gamma^\mu\left(\partial_\mu\psi + u^\nu\partial_\nu\partial_\mu\psi\right) = \frac{mc}{i\hbar}\left(\psi + u^\nu\partial_\nu\psi\right) \ , \tag{6.71}$$

or

$$\partial^\mu\partial_\mu\psi = -\frac{m^2c^2}{\hbar^2}\psi \ , \tag{6.72}$$

which is the Klein-Gordon equation $p_\mu p^\mu\psi = m^2c^2\psi$ (equation (6.55)). This equation describes the motion smoothed over the Zitterbewegung; we note that u is absorbed into ψ, hence the bosonic character of the latter. The ψ in equation (6.72) originates in u multiplied by a bispinor. It is a boson field with four components. By analogy with equation (6.65) we write it as

$$\psi = \frac{1}{\sqrt{V}}\sum_{\mathbf{k}}\beta_k(\psi_{k\alpha})e^{i\mathbf{kr}} \ , \tag{6.73}$$

where

$$(\psi_{\mathbf{k}\alpha}) = \begin{pmatrix} c_{\mathbf{k},+1} \\ c_{\mathbf{k},-1} \\ b^*_{-\mathbf{k},-1} \\ b^*_{-\mathbf{k},+1} \end{pmatrix} \tag{6.74}$$

and

$$\beta_k = c\sqrt{\frac{\hbar}{2\varepsilon_k}} \ . \tag{6.75}$$

This coefficient will be established shortly, by using the energy corresponding to the boson field. Similarly, the time dependence of the operators is $c_{\boldsymbol{k}\sigma}$, $b_{\boldsymbol{k}\sigma} \sim e^{-i\varepsilon_k t}$, where $\sigma = \pm 1$ is the spin label. The c's and the b's operators satisfy usual bosonic commutation relations for four distinct types of bosons. The c-operators correspond to particles (electrons), while the b-operators correspond to antiparticles (positrons). The boson occupation number is the number of electrons, the presence of the cutoff λ_c ensures the exclusion principle.

In the presence of the electromagnetic field we need to use the covariant derivative $D_\mu = \partial_\mu - \frac{e}{ic\hbar}A_\mu$, according to $\psi \longrightarrow \psi + u^\mu D_\mu \psi$; we get the Klein-Gordon equation

$$\left(p_\mu - \frac{e}{c}A_\mu\right)\left(p^\mu - \frac{e}{c}A^\mu\right)\Psi - \frac{ie\hbar}{2c}\sigma^{\mu\nu}F_{\mu\nu}\Psi = m^2c^2\Psi \tag{6.76}$$

with electromagnetic field (equation (6.61)) .

The quantum-mechanical motion of a relativistic particle is effected by bosonic quanta of a field with four components.[8] For a field the localization probability is meaningless. The conservation laws are the energy-momentum conservation, the charge conservation, etc. The avoidance of the self-interaction is achieved by the cutoff λ_c. The quasi-classical limit of the Klein-Gordon equation is meaningless. It is a purely quntum-mechanical equation for a quantum field. With these limitations it can be used for relativistic motion, which is only a change of the reference frame (and the Klein-Gordon equation is relativistically invariant); but it includes an indefinite motion.

6.8 Free bosons

The free equation of motion of the boson field (the Klein-Gordon equation (6.72)) reads

$$\frac{1}{c^2}\frac{\partial^2\psi}{\partial t^2} - \Delta\psi + \frac{m^2c^2}{\hbar^2}\psi = 0 \; . \tag{6.77}$$

If we multiply this equation by $\dot{\overline{\psi}}$, the equation for $\overline{\psi}$ by $\dot{\psi}$, where $\overline{\psi}$ is the adjoint of ψ (transposed conjugate), and add the two equations,

[8] M. Apostol, *A lame Duck: Quantum Electrodynamics*, apoma, Magurele (2020).

we get

$$\frac{\partial}{\partial t}\left(\frac{1}{c^2}\dot{\overline{\psi}}\dot{\psi}+\partial_i\overline{\psi}\partial_i\psi+\frac{m^2c^2}{\hbar^2}\overline{\psi}\psi\right)-\partial_i\left(\dot{\overline{\psi}}\partial_i\psi+\partial_i\overline{\psi}\dot{\psi}\right)=0\ , \quad (6.78)$$

where $i=1,2,3$; hence, we can see that

$$w_e=\frac{1}{c^2}\dot{\overline{\psi}}\dot{\psi}+\partial_i\overline{\psi}\partial_i\psi+\frac{m^2c^2}{\hbar^2}\overline{\psi}\psi \quad (6.79)$$

is the energy density and

$$\mathbf{g}_e=-\frac{1}{c^2}\left(\dot{\overline{\psi}}grad\psi+grad\overline{\psi}\dot{\psi}\right) \quad (6.80)$$

is the momentum density. The total energy is

$$\begin{aligned}W_e&=\int d\mathbf{r}w_e=\sum_{\mathbf{k}\sigma}\beta_k^2\tfrac{2\varepsilon_k^2}{c^2}\left(c^*_{\mathbf{k}\sigma}c_{\mathbf{k}\sigma}+b_{\mathbf{k}\sigma}b^*_{\mathbf{k}\sigma}\right)=\\&=\sum_{\mathbf{k}\sigma}\hbar\varepsilon_k\left(c^*_{\mathbf{k}\sigma}c_{\mathbf{k}\sigma}+b_{\mathbf{k}\sigma}b^*_{\mathbf{k}\sigma}\right)=\\&=\sum_{\mathbf{k}\sigma}\hbar\varepsilon_k\left(c^*_{\mathbf{k}\sigma}c_{\mathbf{k}\sigma}+b^*_{\mathbf{k}\sigma}b_{\mathbf{k}\sigma}+1\right)\ ;\end{aligned} \quad (6.81)$$

this equation establishes the coefficients β_k given by equation (6.75). The time dependence $c_{\boldsymbol{k}\sigma},\,b_{\boldsymbol{k}\sigma}\sim e^{-i\varepsilon_k t}$ is established by the canonical equations of motion of these operators with the hamiltonian W_e. Similarly, the total momentum is

$$\begin{aligned}\mathbf{G}_e&=\int d\mathbf{r}\mathbf{g}_e=\sum_{\mathbf{k}\sigma}\hbar\mathbf{k}\left(c^*_{\mathbf{k}\sigma}c_{\mathbf{k}\sigma}+b_{\mathbf{k}\sigma}b^*_{\mathbf{k}\sigma}\right)=\\&=\sum_{\mathbf{k}\sigma}\hbar\mathbf{k}\left(c^*_{\mathbf{k}\sigma}c_{\mathbf{k}\sigma}+b^*_{\mathbf{k}\sigma}b_{\mathbf{k}\sigma}+1\right)\ .\end{aligned} \quad (6.82)$$

If we multiply the equations for ψ and $\overline{\psi}$ by $\overline{\psi}$ and ψ, respectively, and subtract the two equations from one another, we get another law of conservation, which reads

$$\frac{1}{c^2}\frac{\partial}{\partial t}\left(\overline{\psi}\dot{\psi}-\dot{\overline{\psi}}\psi\right)-\partial_i\left(\overline{\psi}\partial_i\psi-\partial_i\overline{\psi}\psi\right)=0\ ; \quad (6.83)$$

hence,

$$\begin{aligned}Q&=\tfrac{ie}{\hbar c^2}\int d\mathbf{r}\left(\overline{\psi}\dot{\psi}-\dot{\overline{\psi}}\psi\right)=e\sum_{\mathbf{k}\sigma}\left(c^*_{\mathbf{k}\sigma}c_{\mathbf{k}\sigma}-b_{\mathbf{k}\sigma}b^*_{\mathbf{k}\sigma}\right)=\\&=e\sum_{\mathbf{k}\sigma}\left(c^*_{\mathbf{k}\sigma}c_{\mathbf{k}\sigma}-b^*_{\mathbf{k}\sigma}b_{\mathbf{k}\sigma}-1\right)\end{aligned} \quad (6.84)$$

is the electric charge (where e is the particle charge) and

$$\mathbf{J} = -\tfrac{ie}{\hbar}\int d\mathbf{r}\left(\overline{\psi} grad\psi - grad\overline{\psi}\psi\right) = \\ = e\sum_{\mathbf{k}\sigma}\tfrac{c^2\mathbf{k}}{\varepsilon_k}\left(c^*_{\mathbf{k}\sigma}c_{\mathbf{k}\sigma} - b_{\mathbf{k}\sigma}b^*_{\mathbf{k}\sigma}\right) = \\ = e\sum_{\mathbf{k}\sigma}\tfrac{c^2\mathbf{k}}{\varepsilon_k}\left(c^*_{\mathbf{k}\sigma}c_{\mathbf{k}\sigma} - b^*_{\mathbf{k}\sigma}b_{\mathbf{k}\sigma} - 1\right) \tag{6.85}$$

is the electric current. With the notation

$$\rho = \tfrac{ie}{\hbar c^2}\left(\overline{\psi}\dot{\psi} - \dot{\overline{\psi}}\psi\right)\ ,\ j_i = -\tfrac{ie}{\hbar}\left(\overline{\psi}\partial_i\psi - \partial_i\overline{\psi}\psi\right)\ , \tag{6.86}$$

$$j^\mu = (c\rho, \mathbf{j}) = \frac{ie}{\hbar}\left(\overline{\psi}\partial^\mu\psi - (\partial^\mu\overline{\psi})\psi\right)\ , \tag{6.87}$$

equation (6.83) is the continuity equation $\partial_\mu j^\mu = 0$.
The first-order time derivative in the energy (equation (6.79)) requires two types of particles (c and b) and the spin interaction (in equation (6.76)) requires the spin states.

6.9 Interaction

We adopt the energy W_e given by equation (6.81) as the free electron hamiltonian

$$H_e = \sum_{\mathbf{k}\sigma}\hbar\varepsilon_k\left(c^*_{\mathbf{k}\sigma}c_{\mathbf{k}\sigma} + b^*_{\mathbf{k}\sigma}b_{\mathbf{k}\sigma}\right) \tag{6.88}$$

(leaving aside the zero-point energy), where $\varepsilon_k = c\sqrt{k^2+k_0^2}$, $k_0 = mc/\hbar$ being the inverse of the Compton wavelength (Compton wavevector). The interaction of the electrons with the electromagnetic field is derived from equation (6.76); this equation can be written as

$$(p_\mu p^\mu - \tfrac{e}{c}p_\mu A^\mu - \tfrac{e}{c}A_\mu p^\mu + \tfrac{e^2}{c^2}A_\mu A^\mu + \\ + \tfrac{e\hbar}{c}\Sigma\mathbf{H} - \tfrac{ie\hbar}{c}\boldsymbol{\alpha}\mathbf{E} - m^2c^2)\psi = 0\ , \tag{6.89}$$

since

$$-\frac{ie\hbar}{2c}\sigma^{\mu\nu}F_{\mu\nu} = \frac{e\hbar}{c}\Sigma\mathbf{H} - \frac{ie\hbar}{c}\boldsymbol{\alpha}\mathbf{E}\ , \tag{6.90}$$

where $\mathbf{E}$ is the electric field, $\mathbf{H}$ is the magnetic field,

$$\boldsymbol{\alpha} = \begin{pmatrix} 0 & \boldsymbol{\sigma} \\ \boldsymbol{\sigma} & 0 \end{pmatrix} , \quad \boldsymbol{\Sigma} = \begin{pmatrix} \boldsymbol{\sigma} & 0 \\ 0 & \boldsymbol{\sigma} \end{pmatrix} \tag{6.91}$$

and $\boldsymbol{\sigma}$ are the Pauli matrices,

$$\sigma_x = \begin{pmatrix} 0 & 1 \\ 1 & 0 \end{pmatrix} , \quad \sigma_y = \begin{pmatrix} 0 & -i \\ i & 0 \end{pmatrix} ,$$
$$\sigma_z = \begin{pmatrix} 1 & 0 \\ 0 & -1 \end{pmatrix} . \tag{6.92}$$

Hence, we get the equation of motion

$$\tfrac{1}{c^2}\tfrac{\partial^2\psi}{\partial t^2} - \Delta\psi + \tfrac{m^2c^2}{\hbar^2}\psi + \tfrac{e}{c\hbar^2}p_\mu A^\mu\psi + \tfrac{e}{c\hbar^2}A_\mu p^\mu\psi -$$
$$- \tfrac{e^2}{c^2\hbar^2}A_\mu A^\mu\psi - \tfrac{e}{c\hbar}\sum\mathbf{H}\psi + \tfrac{ie}{c\hbar}\boldsymbol{\alpha}\mathbf{E}\psi = 0 . \tag{6.93}$$

The interaction energy densities are

$$v_p = \frac{1}{c}j_\mu A^\mu = \frac{ie}{c\hbar}\left[\overline{\psi}\left(\partial_\mu\psi\right) - \left(\partial_\mu\overline{\psi}\right)\psi\right]A^\mu \tag{6.94}$$

(where the gauge condition $\partial_\mu A^\mu = 0$ is used),

$$v_d = -\frac{e^2}{c^2\hbar^2}\left(\overline{\psi}\psi\right)A_\mu A^\mu \tag{6.95}$$

and

$$v_H = -\tfrac{e}{c\hbar}\overline{\psi}\sum\psi\mathbf{H} ,$$
$$v_E = \tfrac{ie}{c\hbar}\overline{\psi}\boldsymbol{\alpha}\psi\mathbf{E} . \tag{6.96}$$

The interaction is obtained by integrating these densities over the whole space,

$$V_{p,d,H,E} = \int d\mathbf{r}v_{p,d,H,E} . \tag{6.97}$$

The labels p and d are chosen by analogy with the non-relativistic "paramagnetic" and "diamagnetic" contributions. The interactions

v_H and v_E arise from the $\sigma^{\mu\nu}F_{\mu\nu}$-term in equation (6.76). This is known as the "Pauli term".[9]

The interaction v_p can be written as $\rho\Phi - \frac{1}{c}\boldsymbol{j}\boldsymbol{A}$, where the charge density ρ and the current density $\boldsymbol{j}$ are given by equation (6.76), Φ is the scalar potential and $\boldsymbol{A}$ is the vector potential. The vector potential has a longitudinal component $\boldsymbol{A}_l$ and a transverse component $\boldsymbol{A}_t$. The longitudinal component is related to the scalar potential through the Lorenz gauge,

$$\boldsymbol{A}_l = \frac{1}{\sqrt{V}}\sum_{\boldsymbol{k}} \frac{i\boldsymbol{k}}{ck^2}\dot{\Phi}_{\boldsymbol{k}} e^{i\boldsymbol{k}\boldsymbol{r}} \tag{6.98}$$

(V denotes the volume, equation (6.29)) and the scalar potential satisfies the Maxwell equation

$$\frac{1}{c^2}\ddot{\Phi} - \Delta\Phi = 4\pi\rho = \frac{4\pi i e}{\hbar c^2}\left(\overline{\psi}\dot{\psi} - \dot{\overline{\psi}}\psi\right) \tag{6.99}$$

(where equation (6.86) is used). The longitudinal part $\rho\Phi - \frac{1}{c}\boldsymbol{j}\boldsymbol{A}_l$ of the interaction gives minus the Coulomb energy $V_c = \sum_{\mathbf{k}} \frac{2\pi}{k^2}\rho_{\mathbf{k}}\rho_{-\mathbf{k}}$ (equation (6.35)), while the longitudinal part $\frac{1}{8\pi}\int d\mathbf{r}E_l^2$ of the electromagnetic field, where $\boldsymbol{E}_l$ is the longitudinal electric field, gives V_c (equation (6.32)). Therefore, we remain with the transverse interaction

$$v_p = -\frac{1}{c}\mathbf{j}\mathbf{A}_t \tag{6.100}$$

and the transverse part of the energy of the electromagnetic field

$$H_{em} = H_{ph} = \sum_{\mathbf{k}} \hbar\omega_k a^*_{\mathbf{k}} a_{\mathbf{k}} \ , \tag{6.101}$$

which corresponds to photons with energy $\hbar\omega_k = \hbar ck$ (for a fixed polarization, equation (6.46)). The transverse vector potential reads

$$\boldsymbol{A}_t = \frac{1}{\sqrt{V}}\sum_{\boldsymbol{k}} \alpha_k \boldsymbol{e}_{\boldsymbol{k}}\left(a_{\boldsymbol{k}}e^{i\boldsymbol{k}\boldsymbol{r}} + a^*_{\boldsymbol{k}}e^{-i\boldsymbol{k}\boldsymbol{r}}\right) \ , \tag{6.102}$$

where $\alpha_k = c\sqrt{2\pi\hbar/\omega_k}$ and $\boldsymbol{e}_{\boldsymbol{k}}$ is the polarization vector (equation (6.42)). Since the longitudinal field has a vanishing free hamiltonian

[9] W. Pauli, "Relativistic field theories of elementary particles", Revs. Mod. Phys. **13** 203 (1941).

($V_c = \frac{1}{8\pi}\int d\mathbf{r}E_l^2 = 0$ for $\ddot{\Phi}_{\mathbf{k}} + \omega_k^2\Phi_k = 0$ in equation (6.99)), this field cannot be quantized as an independent kind of photons (there is no scalar or longitudinal "photon").

In all other interaction terms where the scalar potential and the longitudinal vector potential occur, we replace them by their expressions given by the equation of motion (6.99):

$$\rho_{\mathbf{k}} = \frac{e}{\hbar c^2\sqrt{V}}\sum_{\mathbf{k}'\sigma}\beta_{\mathbf{k}'+\mathbf{k}}\beta_{\mathbf{k}'}\left(\varepsilon_{\mathbf{k}'+\mathbf{k}} + \varepsilon_{\mathbf{k}'}\right)\cdot$$
$$\cdot\left(c^*_{\mathbf{k}'+\mathbf{k}\sigma}c_{\mathbf{k}'\sigma} - b_{-\mathbf{k}'-\mathbf{k}-\sigma}b^*_{-\mathbf{k}'-\sigma}\right) ,$$
$$\Phi_{\mathbf{k}} = -\frac{4\pi e}{\hbar\sqrt{V}}\sum_{\mathbf{k}'\sigma}\frac{\beta_{\mathbf{k}'+\mathbf{k}}\beta_{\mathbf{k}'}\left(\varepsilon_{\mathbf{k}'+\mathbf{k}}+\varepsilon_{\mathbf{k}'}\right)}{\left(\varepsilon_{\mathbf{k}'+\mathbf{k}}-\varepsilon_{\mathbf{k}'}\right)^2-\omega_k^2}\cdot \tag{6.103}$$
$$\cdot\left(c^*_{\mathbf{k}'+\mathbf{k}\sigma}c_{\mathbf{k}'\sigma} - b_{-\mathbf{k}'-\mathbf{k}-\sigma}b^*_{-\mathbf{k}'-\sigma}\right) ,$$

where $\beta_{\mathbf{k}} = \beta_k = c\sqrt{\frac{\hbar}{2\varepsilon_k}}$,

$$\mathbf{A}_{t\mathbf{k}} = \alpha_k\left(\mathbf{e}_{\mathbf{k}}a_{\mathbf{k}} + \mathbf{e}_{-\mathbf{k}}a^*_{-\mathbf{k}}\right) , \ \alpha_k = c\sqrt{\frac{2\pi\hbar}{\omega_k}} ,$$

$$\mathbf{A}_{l\mathbf{k}} = \frac{4\pi e}{\hbar\sqrt{V}}\sum_{\mathbf{k}'\sigma}\frac{i\mathbf{k}}{k\omega_k}\frac{\beta_{\mathbf{k}'+\mathbf{k}}\beta_{\mathbf{k}'}\left(\varepsilon^2_{\mathbf{k}'+\mathbf{k}}-\varepsilon^2_{\mathbf{k}'}\right)}{\left(\varepsilon_{\mathbf{k}'+\mathbf{k}}-\varepsilon_{\mathbf{k}'}\right)^2-\omega_k^2}\cdot \tag{6.104}$$
$$\cdot\left(c^*_{\mathbf{k}'+\mathbf{k}\sigma}c_{\mathbf{k}'\sigma} - b_{-\mathbf{k}'-\mathbf{k}-\sigma}b^*_{-\mathbf{k}'-\sigma}\right)$$

and

$$\mathbf{E}_{l\mathbf{k}} = -\frac{4\pi i\mathbf{k}}{k^2}\rho_{\mathbf{k}} . \tag{6.105}$$

In addition, from the continuity equation we get the Fourier component of the longitudinal current

$$\mathbf{j}_{l\mathbf{k}} = \frac{e}{\hbar c^2\sqrt{V}}\sum_{\mathbf{k}'\sigma}\frac{\mathbf{k}}{k^2}\beta_{\mathbf{k}'+\mathbf{k}}\beta_{\mathbf{k}'}\left(\varepsilon^2_{\mathbf{k}'+\mathbf{k}} - \varepsilon^2_{\mathbf{k}'}\right)\cdot \tag{6.106}$$
$$\cdot\left(c^*_{\mathbf{k}'+\mathbf{k}\sigma}c_{\mathbf{k}'\sigma} - b_{-\mathbf{k}'-\mathbf{k}-\sigma}b^*_{-\mathbf{k}'-\sigma}\right) .$$

For convenience, also we give below a few bilinear forms:

$$\overline{\psi}\psi = \frac{1}{V}\sum_{\mathbf{k}\mathbf{k}'\sigma}\beta_k\beta_{k'}\left(c^*_{\mathbf{k}\sigma}c_{\mathbf{k}'\sigma} + b_{-\mathbf{k},-\sigma}b^*_{-\mathbf{k}',-\sigma}\right)e^{-i(\mathbf{k}-\mathbf{k}')\mathbf{r}} , \tag{6.107}$$

$$j_0 = \frac{e}{c\hbar V} \sum_{\mathbf{k}\mathbf{k}'\sigma} \beta_k \beta_{k'} \left(\varepsilon_k + \varepsilon_{k'}\right) \cdot \\ \cdot \left(c^*_{\mathbf{k}\sigma} c_{\mathbf{k}'\sigma} - b_{-\mathbf{k},-\sigma} b^*_{-\mathbf{k}',-\sigma}\right) e^{-i(\mathbf{k}-\mathbf{k}')\mathbf{r}} , \tag{6.108}$$

$$\mathbf{j} = \frac{e}{\hbar V} \sum_{\mathbf{k}\mathbf{k}'\sigma} \beta_k \beta_{k'} \left(\mathbf{k} + \mathbf{k}'\right) \cdot \\ \cdot \left(c^*_{\mathbf{k}\sigma} c_{\mathbf{k}'\sigma} + b_{-\mathbf{k},-\sigma} b^*_{-\mathbf{k}',-\sigma}\right) e^{-i(\mathbf{k}-\mathbf{k}')\mathbf{r}} . \tag{6.109}$$

The longitudinal contribution $\Phi^2 - \frac{1}{c^2}(\mathbf{A}_l^2 + 2\mathbf{A}_l\mathbf{A}_t)$ in the interaction v_d (equation (6.95)) brings an interaction of the order e^3, at least; if we limit ourselves to e^2-orders at most, we may use

$$v_d \simeq \frac{e^2}{c^2 \hbar^2} \left(\overline{\psi}\psi\right) \mathbf{A}_t \mathbf{A}_t . \tag{6.110}$$

Similarly, only the transverse field contributes to v_H. It is worth noting that the longitudinal degrees of freedom of the electromagnetic field are removed from the problem, by means of equations (6.103)-(6.106); they are replaced by the degrees of freedom of the charges. Similar interaction contributions arise from an external (transverse, purely radiation field).

For convenience, the interactions $v_{H,E}$ (equations (6.96)) can be written as

$$v_H = -\frac{e}{2c\hbar} \overline{\psi} \left(\sum + \sum^*\right) \psi \mathbf{H} , \\ v_E = \frac{ie}{2c\hbar} \overline{\psi} \left(\boldsymbol{\alpha} - \boldsymbol{\alpha}^*\right) \psi \mathbf{E} . \tag{6.111}$$

In the first equation (6.111) only the matrices $\sigma_{x,z}$ remain, while in the second equation (6.111) only the matrix σ_y remains.

6.10 Classical electromagnetic radius

The representation of the pointlike charges by the Dirac δ-function leads to singular classical electromagnetic potentials at the position of the charge. The partial differential equations of the classical electromagnetism require to view the pointlike charges as having a radius, at least infinitesimal. In order to avoid the self-interaction the same Maxwell equations requires a finite classical electromagnetic radius for "pointlike" charges. We recall here the classical electromagnetic radius of a charge.

A charge in motion generates a propagating electromagnetic field. The charge moves in this field, so it interacts with its own field, as a consequence of the retardation. This is a self-interaction. Since we are interested in the field near the charge, we may leave aside the scalar potential and, according to the Lorenz gauge, the longitudinal part of the vector potential. The transverse vector potential $\boldsymbol{A}$ is generated according to the equation

$$\frac{1}{c^2}\frac{\partial^2 \boldsymbol{A}}{\partial t^2} - \Delta \boldsymbol{A} = \frac{4\pi}{c}\boldsymbol{j} \ , \tag{6.112}$$

where $\boldsymbol{j}$ is the transverse current density. Let us assume a pointlike charge q with trajectory $\boldsymbol{r}_0(t)$, acted by an external force; its charge density is $\rho = q\delta(\boldsymbol{r} - \boldsymbol{r}_0)$ and its current density is $\boldsymbol{j} = q\boldsymbol{v}\delta(\boldsymbol{r} - \boldsymbol{r}_0)$, where $\boldsymbol{v} = \dot{\boldsymbol{r}}_0$. Equation (6.112) becomes

$$\frac{1}{c^2}\frac{\partial^2 \boldsymbol{A}}{\partial t^2} - \Delta \boldsymbol{A} = \frac{4\pi q}{c}\boldsymbol{v}\delta(\boldsymbol{r} - \boldsymbol{r}_0) \ . \tag{6.113}$$

The field $\boldsymbol{A}$ acts upon the charge. Consequently, the charge with mass m acquires an additional velocity $\boldsymbol{v}'$, whose (classical, non-relativistic) equation of motion is

$$m\frac{d\boldsymbol{v}'}{dt} = q\boldsymbol{E}_0 + \frac{1}{c}(\boldsymbol{v} + \boldsymbol{v}') \times \boldsymbol{H}_0 \ , \tag{6.114}$$

where $\boldsymbol{E}_0 = -\frac{1}{c}\frac{\partial \boldsymbol{A}_0}{\partial t}$ is the electric field and $\boldsymbol{H}_0 = curl\boldsymbol{A}_0$ is the magnetic field. The suffix 0 indicates that the fields are taken at position $\boldsymbol{r}_0$. This is the Lorentz force. Since we are interested in the region close to the charge, we may neglect the second term on the right in equation (6.114), and get

$$\frac{d\boldsymbol{v}'}{dt} = -\frac{q}{mc}\frac{d\boldsymbol{A}_0}{dt} \tag{6.115}$$

and $\boldsymbol{v}' = -\frac{q}{mc}\boldsymbol{A}_0$ (the relativistic equation of motion does not change these things). It follows that the self-interaction produces a reaction current density

$$\boldsymbol{j}' = q\boldsymbol{v}'\delta(\boldsymbol{r}-\boldsymbol{r}_0) = -\frac{q^2}{mc}\boldsymbol{A}\delta(\boldsymbol{r}-\boldsymbol{r}_0) \ , \tag{6.116}$$

which, in turn, determines a field given by

$$\frac{1}{c^2}\frac{\partial^2 \boldsymbol{A}'}{\partial t^2} - \Delta\boldsymbol{A}' = -\frac{4\pi q^2}{mc^2}\boldsymbol{A}\delta(\boldsymbol{r}-\boldsymbol{r}_0) \ . \tag{6.117}$$

For $\boldsymbol{r}$ close to $\boldsymbol{r}_0$ we may limit ourselves to the static part of this field, which is $\boldsymbol{A}' = -\frac{q^2}{mc^2}\frac{\boldsymbol{A}_0}{|\boldsymbol{r}-\boldsymbol{r}_0|}$. The total field at the position of the charge is $\boldsymbol{A}_0 - \frac{q^2\boldsymbol{A}_0}{mc^2}\lim|_{\boldsymbol{r}\to\boldsymbol{r}_0}\frac{1}{|\boldsymbol{r}-\boldsymbol{r}_0|}$. This shows that the charge must be viewed as having an undetermined position inside a region with dimension of order

$$a = \frac{q^2}{mc^2} \ ; \tag{6.118}$$

the field at the undetermined position of the charge is zero and the force acting on the charge is vanishing; this way, the self-interaction is avoided. Therefore, we must limit ourselves to a cutoff length of the order $a = q^2/mc^2$, a cutoff wavevector of the order $1/a$ and a cutoff frequency of the order c/a. In addition, if we see the charge as being pointlike and moving indefinitely inside a region with radius of the order a, its self-energy is of the order $q^2/a = mc^2$. The parameter a is the classical electromagnetic radius of the charge q with mass m. It sets the limits of the classical electromagnetism. The fields are limited by $E_s = H_s = q/a^2 = m^2c^4/q^3$ (or $qE_sa = mc^2$, $m\omega^2 = q\omega H_s/c$, with $\omega = c/a$). For electrons $q = -e = -4.8 \times 10^{-10} esu$, $a \simeq 2.8 \times 10^{-13} cm^{-1}$ ($c/a \simeq 10^{23} s^{-1}$, $m \simeq 10^{-27} g$, $mc^2 \simeq 0.5 MeV$, $1eV = 1.6 \times 10^{-12} erg$), $E_s = H_s \simeq 6 \times 10^{15} esu$ ($1esu = 3 \times 10^4 V/m$). The Lorentz contraction along the direction of motion makes a radius $q^2/\mathcal{E}$ along this direction, where $\mathcal{E}$ is the charge energy.

The undefined motion of a pointlike charge inside of a region with dimensions of the order a has not very relevant consequences in the classical electromagnetism, except for its governing the Lorenz damping and the natural breath of the spectroscopic line. The radius a may be viewed as an uncertainty in the charge position, arising from its electromagnetic field.

6.11 Relativistic quantum uncertainty

A relativistic quantum particle exhibits a special uncertainty, arising from both the quantum-mechanical and relativistic character of its motion. The quantum-mechanical motion implies variations of the mechanical action of the order of the Planck constant $\hbar$. This motion is obtained by replacing the energy by $\mathcal{E} \rightarrow i\hbar\frac{\partial}{\partial t}$ and the momentum by $\boldsymbol{p} \rightarrow -i\hbar grad$, and applying these operators to a wavefunction. It follows immediately that there exists a momentum and distance scale variations Δp and Δl, such that $\Delta p \Delta l \simeq \hbar$ and, similarly, an energy and time scale variations $\Delta\mathcal{E}$ and Δt, such that $\Delta\mathcal{E}\Delta t \simeq \hbar$. This is valid for non-relativistic particles with mass m, where the energy is $\mathcal{E} = \frac{1}{2m}p^2$, or for ultra-relativistic particles with energy $\mathcal{E} = cp$. The energy of a relativistic particle is given by $\mathcal{E}^2 = c^2p^2 + m^2c^4$, and the quantization leads to a Klein-Gordon-type equation

$$\hbar^2\frac{\partial^2\psi}{\partial t^2} - c^2\hbar^2\Delta\psi + m^2c^4\psi = 0 \ , \tag{6.119}$$

whatever ψ may mean. We can see that the time scale is $\tau = \frac{\hbar}{mc^2}$ and the distance scale is $l = \frac{\hbar}{mc}$. The meaning of these equalities is obvious. These are quantum-mechanical uncertainties, for a quantum-mechanical motion, where the durations cannot be shorter than τ and the distances cannot be shorter than l; if they were shorter, the quantum mechanical motion would be undefined, because variations smaller than $\hbar$ of the mechanical action are meaningless. A similar conclusion holds for the Dirac equation, and even for $m = 0$ (*e.g.*, for photons), where τ and l take on their relativistic form $\tau = \hbar/\mathcal{E}$ and $l = \hbar c/\mathcal{E}$; for photons l is their wavelength λ, for massive particles l is their Compton wavelength λ_c. The reason is the square relativistic dependence of the energy $\mathcal{E}^2 = c^2p^2 + m^2c^4$, $\mathcal{E}^2 = c^2p^2$.

The relativistic energy $\mathcal{E}^2 = c^2p^2 + m^2c^4$, $\mathcal{E}^2 = c^2p^2$ implies both a positive energy $\mathcal{E} = c\sqrt{p^2 + m^2c^2}$, $\mathcal{E} = cp$ and a negative energy $\mathcal{E} = -c\sqrt{p^2 + m^2c^2}$, $\mathcal{E} = -cp$. The corresponding presence of the second-order time derivative in the quantization equation requires the presence of both positive and negative frequencies $\varepsilon_{\boldsymbol{k}} = \pm c\sqrt{k^2 + k_0^2}$, $\omega_{\boldsymbol{k}} = \pm ck$ in the solution of this equation, where $\boldsymbol{k} = \boldsymbol{p}/\hbar$ is the wavevector and $k_0 = mc/\hbar$ is the Compton wavevector. A negative energy (frequency) means the absence of a positive energy (frequency). It follows that, apart from particles in the quantized relativistic motion, we have also antiparticles, even if they may coincide with the particles. Therefore, over the indefinite motion within a region with dimension of the order of the Compton wavelength λ_c (or wavelength λ) we have an indefinite superposition of particles and antiparticles, both states of a generic particle. This indefiniteness precludes a quantum-mechanical interpretation of the quantization of the relativistic energy, which is replaced by the quantum-field picture. Indeed, the Klein-Gordon equation has not a definite probability density (electromagnetic field included), while the Dirac equation exhibits an indefinite velocity $c\boldsymbol{\gamma}$ of the probability current, where $\boldsymbol{\gamma}$ are the Dirac matrices. The fields which enter the quantum-field equations are amplitudes of creation or destruction of particles (antiparticles).

The relativistic quantum uncertainty described above holds both for bosons and fermions. In particular, this uncertainty is visible in the Dirac equation

$$\gamma^{\mu} p_{\mu} \psi = mc\psi \tag{6.120}$$

by the presence of the Dirac matrices γ^{μ} ($\mu = 0, 1, 2, 3$), which mix up the coordinates and the time over a length scale $\lambda_c = \hbar/mc$ and over durations $\tau = \hbar/mc^2$. This indefinite motion is the fermion Zitterbewegung.

It follows that in treating the quantum-field interaction, as long as we are interested in quantum-mechanical motions, we need to use a cutoff length of the order of the Compton wavelength $\lambda_c = \hbar/mc$ (and a cutoff time of the order $\tau = \hbar/mc^2$, a cutoff energy mc^2 and a cutoff momentum mc). In particular, we emphasize that the quantum-field equations have not a (quasi-) classical limit. The quasi-classical limit of the Klein-Gordon equation is both the Hamilton-Jacobi equation and a free wave equation, which is contradictory, while the Dirac equation in the quasi-classical limit $\mathcal{E} - c\boldsymbol{\alpha}\boldsymbol{p} = mc\beta$, where $\boldsymbol{\alpha} = \gamma^0\boldsymbol{\gamma}$, is not a valid classical equation.

The Compton wavelength sets a quantum limit for the motion of the relativistic fields and for the fields themselves. The latter is the Schwinger limit $E_s = H_s = m^2c^3/\hbar \mid q \mid$. For electrons $\lambda_c = \hbar/mc \simeq 3.8 \times 10^{-11} cm$ and $E_s = H_s \simeq 4 \times 10^{13} esu$. The ratio a/λ_c is

$\alpha = a/\lambda_c = q^2/\hbar c$. For electrons $\alpha = e^2/\hbar c \simeq \frac{1}{137}$ is the fine-structure constant.

7 Electromagnetic Quantum Effects

7.1 Perturbation theory

The state vectors $\mid v>$ are obtained by the action of the electron and photon creation operators upon the vacuum ($c^*_{\mathbf{k}\sigma}$, $b^*_{\mathbf{k}\sigma}$, $a^*_{\mathbf{k}}$). The time evolution of the state vectors is given by Schroedinger equation

$$i\hbar\frac{\partial}{\partial t} \mid v >= (H_0 + V) \mid v > \ , \tag{7.1}$$

where $H_0 = H_e + H_{ph}$ and $V = V_p + V_d + V_H + V_E$ (given in the chapter "Electron-Photon Interaction"). The electron hamiltonian is

$$H_e = \sum_{\mathbf{k}\sigma} \hbar\varepsilon_k \left(c^*_{\mathbf{k}\sigma} c_{\mathbf{k}\sigma} + b^*_{\mathbf{k}\sigma} b_{\mathbf{k}\sigma}\right) \ , \tag{7.2}$$

where $\varepsilon_k = c\sqrt{k^2 + k_0^2}$, $k_0 = mc/\hbar$ being the inverse of the Compton wavelength. The spin label takes two values ± 1. The photon hamiltonian is

$$H_{ph} = \sum_{\mathbf{k}} \hbar\omega_k a^*_{\mathbf{k}} a_{\mathbf{k}} \ , \tag{7.3}$$

where $\hbar\omega_k = \hbar ck$ (for a fixed polarization). The interaction is given by

$$V_{p,d,E,H} = \int d\boldsymbol{r} v_{p,d,E,H} \ , \tag{7.4}$$

where the interaction densities $v_{p,d,E,H}$ are given in chapter "Electron-Phonon Interaction".

We adopt the interaction representation, where $\mid v >= e^{-\frac{i}{\hbar}H_0 t}\widetilde{\mid v >}$, which means that all the operators have the form

$$\widetilde{O}=e^{\frac{i}{\hbar}H_0t}Oe^{-\frac{i}{\hbar}H_0t} \ , \tag{7.5}$$

such that their equation of motion is

$$\dot{\widetilde{O}}=\frac{i}{\hbar}[H_0,\widetilde{O}] \ ; \tag{7.6}$$

we can see that $\widetilde{a}_{\mathbf{k}}=e^{-i\omega_k t}a_{\mathbf{k}}$, $\widetilde{c}_{\mathbf{k}\sigma}=e^{-i\varepsilon_k t}c_{\mathbf{k}\sigma}$ and $\widetilde{b}_{\mathbf{k}\sigma}=e^{-i\varepsilon_k t}b_{\mathbf{k}\sigma}$. Equation (7.1) becomes

$$i\hbar\frac{\partial}{\partial t}\widetilde{\mid v>}=e^{\frac{i}{\hbar}H_0t}Ve^{-\frac{i}{\hbar}H_0t}\widetilde{\mid v>} \ ; \tag{7.7}$$

we denote all the $\widetilde{O}$-operators by $O(t)$ and all the states $\widetilde{\mid v>}$ by $\mid v(t)>$, and write

$$i\hbar\frac{\partial}{\partial t}\mid v(t)>=V(t)\mid v(t)> \ . \tag{7.8}$$

The solution of this equation is

$$\mid v(t)>=\mid v(t_i)>-\frac{i}{\hbar}\int_{t_i}^{t}dt^{'}V(t^{'})\mid v(t^{'})> \ , \tag{7.9}$$

or

$$\begin{gathered}\mid v(t)>=\mid v(t_i)>-\tfrac{i}{\hbar}\int_{t_i}^{t}dt^{'}V(t^{'})\mid v(t_i)>+\\ +\left(-\tfrac{i}{\hbar}\right)^2\int_{t_i}^{t}dt^{'}V(t^{'})\int_{t_i}^{t^{'}}dt^{''}V(t^{''})\mid v(t_i)>+... \ ,\end{gathered} \tag{7.10}$$

where t_i is the initial moment of time. The initial time is $t_i=0$ and the final time is $t_f=t$. The product $<v_f\mid v_i>$, where $\mid v_{i,f}>=\mid v(t_{i,f})>$, is the amplitude of transition from the initial to the final state (with normalized states). We note that the Schroedinger evolution equation is not relativistically invariant. However, in computing the transition probabilities we let the time go to infinity (and the initial moment to minus infinity), such that the solution gets a constant coefficient (the scattering matrix), which is irrelevant for the relativistic invariance. The interaction includes a position integral over the whole space. The relativistic invariance is improper for an evolution equation which does not include a position and a moment of time.

In an elementary act of interaction there exists a photon with energy $\hbar\omega_k = \hbar ck$ and momentum $\hbar\boldsymbol{k}$ and an electron with energy e^2k, such that the ratio of these two energies gives the coupling constant $\alpha = e^2/\hbar c$. For electrons $\alpha = 1/137$ is the fine structure constant. It follows that calculations up to the second-order of the perturbation theory suffice. In addition, whenever divergent integrals occur we use the cutoff wavevector $k_0 = mc/\hbar$ (the inverse of the Compton wavelength). Being associated with an unphysical, indefinite motion region, it is not subject to the relativistic invariance requirements.

7.2 Interacting electron

The initial state of an electron is $| v_i >= c^*_{\mathbf{k}\sigma} | 0 >$. The first-order perturbation state produced by the interaction v_p is

$$| v >^{(1)}_p = \frac{2e}{c\hbar^2} \sum_{\mathbf{k}'} \alpha_{k'} \beta_k \beta_{\mathbf{k}'+\mathbf{k}} \mathbf{k} \mathbf{e}_{-\mathbf{k}'} S_{\Delta\varepsilon} a^*_{-\mathbf{k}'} c^*_{\mathbf{k}'+\mathbf{k}\sigma} | 0 > , \qquad (7.11)$$

where

$$S_{\Delta\varepsilon} = \frac{e^{i(\varepsilon_{\mathbf{k}'+\mathbf{k}} + \omega_{k'} - \varepsilon_k)t} - 1}{\varepsilon_{\mathbf{k}'+\mathbf{k}} + \omega_{k'} - \varepsilon_k} , \ \Delta\varepsilon = \varepsilon_{\mathbf{k}'+\mathbf{k}} + \omega_{k'} - \varepsilon_k \qquad (7.12)$$

and $\alpha_k = c\sqrt{2\pi\hbar/\omega_k}$, $\beta_k = \beta_{\boldsymbol{k}} = c\sqrt{\hbar/2\varepsilon_k}$. The electron changes its state and emits a photon. The second-order perturbation state produced by the same interaction v_p is

$$\begin{aligned} | v >^{(2)}_p = & \tfrac{e^2}{c^2\hbar^4} \textstyle\sum_{\mathbf{k}'} \alpha_k \alpha_{\mathbf{k}'-\mathbf{k}} \beta_k \beta_{\mathbf{k}'+\mathbf{k}} \beta^2_{k'} \left[2\mathbf{k}'\mathbf{e}_{-\mathbf{k}} \right] \cdot \\ & \cdot \left[(\mathbf{k}' + \mathbf{k}) \mathbf{e}_{\mathbf{k}-\mathbf{k}'} \right] \tfrac{S_{\Delta\varepsilon} + S_{\Delta\varepsilon_2}}{\Delta\varepsilon_1} a^*_{-\mathbf{k}} a^*_{\mathbf{k}-\mathbf{k}'} c^*_{\mathbf{k}'+\mathbf{k}\sigma} | 0 > , \end{aligned} \qquad (7.13)$$

where

$$\begin{aligned} & \Delta\varepsilon_1 = \varepsilon_{k'} + \omega_{\mathbf{k}'-\mathbf{k}} - \varepsilon_k , \ \Delta\varepsilon_2 = \varepsilon_{k'} - \omega_k - \varepsilon_{\mathbf{k}'+\mathbf{k}} , \\ & \Delta\varepsilon = \Delta\varepsilon_1 - \Delta\varepsilon_2 ; \end{aligned} \qquad (7.14)$$

the electron emits two photons (an additional term is present for bound states). The first non-vanishing contribution of the interaction v_d is of the order e^2:

$$| v >_d^{(2)} = -\frac{ie^2}{c^2\hbar^3} t\beta_k^2 \sum_{\mathbf{k}'} \alpha_{\mathbf{k}'}^2 c_{\mathbf{k}\sigma}^* \mid 0 > -$$

$$-\frac{ie^2}{c^2\hbar^3} \sum_{\mathbf{k}'\mathbf{q}} \alpha_{k'} \alpha_{\mathbf{k}'-\mathbf{q}} \beta_k \beta_{\mathbf{k}'+\mathbf{q}} \cdot \tag{7.15}$$

$$\cdot \left(\mathbf{e}_{-\mathbf{k}'} \mathbf{e}_{\mathbf{k}'-\mathbf{q}}\right) S_{\Delta\varepsilon} a_{-\mathbf{k}'}^* a_{\mathbf{k}'-\mathbf{q}}^* c_{\mathbf{k}+\mathbf{q}\sigma}^* \mid 0 > ,$$

where

$$\Delta\varepsilon = \omega_{k'} + \omega_{\mathbf{k}'-\mathbf{q}} - \varepsilon_k \ . \tag{7.16}$$

The contributions of the other interactions (v_H, v_E) are estimated in the same way.

7.3 Spontaneous emission

Equation (7.11) gives the amplitude of spontaneous emission of a photon,

$$f = ec\sqrt{\frac{2\pi E_{ph}}{E_i E_f}} \sin\theta \cdot S_{\Delta\varepsilon} \ , \tag{7.17}$$

where $E_{ph} = \hbar\omega$ is the energy of the photon, $E_{i,f}$ are the initial and final energies of the electron and θ is the angle between the direction of propagation of the electron and the direction of propagation of the photon. For the probability of emission (polarized photon, per unit volume),

$$|S_{\Delta\varepsilon}|^2 = 2\pi t\delta(\Delta\varepsilon) \ . \tag{7.18}$$

If there is a width of the energy levels, the factor $S_{\Delta\varepsilon}$ becomes

$$S_{\Delta\varepsilon} = \frac{e^{i\Delta\varepsilon t} e^{-\gamma t} - 1}{\Delta\varepsilon + i\gamma} \tag{7.19}$$

and $|S_{\Delta\varepsilon}|^2 \longrightarrow 1/\left(\Delta\varepsilon^2 + \gamma^2/4\right)$ for $t \longrightarrow \infty$ (a natural line breadth is caused by the field emitted by the charge).[1] On the other hand,

[1] V. Weisskopf and E. Wigner, "Berechnung der natuerlichen Linienbreite auf Grund der Diracschen Lichttheorie", Z. Phys. **63** 54 (1930); "Ueber die

$\pi\delta(\Delta\varepsilon) \longleftarrow (\gamma/2)/\left(\Delta\varepsilon^2+\gamma^2/4\right)$ for $\gamma \ll \Delta\varepsilon$; it follows that the relevant time is of the order $t \simeq 1/\gamma$, as expected. This is the typical result for photon emission or absorption, dipole (multipole) radiation (from bound states), Zeeman, Stark and photoelectric effects. The results are similar with those of the radiation theory.[2]

7.4 Diamagnetic self-energy: a mass renormalization

The first term in equation (7.15) gives an interacting state

$$\mid v>=\mid v_i> -\frac{ie^2}{c^2\hbar^3}t\beta_k^2\sum_{\mathbf{k}'}\alpha_{\mathbf{k}'}^2\mid v_i> \ , \tag{7.20}$$

which indicates a change

$$\Delta E_d=\frac{e^2}{c^2\hbar^2}\beta_k^2\sum_{\mathbf{k}}\alpha_{\mathbf{k}'}^2 \tag{7.21}$$

in the energy of the electron; or

$$\Delta E_d=\frac{e^2c\hbar}{2\pi E}\int dk'\cdot k' \ , \tag{7.22}$$

where $E=\sqrt{p^2c^2+m^2c^4}$ is the original energy of the electron. This self-energy is infinite. It is associated with photon fluctuations of the vacuum: the electron generates and absorbs a photon, according to the diamagnetic interaction $\sim(\overline{\psi}\psi)A^2$ (a factor 2 should be included for the two photon polarizations).

naturliche Linienbreite in der Strahlung des harmonischen Oszillators", Z. Phys. **65** 18 (1930). See also W. Heitler, *The Quantum Theory of Radiation*, Dover (1984).

[2]P. A. M. Dirac, "The quantum theory of the emission and absorption of radiation", Proc. Roy. Soc. **A114** 243 (1927); "The quantum theory of dispersion", Proc. Roy. Soc. **A114** 710 (1927). See also E. Fermi, "Quantum theory of radiation", Revs. Mod. Phys. **4** 87 (1932); W. Heitler, *The Quantum Theory of Radiation*, Dover (1984).

Such divergences are typical for higher-order terms in the perturbation series, including even some second-order contributions. Their origin is twofold. On one hand, they arise from the pointlike nature of the electron, which generates a Coulomb static potential (related to the integration extended to infinity in equation (7.22)). On the other hand, they arise in higher-orders of the perturbation series as a consequence of the deficient formulation of the interaction problem. Indeed, the interaction problem is based on two equations. First, the electromagnetic field satisfies the Maxwell equations

$$\frac{1}{c^2}\frac{\partial^2 A_\mu}{\partial t^2} - \Delta A_\mu = \frac{4\pi}{c} j_\mu \ , \tag{7.23}$$

where the current j_μ is a bilinear form of ψ and $\overline{\psi}$. These later quantities are expressed with creation and destruction operators, which obey the Schroedinger equation of motion; for an operator O, this equation reads $\dot{O} = \frac{i}{\hbar}[H, O]$, where H is the hamiltonian which includes the interaction with the electromagnetic field. Starting with the second order of the perturbation theory, we have a contribution to j_μ from the field, by using Schroedinger equation; introducing this contribution in equation (7.23), it gives a field-field self-interaction, which is unphysical. Similarly, we may express the field by means of the current density j_μ from equation (7.23) and introduce it into the Schroedinger equation; then, starting with the second-order of the perturbation theory, we have an interaction of the particle (charge) with itself, which again is unphysical. In fact, the separation of the charges from their fields is impossible in higher orders of the perturbation theory, as it is known from the impossibility of eliminating the fields from the classical equation of motion of the charges in higher-orders of relativistic corrections. Ultimately, this entanglement of the charges and their fields arises from the formulation of the interaction problem; it manifests itself as an indeterminacy of the motion over short distances. In relativistic Quantum Mechanics the motion is not determined over distances of the order of the Compton wavelength.

It is worthwhile recalling in this context the relativistic equation of motion

$$\mu c \frac{du^\mu}{dt} = \frac{1}{c} F^{\mu\nu} j_\nu \tag{7.24}$$

for a mass density μ, where u^μ is the four-velocity ($u^\mu = dx^\mu/ds$)

and $F_{\mu\nu} = \partial_\mu A_\nu - \partial_\nu A_\mu$ is the electromagnetic field (the change of μ and t makes this a relativistic invariant equation); the current density is $j_\mu = \rho dx_\mu/dt = \rho u_\mu(ds/dt)$, where ρ is the charge density (which changes like μ). We may solve this equation for j^μ (u^μ) as a function of $F^{\mu\nu}$, introduce this j^μ in equation (7.23) and get fields determined by themselves. Similarly, we may solve equation (7.23) for fields as functions of u^μ (j^μ), introduce these fields in equation (7.24) and get four-velocities determined by four-velocities. In both cases we get field or particle (charge) self-interaction, which is unphysical. We cannot disentangle the fields from the charges. This situation is well known from the relativistic corrections to the classical lagrangian of the charges, where, in higher-orders of relativistic corrections, the charges cannot be separated from fields. Similarly, the Dirac equation

$$\gamma^\mu \left(p_\mu - \frac{e}{c}A_\mu\right)\psi = mc\psi \tag{7.25}$$

is entangled with Maxwell equations (7.23), the indeterminacy of the motion over short distances being embodied in the Dirac matrices. A perturbation-theory approach to (quantized) Maxwell equations coupled with Dirac equation has been done by Stueckelberg.[3]

It follows that in computing higher-order contributions to the perturbation series of the relativistic quantum-mechanical quantities one should be content with the approximation scheme provided by averaging the Dirac equation over distances of the order of the Compton wavelength, *i.e.* with the approximation scheme provided by the boson field; this scheme gives only orders of magnitude. The fact that in Dirac equation the indeterminacy of the short-distance motion is associated with the Dirac matrices may support the attempt of disentangling the fields from the charges, by means, for instance of the renormalization technique, carried out in a covariant fashion. Though a reasonable idea, the renormalization implies regularization techniques which are improper. The standard renormalization scheme forces the separation of the particles from the fields; and it implies a subtraction of infinities in the regularization of the Feynman integrals which is arbitrary. The experimental measurements force, indeed, the

[3]E. C. G. Stueckelberg, "Relativistisch invariante Stoerungstheorie des Diracschen Elektrons. I. Teil: Streustrahlung und Bremsstrahlung", Ann. Phys. 5. Folge, **21** 367 (1934) (Berichtigung, p. 744).

particle-field separation, and we may admit that the two separations, the theoretical one and the experimental one, may coincide. However, the subtraction of the infinities remains arbitrary, undetermined. In addition, the existence of competing interaction with other charges (like mesons), makes any agreement problematic, and, ultimately, the agreement with the experiment, if it exists, is accidental. In fact, as it is well known, the consistent application of the separation procedure in all orders of the perturbation theory, makes the interaction infinite or zero.[4] The renormalization and the regularization techniques are inconsistent.

The cutoff in equation (7.22) is the inverse of the Compton wavelength $\lambda_c = \frac{\hbar}{mc}$; by using it, we get

$$\Delta E_d = \frac{e^2 m^2 c^3}{4\pi\hbar E} \ , \tag{7.26}$$

which, in the non-relativistic limit, becomes $\Delta E_d \simeq e^2/4\pi\lambda_c = (1/4\pi)\cdot$ $\cdot(e^2/\hbar c)mc^2$. This is a mass renormalization, due to photon fluctuations ($\alpha = e^2/\hbar c = 1/137$ is the fine structure constant). Similar results are obtained from second-order contributions.

7.5 Lamb shift

Let us consider a set of electron bound states denoted by n, with orthonormalized wavefunctions φ_n (instead of plane waves $e^{i\mathbf{kr}}$). The first-order interacting state generated by the interaction v_p is

$$\mid v >_p^{(1)} = -\frac{ie}{c\hbar^2}\sum_{n_1\mathbf{k}} \alpha_k \beta_n \beta_{n_1} \left[\mathbf{G}_{n_1 n}(\mathbf{k})\mathbf{e}_{-\mathbf{k}}\right] S_{\Delta\varepsilon} a^*_{-\mathbf{k}} c^*_{n\sigma} \mid 0 > \ , \tag{7.27}$$

where $\Delta\varepsilon = \omega_k - \varepsilon_n$ and

$$\mathbf{G}_{nn'}(\mathbf{k}) = \int d\mathbf{r}\,(\varphi_n^* grad\varphi_{n'} - grad\varphi_n^* \cdot \varphi_{n'})\, e^{i\mathbf{kr}} \tag{7.28}$$

[4]L. Landau, "On the Quantum Theory of Fields", in *Niels Bohr and the Development of Physics*, ed. W. Pauli, Pergamon Press (1955) and references therein; L. Landau and E. Lifshitz, *Course of Theoretical Physics*, vol. 4, *Quantum Electrodynamics* (V. Berestetskii, E. Lifshitz, L. Pitaevskii), Butterworth-Heinemann (1971); M. Gell-Mann and F. E. Low, "Quantum Electrodynamics at small distances", Phys. Rev. **95** 1300 (1954).

(without the b-electrons). The second-order interacting state includes, apart from two photons, a contribution arising from the photon fluctuations, given by

$$\mid v>_p^{(2)}=\frac{e^2}{c^2\hbar^4}\sum_{n_1n_2\mathbf{k}}\alpha_k^2\beta_n\beta_{n_1}\beta_{n_2}^2\cdot$$

$$\cdot\left[\mathbf{G}_{n_1n_2}(\mathbf{k})\mathbf{e_k}\right]\left[\mathbf{G}_{n_2n}(-\mathbf{k})\mathbf{e_k}\right]\cdot \tag{7.29}$$

$$\cdot\int_0^t dt_1e^{-i(\omega_k+\varepsilon_{n_2})t_1}\int_0^{t_1}dt_2e^{i(\omega_k-\varepsilon_n+\varepsilon_{n_2})t_2}c_{n_1\sigma}^*\mid 0> \ .$$

We can see that the electron emits and absorbs a photon and changes its state. Noteworthy, for free electrons this contribution is zero. A self-interacting contribution corresponds to $n_1=n$. In addition, the main contribution arises from a degenerate state $n_2=n'$, if it exists. In these circumstances, equation (7.29) becomes

$$\mid v>_p^{(2)}=\frac{e^2}{c^2\hbar^4}\sum_{\mathbf{k}}\alpha_k^2\beta_n^4\left[\mathbf{G}_{nn'}(\mathbf{k})\mathbf{e_k}\right]\left[\mathbf{G}_{n'n}(-\mathbf{k})\mathbf{e_k}\right]\cdot$$

$$\cdot\int_0^t dt_1e^{-i(\omega_k+\varepsilon_n)t_1}\int_0^{t_1}dt_2e^{i\omega_kt_2}c_{n\sigma}^*\mid 0> \ , \tag{7.30}$$

or

$$\mid v>_p^{(2)}=\frac{e^2}{c^2\hbar^4}\sum_{\mathbf{k}}\alpha_k^2\beta_n^4\cdot$$

$$\cdot\left[\mathbf{G}_{nn'}(\mathbf{k})\mathbf{e_k}\right]\left[\mathbf{G}_{n'n}(-\mathbf{k})\mathbf{e_k}\right]Sc_{n\sigma}^*\mid 0> \ , \tag{7.31}$$

where

$$S=\frac{1}{\omega_k}\left[\frac{e^{-i\varepsilon_nt}-1}{\varepsilon_n}-\frac{e^{-i(\varepsilon_n+\omega_k)t}-1}{\varepsilon_n+\omega_k}\right] \ . \tag{7.32}$$

For atomic bound states $\mathbf{G}_{nn'}(\mathbf{k})\mathbf{e_k}$ and the range of k are of the order $1/a$, where a is the dimension of the atomic state. In the limit $k\longrightarrow 0$ equation (7.32) becomes

$$S=-\frac{\partial}{\partial\varepsilon_n}\frac{e^{-i\varepsilon_nt}-1}{\varepsilon_n} \ ; \tag{7.33}$$

in the limit of large t we get $S\simeq it/\varepsilon_n$, such that equation (7.31) becomes

$$\mid v>_p^{(2)}=it\frac{e^2\lambda_c^3}{8\pi\hbar a^4}\mid v_i> \ , \tag{7.34}$$

where $\varepsilon_n \simeq c/\lambda_c$. Therefore, we get an energy shift

$$\Delta E_p \simeq -E_n \left(\frac{e^2}{8\pi a E_n}\right)\left(\frac{\lambda_c}{a}\right)^3 ; \tag{7.35}$$

since E_n is of the order e^2/a, this shift is $\Delta E_p/E_n \simeq \frac{1}{8\pi}(\lambda_c/a)^3$. It is the splitting of the two degenerate states. The result is smaller by one order of magnitude than the standard result.[5] The standard result is $\Delta E \simeq 10^3 MHz$ ($1eV \simeq 2.4 \times 10^{14} Hz$), $\Delta E/E \simeq 3 \times 10^{-7}$ ($E = 13.6eV$), while $\Delta E_p/E_n \simeq 2 \times 10^{-8}$ (equation (7.35) with $\lambda_c = 3.8 \times 10^{-11} cm$ and $a = 0.53$Å). The main source of errors is the restriction to long-wavelength photons in equation (7.33) and the approximation $\varepsilon_n \simeq c/\lambda_c$. A slight variation in the cutoff λ_c may compensate the difference. The "mass renormalization" applied by Bethe is improper both in the non-relativistic approximation and the cutoff $242eV$.[6]

7.6 Spin-flip photon emission

In the interaction $v_H = -\frac{e}{2c\hbar}\overline{\psi}(\boldsymbol{\Sigma} + \Sigma^*)\psi\mathbf{H}$ given in the chapter "Electron-Photon Interaction" it is convenient to use the spin wavefunctions φ_σ in writing the field ψ. The interacting-electron initial state $c^*_{\mathbf{k}\sigma} \mid 0 >$ becomes

$$\begin{gathered} \mid v >_H^{(1)} = \frac{e}{c\hbar^2} \sum_{\mathbf{k}'} \sin\varphi(\mathbf{k}') k' \alpha_{k'} \beta_k \beta_{\mathbf{k}'+\mathbf{k}} \cdot \\ \cdot S_{\Delta\varepsilon} a^*_{-\mathbf{k}'} c^*_{\mathbf{k}'+\mathbf{k},-\sigma} \mid 0 > , \end{gathered} \tag{7.36}$$

[5] W. E. Lamb and R. C. Retherford, "Fine structure of the hydrogen atom by a microwave method", Phys. Rev. **72** 241 (1947); H. Bethe, "The electromagnetic shift of the energy levels", Phys. Rev. **72** 339 (1947); J. B. French and V. F. Weisskopf, "The electromagnetic shift of the energy levels", Phys. Rev. **75** 1240 (1949); R. Feynman, "Relativistic cut-off for Quantum Electrodynamics", Phys. Rev. **74** 1430 (1948) (correction in "Space-time approach to Quantum Electrodynamics", Phys. Rev. **76** 769 (1949)); J. Schwinger, "Quantum Electrodynamics. III. The electromagnetic properties of the electron-Radiative corrections to scattering", Phys. Rev. **76** 790 (1949).

[6] H. Bethe, "The electromagnetic shift of the energy levels", Phys. Rev. **72** 339 (1947).

where $\varphi(\mathbf{k}')$ is the angle in the parametrization

$$\begin{aligned} \mathbf{k}' &= k'(\sin\theta\cos\varphi, \sin\theta\sin\varphi, \cos\theta) \ , \\ \mathbf{e}_{\mathbf{k}'} &= \mathbf{e}_\theta = (\cos\theta\cos\varphi, \cos\theta\sin\varphi, -\sin\theta) \end{aligned} \tag{7.37}$$

and

$$\mathbf{k}' \times \mathbf{e}_{\mathbf{k}'} = \mathbf{e}_\varphi = (-\sin\varphi, \cos\varphi, 0) \tag{7.38}$$

(actually, related to the angle between the spin and the propagation wavevector of the photon); the spin operator $\boldsymbol{\sigma}$ is directed along the vector $\mathbf{e}_\varphi$ (where only the σ_x-component remains), such that, on emitting a photon, the magnetic field changes and the spin is reversed. Equation (7.36) gives the amplitude of spin-flip photon emission ($\Delta\varepsilon = \varepsilon_{\mathbf{k}'+\mathbf{k}} + \omega_{k'} - \varepsilon_k$). The second-order correction, as well as the contributions of the interaction v_E can be estimated in the same way; the latter implies vacuum polarization and photon fluctuations.

7.7 Anomalous magnetic moment

Let us assume an external, uniform and constant, magnetic field $\mathbf{H}_0$. The first contribution of v_H with $H_\mathbf{0}$ to the interacting state (apart from the zeroth order contribution) arises in the third-order of the perturbation theory. The structure of this contribution is

$$-\left(\frac{e}{c\hbar}\right)^3 \left[\overline{\psi}\left(\boldsymbol{\Sigma}\psi\right)\mathbf{H}\right]_1 \left[\overline{\psi}\left(\boldsymbol{\Sigma}\psi\right)\mathbf{H}\right]_2 \left[\overline{\psi}\left(\boldsymbol{\Sigma}\psi\right)\right]_3 \mathbf{H}_0 c^*_{\mathbf{k}\sigma} \mid 0> \ , \tag{7.39}$$

where the magnetic field is

$$\mathbf{H} = \sum_{\mathbf{q}} i\alpha_q \left(\mathbf{q}\times\mathbf{e}_\mathbf{q}\right)\left(a_\mathbf{q} - a^*_{-\mathbf{q}}\right) e^{i\mathbf{q}\mathbf{r}} \tag{7.40}$$

and the suffixes $1, 2, 3$ denote the times; in equation (7.39) the space integration is included. Spin wavefunctions should be included, which amounts to products of spin operators. The full contribution to the interacting state is obtained by inserting the time integrations

$$\left(-\frac{i}{\hbar}\right)^3 \int_0^t dt_1 \cdot \int_0^{t_1} dt_2 \cdot \int_0^{t_2} dt_3 \ ; \tag{7.41}$$

a factor 3 is included for the three positions of $\mathbf{H}_0$ in equation (7.39). Obviously, only the c-operators contribute.

We can see that, in fact, the time t_3 does not appear in equation (7.39), as expected. The external field plays the role of a probe, which lasts a short time Δt; consequently, we replace the t_3-integration by

$$-\frac{i}{\hbar}\int_0^{\Delta t} dt_3 = -\frac{i\Delta t}{\hbar} = \frac{\delta t}{\hbar} = \frac{1}{mc^2} \ . \tag{7.42}$$

The remaining time integrations are

$$S = \int_0^t dt_1 e^{-i\varepsilon_{\mathbf{k}+\mathbf{q}}t_1} \cdot \int_0^{t_1} dt_2 e^{i(\varepsilon_{\mathbf{k}+\mathbf{q}}-\varepsilon_k)t_2} \ . \tag{7.43}$$

In this equation the wavevector $\mathbf{q}$ of the (emitted and absorbed) photon brings a comparatively small contribution, such that

$$S = -\frac{\partial}{\partial\varepsilon_k}\frac{e^{-i\varepsilon_k t}-1}{\varepsilon_k} \simeq \frac{it}{\varepsilon_k} \ . \tag{7.44}$$

Taking into account the spin contribution

$$[\boldsymbol{\sigma}_y(\mathbf{q}\times\mathbf{e}_{\mathbf{q}})_y]^2 = q^2\cos^2\varphi \tag{7.45}$$

(we recall the relationship $\sigma_i\sigma_j = \delta_{ij} + i\varepsilon_{ijk}\sigma_k$ for the Pauli matrices), we get the energy change

$$\Delta E_H = -\frac{3}{\hbar}\left(\frac{e}{c\hbar}\right)^3 \sum_{\mathbf{q}} \frac{\cos^2\varphi(\mathbf{q})}{\varepsilon_k mc^2} q^2\alpha_q^2\beta_k^4\beta_{\mathbf{k}+\mathbf{q}}^2 (\mathbf{H}_0\boldsymbol{\sigma}) \ . \tag{7.46}$$

In this expression we may approximate ε_k by ck_0 ($k_0 = mc/\hbar$) and $\beta_{\mathbf{k}+\mathbf{q}} \simeq \beta_k$ (and average over directions); we get the relative change in the magnetic moment (the Bohr magneton $\mu =\mid e \mid \hbar/2mc$)

$$\Delta\mu/\mu \simeq \frac{3}{32\pi}\cdot\frac{e^2}{\hbar c} \tag{7.47}$$

(a factor 2 should be included for the two polarizations). The result is comparable with Schwinger's standard result ($\Delta\mu/\mu = \alpha/2\pi$).[7] The

[7] J. Schwinger, "On Quantum-Electrodynamics and the magnetic moment of the electron", Phys. Rev. **73** 416 (1948); "Quantum Electrodynamics. III. The electromagnetic properties of the electron-Radiative corrections to scattering", Phys. Rev. **76** 790 (1949).

result should be multiplied by a factor $(k_c/k_0)^3$, where k_c is the cut-off. The standard result is obtained for $k_c \simeq k_0/\sqrt{2}$, characteristic for a gaussian distribution. The standard result implies an improper regularization of a divergent integral.[8]

7.8 Pair creation: photon annihilation

The interaction $v_E = \frac{ie}{2c\hbar}\overline{\psi}\left(\boldsymbol{\alpha}-\boldsymbol{\alpha}^*\right)\psi\mathbf{E}$ is responsible for pair creation (destruction). The electric field is

$$\mathbf{E} = \sum_{\mathbf{q}} iq\alpha_q\left(\mathbf{e}_{\mathbf{q}}a_{\mathbf{q}} - \mathbf{e}_{-\mathbf{q}}a^*_{-\mathbf{q}}\right)e^{i\mathbf{qr}} \ , \tag{7.48}$$

after the spatial integration the interaction becomes

$$\begin{gathered} V_E = -\frac{e}{c\hbar}\sum\nolimits_{\mathbf{kq}} q\alpha_q\beta_k\beta_{\mathbf{k}-\mathbf{q}}\left(c^*_{\mathbf{k}}\boldsymbol{\sigma}_y b^*_{-\mathbf{k}+\mathbf{q}} + b_{-\mathbf{k}}\boldsymbol{\sigma}_y c_{\mathbf{k}-\mathbf{q}}\right)\cdot \\ \cdot\left(\mathbf{e}_{\mathbf{q}}a_{\mathbf{q}} - \mathbf{e}_{-\mathbf{q}}a^*_{-\mathbf{q}}\right) \ , \end{gathered} \tag{7.49}$$

where the spin suffixes are included. We can see that an external photon is absorbed and generates an electron pair, with opposite spins. The interacting state is given by

$$-\frac{e}{c\hbar}\sum_{\mathbf{k}} q\alpha_q\beta_k\beta_{\mathbf{k}-\mathbf{q}}c^*_{\mathbf{k}}\sigma_y e_{\mathbf{q}y}b^*_{-\mathbf{k}+\mathbf{q}}e^{-i\omega_q t} \mid 0> \ , \tag{7.50}$$

where the time integration should be included. We get the amplitude

$$f = -\frac{ie}{c\hbar^2}q\alpha_q\beta_k\beta_{\mathbf{k}-\mathbf{q}}\cos\varphi\sin\varphi S_{\Delta\varepsilon} \tag{7.51}$$

for the pair $c^*_{\mathbf{k}+}b^*_{-\mathbf{k}+\mathbf{q},-}$, where $\Delta\varepsilon = \varepsilon_k + \varepsilon_{\mathbf{k}-\mathbf{q}} - \omega_q$ ($\left|S_{\Delta\varepsilon}\right|^2 = 2\pi t\delta(\Delta\varepsilon)$) and $\varphi = \varphi(\mathbf{q})$ is given by equation (7.37). We can see that the momentum is conserved, but the energy cannot be conserved, since the equation

$$\sqrt{k^2+k_0^2} + \sqrt{(\mathbf{k}-\mathbf{q})^2+k_0^2} = q \tag{7.52}$$

has no solutions (as it is well known).

[8]See, for instance, W. Heitler, *The Quantum Theory of Radiation*, Dover (1984).

7.9 Pair creation: external field

The coupling of v_E with v_p or v_H leads to pair creation; v_H can be generated by an external magnetic field. Similarly, we can consider an interaction v_E generated by an external (static) electric field, like the field $grad(Ze/r)$ of a nucleus with charge Ze. This electric field is

$$\mathbf{E} = -\sum_{\mathbf{k}} \frac{4\pi i Ze\mathbf{k}}{k^2} e^{-i\mathbf{kr}} \; ; \tag{7.53}$$

it generates an interaction

$$\begin{gathered} V_0 = \frac{4\pi e^2 Z}{c\hbar} \sum_{\mathbf{kk}_1} \frac{1}{k^2} \beta_{k_1} \beta_{\mathbf{k}_1+\mathbf{k}} \cdot \\ \cdot \left(c^*_{\mathbf{k}_1} \boldsymbol{\sigma}_y k_y b^*_{-\mathbf{k}_1-\mathbf{k}} + b_{-\mathbf{k}_1} \boldsymbol{\sigma}_y k_y c_{\mathbf{k}_1+\mathbf{k}} \right) \end{gathered} \tag{7.54}$$

(of the v_E-type). We couple this interaction with

$$\begin{gathered} V_p = -\frac{e}{c\hbar} \sum_{\mathbf{kq}} \alpha_q \beta_k \beta_{\mathbf{k}-\mathbf{q}} (2\mathbf{k} - \mathbf{q}) \cdot \\ \cdot \left(c^*_{\mathbf{k}} c_{\mathbf{k}-\mathbf{q}} + b^*_{-\mathbf{k}+\mathbf{q}} b_{-\mathbf{k}} \right) \left(\mathbf{e}_{\mathbf{q}} a_{\mathbf{q}} - \mathbf{e}_{-\mathbf{q}} a^*_{-\mathbf{q}} \right) \; . \end{gathered} \tag{7.55}$$

The interaction V_0 does not depend on time; we use for it the integral given by equation (7.42). We apply

$$\frac{-2i}{\hbar mc^2} \int_0^t dt_1 V_p(t_1) V_0 \tag{7.56}$$

to the state $a^*_{\mathbf{q}} \mid 0 >$ (the factor 2 in equation (7.56) arises from the product $(V_p + V_0)(V_p + V_0)$). The temporal factor $S_{\Delta\varepsilon}$ implies

$$\varepsilon_{\mathbf{k}+\mathbf{q}} - \varepsilon_q = \omega_q \; , \tag{7.57}$$

an equation which is only approximately satisfied for small $\mathbf{k}$ and q of the order k_0. The momentum is not conserved, since the (large) nucleus is static. There are two amplitudes for creation of pairs $(\mathbf{k}, -\mathbf{k} + \mathbf{q} - \mathbf{k}')$ and $(-\mathbf{k} + \mathbf{q}, \mathbf{k} - \mathbf{k}')$, with opposite spins and undetermined $\mathbf{k}'$. The order of magnitude of these amplitudes is

$$f \simeq \pm \frac{4\pi i e^3 Z}{\hbar mc^2 k_0^2} \sqrt{\frac{2\pi c\hbar}{k_0}} \frac{k}{k'} \sin\theta \sin\theta' \cos\varphi' S_{\Delta\varepsilon} \; , \tag{7.58}$$

where θ is the angle between $\mathbf{k}$ and $\mathbf{q}$ and θ', φ' are the angles of $\mathbf{k}'$. This amplitude has the dimension $\sqrt{\alpha}\cdot vol^{3/2}$, where $\alpha = e^2/c\hbar$ is the fine-structure constant. The factor $vol^{3/2}$ is reduced by the density of states. Also, the amplitude given by equation (7.58) can be written as $f \simeq \sqrt{\alpha}Zr_e\lambda_c^{7/2}$, where $r_e = e^2/mc^2$ is the classical radius of the electron. Electron-positron pairs can also be created by polarizing the vacuum by an external electromagnetic field.[9]

The uncertainty implied by the second-order perturbation theory requires a density of states $\sim 1/\lambda_c^3$ for each electron; and a factor $\sim 1/\lambda_c$ for the photon; it follows a cross-section $\sigma \simeq \alpha Z r_e^2$, which has the order of magnitude of the standard result.[10] This is the Bethe-Heitler process. The coupling of V_E with V_0 leads to the Bremsstrahlung of an electron ($c^*_{\mathbf{k}\sigma} \mid 0>$) in an external field (and synchroton radiation).

7.10 Pair creation: two-photon annihilation

The coupling of V_E (equation (7.49)) with V_p (equation (7.55)) leads to pair creation by the annihilation of two photons. We apply the second-order perturbation operator (V_pV_E) to the intial state $a^*_{\mathbf{q}}a^*_{-\mathbf{q}} \mid 0>$. The time integration leads to $S_{\Delta\varepsilon}$, where $\Delta\varepsilon = 2\varepsilon_k - 2\omega_q$, for the pair ($\mathbf{k}$, $-\mathbf{k}$) with opposite spins, and to $S_{\Delta\varepsilon}$, $\Delta\varepsilon = \varepsilon_k - \varepsilon_{\mathbf{k}-\mathbf{q}} - \omega_q$, an approximate energy conservation. This is the Breit-Wheeler process. The energy conservation $2\varepsilon_k - 2\omega_q$ is satisfied for $\mathbf{k} \longrightarrow 0$ and $q \simeq k_0$. The amplitude of formation of the pair ($\mathbf{k}$, $-\mathbf{k}$) is

$$f \simeq \mp\frac{4i}{\hbar^2}\left(\frac{e}{c\hbar}\right)^2 q^2\alpha_q^2\beta_0^2\frac{\beta_q^2}{\varepsilon_q}\sin\Theta\cos\theta\sin\varphi\cdot S_{\Delta\varepsilon}\ , \tag{7.59}$$

where Θ is the angle between $\mathbf{k}$ and $\mathbf{q}$ and θ, φ are the angles of the wavevector $\mathbf{q}$ (with respect to the electron spin). For $q \simeq k_0$ the order of magnitude of the amplitude f is $f \simeq \alpha\lambda_c^3$ (where $S \simeq \tau \simeq \lambda_c/c$), in

[9] M. Apostol, "Dynamics of electron-positron pairs in a vacuum polarized by an external electromagnetic field", J. Mod. Opt. **58** 611 (2011).

[10] H. Bethe and W. Heitler, "On the stopping of fast particles and on the creation of positive electrons", Proc. Roy. Soc. **A146** 83 (1934).

agreement with the standard result (we note that $r_e = \alpha\lambda_c$).[11] The process of pair formation by annihilation of two photons is related to the process of pair annihilation with the formation of two photons.[12]

7.11 Charge renormalization

Let us consider the interaction of a charge e with a static charge Q. The latter generates an interaction V_0 given by equation (7.54). The first non-vanishing correction to the initial state $c^*_{\mathbf{k}\sigma} \mid 0>$ appears in the second order of the perturbation theory; it is due to vacuum polarization. Before effecting the time integration, this contribution reads

$$\begin{gathered}\left(\tfrac{4\pi eQ}{c\hbar}\right)^2 \sum_{\mathbf{q},\mathbf{q}'} \tfrac{1}{q^2q'^2}\beta_k\beta^2_{\mathbf{k}-\mathbf{q}'}\beta_{\mathbf{k}-\mathbf{q}-\mathbf{q}'}(\sigma_y q_y)(\sigma_y q'_y)\cdot \\ \cdot e^{i(\varepsilon_{\mathbf{k}-\mathbf{q}'}-\varepsilon_k)t_1} e^{i\varepsilon_{\mathbf{k}-\mathbf{q}'}t_2} c^*_{\mathbf{k}-\mathbf{q}-\mathbf{q}'} \mid 0> \ .\end{gathered} \tag{7.60}$$

The main contribution to the state $\mathbf{k}\sigma$, with small $\mathbf{k}$, comes from small $\mathbf{q} = -\mathbf{q}'$. This is a quasi-static interaction; instead of two V_0 generated by Q, we may use only one generated by Q and another generated by e; then $(eQ)^2$ is replaced by e^3Q. The integration over $\mathbf{q}'$ gives $q^2\Delta q \simeq 1/\lambda_c^3$. The main contribution of the time integration is $S_{\Delta\varepsilon}/\varepsilon_k$,

$$S_{\Delta\varepsilon} = \frac{e^{i(\varepsilon_{\mathbf{k}+\mathbf{q}}-\varepsilon_k)t} - 1}{\varepsilon_{\mathbf{k}+\mathbf{q}} - \varepsilon_k} \simeq it \ . \tag{7.61}$$

After averaging over angles we get a change

$$\Delta E \simeq \frac{1}{3\hbar}\left(\frac{4\pi eQ}{c\hbar}\right)^2 \frac{\beta_0^4}{\varepsilon_0}\lambda_c^3 \sum_{\mathbf{k}} \frac{1}{k^2} \tag{7.62}$$

in energy, which should be compared with the Coulomb interaction $4\pi eQ\sum_{\mathbf{k}}\frac{1}{k^2}$. It follows a charge renormalization of the order $\delta e/e \simeq \frac{\pi}{6}\alpha$, a result comparable with the standard results.[13]

[11] G. Breit and J. A. Wheeler, "Collision of two light quanta", Phys. Rev. **46** 1087 (1934).

[12] P. A. M. Dirac, "On the annihilation of electrons and protons", Proc. Cambr. Phil. Soc. **26** 361 (1930).

[13] J. Schwinger, "Quantum Electrodynamics. II. Vacuum polarization and self-energy", Phys. Rev. **75** 651 (1949); R. P. Feynman, "Space-time approach to

7.12 Photon mass

The vacuum polarization accompanied by photon fluctuations in the perturbation series does not change the energy of the photon. However, the interaction with a "real" electron gives mass to the photon. Indeed, the interaction leads to

$$\frac{1}{c^2}\frac{\partial^2 A^\mu}{\partial t^2} - \Delta A^\mu + \frac{8\pi e^2}{c^2\hbar^2}\left(\overline{\psi}\psi\right)A^\mu = \\ = \frac{4\pi}{c}j^\mu - \frac{4\pi i e\hbar}{c}\partial_\nu\overline{\psi}\left(\sigma^{\mu\nu} - \sigma^{\mu\nu *}\right)\psi \ , \tag{7.63}$$

which is a wave equation with sources and mass. For an electron (both spin orientations) we get a photon mass given by

$$m_{ph}^2 = \frac{e^2\hbar^2}{Ec^2}\cdot\frac{1}{\lambda_c^3} = \alpha\left(mc^2/E\right)m^2 \ , \tag{7.64}$$

where E is the energy of the electron. The photons interacting with the electrons are similar to radiation propagating in matter (polaritons). The photon mass leads to a screened Coulomb interaction $\frac{1}{r}e^{-\frac{m_{ph}c}{\hbar}r}$.

7.13 "The meaning of it all"

Apart from regular processes of emission of their own electromagnetic field (of one or several photons), the charges undergo self-interaction with their own field. This self-interaction proceeds by vacuum polarization and photon fluctuations. Such processes can be seen not only in the mass and charge renormalization, but also in the interaction of the charges with an external electromagnetic field (*e.g.* Lamb shift, anomalous magnetic moment), or other processes, like pair creation, etc. It is in such processes where the Compton cutoff appears, not always as a necessary tool of eliminating divergences, but rather

---

Quantum Electrodynamics", Phys. Rev. **76** 769 (1949); R. Serber, "Linear modifications in the Maxwell field equations", Phys. Rev. **48** 49 (1935); E. A. Uehling, "Polarization effects in the positron theory", Phys. Rev. **48** 55 (1935).

as a characteristic momentum (wavelength) which controls the process. Such processes are governed by the Zitterbewegung. Although a cutoff, the Compton wavelength $\lambda_c = \hbar/mc$ may be viewed as a satisfactory estimate of the dimension of the spatial extension of the Zitterbewegung. The mass and charge renormalization should be effected everywhere. For instance, the renormalized coupling constant is $\overline{\alpha} \simeq \alpha + \frac{\pi}{3}\alpha\overline{\alpha}$ (*i.e.* $\overline{\alpha} = 1/137 \simeq \alpha/(1-\pi\alpha/3)$), from charge renormalization, and the renormalized electron mass is $\overline{m}\,(\simeq 10^{-27}g) \simeq \overline{\alpha}m$. All these (relativistic) quantum-mechanical corrections are small (they are observable for a large number of particles). More important, according to the manner they are derived (and defined) they have the status of uncertainties produced by (photon and charge) fluctuations. Consequently, an accurate value for such quantities is not worth much. The cutoff used in deriving them may be viewed as a fitting parameter, for each interaction process. The effects of the self-interaction depend on the interaction process. The mass and the charge renormalization in the standard Quantum Electrodynamics is an arbitrary scheme of calculation, leading to undetermined results; it is an arbitrary, prescribed calculation recipe. The "excellent agreement" with experiment is only the result of such an arbitrary choice. In this context it is worth emphasizing that the boson theory of the Dirac equation is different. While, basically, the standard Quantum Electrodynamics is a quantum-mechanical theory of Dirac equation, the boson theory of the Dirac equation is a field theory. The former is the theory of the quantum-mechanical motion of the particle (charge; in relativistic form, by using Dirac equation), while the latter is the theory of the motion of the Zitterbewegung of the Dirac particle (the motion of a motion). The former includes typical logarithmic infinities, while the divergences appearing in the latter for a vanishing wavelength cutoff are power-like (*e.g.*, quadratic). The cutoff method implies an uncertainty, such that an "excellent" agreement cannot be expected from the boson theory. The "agreement" with the experiment of the standard theory is not a "fluke",[14] but a fake. The standard theory is a failure, while the boson theory is a theory.

The boson theory looks like an effective field theory, inasmuch as its

[14]P. A. M. Dirac, "The evolution of the physicist's picture of Nature", Am. Sci. **208** 45 (1963).

basic elements are quanta of interaction fields governed by a cutoff.[15] The bosons represent changes in the fermionic states of the particles (electrons), similar to the photons which change the electromagnetic field. In the standard theory, the quanta are the electrons themselves, and even the photons are viewed more as quantum-mechanical particles;[16] the latter theory resembles much a quantum-mechanical theory of transitions, similar to the second-quantization radiation theory of Dirac, which later became the "field-theory" of condensed matter;[17] (where it is reasonable).

Both the Dirac equation and the Klein-Gordon equation on which the boson theory is based are relativistically invariant. Consequently, with the limitations described above, both may be used for quantum-mechanical effects of relativistic particles, since the latter imply only a change of the reference frame. The cutoff length is not a physical length, it is a fictitious length which is not subject to the relativistic invariance.

[15] P. Jordan, "Ueber Wellen und Korpuskeln in der Quantenmechanik", Z. Phys. **45** 765 (1927).

[16] R. P. Feynman, *Quantum Electrodynamics*, Benjamin, NY (1961).

[17] P. A. M. Dirac, "The quantum theory of the emission and absorption of radiation", Proc. Roy. Soc. **A114** 243 (1927); P. Jordan and E. Wigner, "Ueber das Paulische Aequivalenzverbot", Z. Phys. **47** 631 (1928); V. M. Galitskii and A. B. Migdal, "Application of quantum field theory methods to the many body problem", Sov. Phys.-JETP **7** 96 (1958) (ZhETF **34** 139 (1958)); A. Klein and R. Prange, "Perturbation theory for an infinite medium of fermions", Phys. Rev. **112** 994 (1958); P. C. Martin and J. Schwinger, "Theory of many-particle systems. I.", Phys. Rev. **115** 1342 (1959).

8 Quasi-classical relativistic electron

8.1 Relativistic notations

We use the contravariant coordinates $x^\mu = (ct, \boldsymbol{r})$ and the covariant coordinates $x_\mu = (ct, -\boldsymbol{r})$, where t is the time, $\boldsymbol{r}$ is the position, c $(3 \times 10^{10} cm/s)$ is the speed of light in vacuum, $\mu = 0, 1, 2, 3$ and $x^\mu x_\mu = c^2t^2 - \boldsymbol{r}^2$. The metric is $g^{\mu\nu} = (1, -1, -1, -1)$. The wavevector is $k^\mu = (\omega/c, \boldsymbol{k})$, $k_\mu = (\omega/c, -\boldsymbol{k})$, where ω is the frequency and $\boldsymbol{k}$ is the spatial wavevector. The derivatives are $\partial^\mu = (\frac{\partial}{c\partial t}, -grad)$, $\partial_\mu = (\frac{\partial}{c\partial t}, grad)$, such that $\partial^\mu(k^\nu x_\nu) = k^\mu$, $\partial_\mu(k_\nu x^\nu) = k_\mu$ ($k^\nu x_\nu = k_\nu x^\nu$), The momentum is $p^\mu = i\hbar\partial^\mu$, $p_\mu = i\hbar\partial_\mu$. The electromagnetic potentials are $A^\mu = (\Phi, \boldsymbol{A})$, $A_\mu = (\Phi, -\boldsymbol{A})$, where Φ is the scalar potential and $\boldsymbol{A}$ is the vector potential. The electric current density is $j^\mu = (c\rho, \boldsymbol{j})$, $j_\mu = (c\rho, -\boldsymbol{j})$, where ρ is the charge density and $\boldsymbol{j}$ is the spatial current density.

We use Fourier transforms of the form

$$\begin{aligned} f(t, \boldsymbol{r}) = f(x) = \tfrac{1}{2\pi\sqrt{V}} \int d\omega \textstyle\sum_{\boldsymbol{k}} e^{-i\omega t + i\boldsymbol{k}\boldsymbol{r}} f(\omega, \boldsymbol{k}) = \\ = \tfrac{1}{2\pi\sqrt{V}} \int d\omega \textstyle\sum_{\boldsymbol{k}} e^{-ik_\mu x^\mu} f(k) = \sum_k e^{-ikx} f(k) \end{aligned} \tag{8.1}$$

and

$$f(k) = \frac{1}{\sqrt{V}} \int dx f(x) e^{ikx} \ , \tag{8.2}$$

where V is the volume. We have

$$f(x)g(x) = \sum_k \left(\sum_{k_1} f(k-k_1)g(k_1)\right) e^{-ikx} \ ,$$
$$f^*(x)g(x) = \sum_k \left(\sum_{k_1} f^*(k_1-k)g(k_1)\right) e^{-ikx} \ ,$$
$$f(x)g(x)hx) = \sum_k \{\sum_{k_1,k_2} f(k-k_1-k_2)\cdot$$
$$\cdot g(k_1)h(k_2)\}e^{-ikx} \ . \tag{8.3}$$

$\sum_k$ is a notation for $\frac{1}{2\pi\sqrt{V}}\int d\omega \sum_{\boldsymbol{k}}$.

8.2 Relativistic electron

A relativistic particle with mass m has the energy $\mathcal{E}$ given by $\mathcal{E}^2 = c^2\left(\boldsymbol{p}^2 + m^2c^2\right)$, where $\boldsymbol{p}$ is the momentum. If we admit only positive energies, this equation becomes $\mathcal{E} = c\sqrt{\boldsymbol{p}^2 + m^2c^2}$. In the non-relativistic limit we may write

$$\mathcal{E} = mc^2 + \frac{1}{2m}\boldsymbol{p}^2 + ... \ . \tag{8.4}$$

We may leave aside the constant term mc^2 and perform the quantization according to the rules $\mathcal{E} \rightarrow i\hbar\frac{\partial}{\partial t}$, $\boldsymbol{p} \rightarrow -i\hbar grad$, where $\hbar$ ($\simeq 10^{-27} erg \cdot s$) is Planck's constant; we get the Schroedinger equation

$$i\hbar\frac{\partial\psi}{\partial t} = -\frac{\hbar^2}{2m}\Delta\psi + ... \tag{8.5}$$

with relativistic corrections. Interaction can be included, *e.g.* by $\mathcal{E} \rightarrow \mathcal{E} - \Phi$, $\boldsymbol{p} \rightarrow \boldsymbol{p} - \frac{e}{c}\boldsymbol{A}$ for electromagnetic potentials Φ and $\boldsymbol{A}$ and a charge e (for electrons $e = -4.8 \times 10^{-10} esu$). This provides a consistent quantum-mechanical description, with ψ the wavefunction whose square is the localization probability. Noteworthy, the quantization is meaningful due to the fact that it is done on a classical, non-relativistic equation $\mathcal{E} = \frac{1}{2m}\boldsymbol{p}^2 + ...$; the Schroedinger equation has a (quasi)-classical limit (for $\hbar \rightarrow 0$ the derivatives $\rightarrow \infty$). The quantum-mechanical picture describes motions which imply small amounts of action, of the order of $\hbar$. For larger amounts, the Schroedinger equation passes into the classical equation of motion (Hamilton-Jacobi equation), including a meaningful current conservation.

The Schroedinger equation is not relativistic invariant.[1] Although a relativistic particle has a large momentum, it may undergo small quantum-mechanical movements. In order to quantize the relativistic expression $\mathcal{E}^2 = c^2\left(\boldsymbol{p}^2 + m^2c^2\right)$ we need to include the negative energies too, which leads to $\mathcal{E} = \pm c\sqrt{\boldsymbol{p}^2 + m^2c^2}$. This requires to give up the current quantum-mechanical description and to view a positive energy as the creation of a particle with that energy and a negative energy as the destruction of a particle with that positive energy. This is a quantum-field picture. For particles with electric charge, like electrons, the destruction of a charge e means the creation of a charge $-e$. Thus, we are lead to admit the existence of the antiparticles. The particle field has then (at least) two distinct states, particle and antiparticle. A similar situation occurs for other intrinsic characteristics of the relativistic particles. The existence of two distinct states (at least) requires the existence of a non-trivial matrix in the quantized version of the relativistic equation $\mathcal{E}^2 = c^2\left(\boldsymbol{p}^2 + m^2c^2\right)$, which is absent. The occurrence of such matrices is only possible in a linearized version of this equation, which is the Dirac equation

$$\gamma^\mu p_\mu \psi = mc\psi \ ; \tag{8.6}$$

γ^μ are (four times four) matrices, which, apart from the antiparticle states includes also the 1/2-spin states; ψ is a bispinor corresponding to all these states. By applying twice this equation we get the Klein-Gordon equation

$$\partial^\mu \partial_\mu \psi = -\frac{m^2c^2}{\hbar^2}\psi \ , \tag{8.7}$$

or

$$\hbar^2 \frac{\partial^2 \psi}{\partial t^2} - \hbar^2 c^2 \Delta \psi + m^2 c^4 \psi = 0 \ , \tag{8.8}$$

which we would get by quantizing the expresion $\mathcal{E}^2 = c^2\left(\boldsymbol{p}^2 + m^2c^2\right)$ (we note that $\{\gamma^\mu, \gamma^\nu\} = 2g^{\mu\nu}$). The only difference is that ψ is a

[1] A quantization of the expression $E = c\sqrt{\boldsymbol{p}^2 + m^2c^2}$ amounts to a quantization of the series expansion of this expression.

bispinor for four states. The 1/2-spin appears explicitly in the presence of the electromagnetic field. Besides, the energy of the field is meaningful only if we accept that the creation and destruction operators (field operators) obey anticommutation rules, *i.e.* these particles are fermions. The Dirac field describes the electron, according to the experimental evidence.

The Dirac equation has a few particularities. First, we note that it has no (quasi-) classical limit. Indeed, in this limit it becomes $\gamma^0\mathcal{E} - c\boldsymbol{\gamma}\boldsymbol{p} = mc^2$, which is not a valid classical equation. The Dirac equation is a quantum-field equation, which is not applicable in the quasi-classical limit. The same is true for its Klein-Gordon equivalent, which, in the quasi-classical limit $\psi \sim Ae^{\frac{i}{\hbar}S}$, with a slowly-varying amplitude A, leads to a meaningless free wave equation for the classical action S.

Second, the Dirac equation leads to a continuity equation

$$\frac{\partial}{\partial t}(\psi^*\psi) + c \cdot grad(\psi^*\boldsymbol{\alpha}\psi) = 0 \ , \tag{8.9}$$

where ψ^* is the transposed conjugate bispinor and $\boldsymbol{\alpha} = \gamma^0\boldsymbol{\gamma}$. This equation shows that the localization probability of a particle, irrespective of its state, moves with an undefined velocity $c\boldsymbol{\alpha}$;[2] such an equation makes no sense. Finally, the Dirac equation mixes up states over distances of the order of the Compton wavelength $\lambda_c = \hbar/mc$ and times of the order $\hbar/mc^2$. Therefore, over such distances and times the Dirac equation describes an unphysical, undefined motion, known as the Zitterbewegung. The quantum-mechanical effects of a relativistic particle should imply a cutoff momentum of the order mc, a cutoff time $\hbar/mc^2$ and a corresponding cutoff energy mc^2.[3] Beyond these limits the interaction is quantum-mechanical and, in the extreme limit, classical. These limits may be taken as the quantum-mechanical limits of the Electromagnetism (for fields they are known as the Schwinger limits). For electrons $\lambda_c \simeq 3.8 \times 10^{-11}cm$ and the

[2]G. Breit, "An interpretation of Dirac's theory of the electron", Proc. Nat. Acad. Sci. **14** 553 (1928).

[3]L. Landau and R. Peierls, "Erweiterung des Unbestimmtheitsprinzips fur die relativistische Quantentheorie", Z. Phys. **69** 56 (1931); L. Landau and E. Lifshitz, *Course of Theoretical Physics*, vol. 4, *Quantum Electrodynamics* (V. Berestetskii, E. Lifshitz, L. Pitaevskii), Butterworth-Heinemann (1971).

frequency $\omega = mc^2/\hbar \simeq 10^{21}s^{-1}$. The relativistic uncertainty given by λ_c arises from the existence of the antiparticles, *i.e.* states with negative energy, and, ultimately, in the relativistic energy-momentum relationship, which implies the squared energy. For fermions, this uncertainty acquires the aspect of the Zitterbewegung, originating in the Dirac matrices.

8.3 Bosons

The measurable quantum-field motion of a relativistic electron is the quantum motion of the Zitterbewegung. To this end, we give a variation δx^μ to the coordinates x^μ, and denote it by u^μ, according to the scheme $x^\mu \longrightarrow x^\mu + \delta x^\mu$, $\delta x^\mu = u^\mu$, $x^\mu \longrightarrow x^\mu + u^\mu$. We take the first-order variations with respect to u^μ of the Dirac equation (8.6)

$$\gamma^\mu \partial_\mu \psi = \frac{mc}{i\hbar}\psi \ . \tag{8.10}$$

We write $x^\mu = s^\mu \cdot 1$, where $s^0 = ct$, $\mathbf{s} = \mathbf{r}$ and 1 denotes the unit matrix; we have $x_\mu x^\mu = s_\mu s^\mu \cdot 1 = s^2 \cdot 1$, where $s^2 = c^2t^2 - \mathbf{r}^2$. For δx^μ we need $\delta x_\mu \delta x^\mu = u_\mu u^\mu = u^2 = ds^2$; the (non-trivial) solution of this equation is

$$\delta x^\mu = u^\mu = \frac{1}{2} u \gamma^\mu \tag{8.11}$$

(since $\gamma_\mu \gamma^\mu = 4$). The first-order expansion of the Dirac equation is

$$\gamma^\mu \left(\partial_\mu \psi + u^\nu \partial_\nu \partial_\mu \psi\right) = \frac{mc}{i\hbar}\left(\psi + u^\nu \partial_\nu \psi\right) \ , \tag{8.12}$$

or

$$\partial^\mu \partial_\mu \psi = -\frac{m^2c^2}{\hbar^2}\psi \ , \tag{8.13}$$

which is the Klein-Gordon equation (8.7). We note that u is absorbed into ψ, hence the bosonic character of the latter. The bosonic field thus obtained has a density, the presence of an electron means a boson quantum, more boson quanta means more electrons. The boson field should be used with the cutoff Compton wavelength, like an electron radius, so the exclusion principle is satisfied. The ψ in equation (8.13) originates in u multiplied by a bispinor. It is a boson field with four components. We write it as

$$\psi = \frac{1}{\sqrt{V}} \sum_{\mathbf{k}} \beta_k (\psi_{k\alpha}) e^{i\mathbf{kr}} \ , \tag{8.14}$$

where

$$(\psi_{\mathbf{k}\alpha}) = \begin{pmatrix} c_{\mathbf{k},+1} \\ c_{\mathbf{k},-1} \\ b^*_{-\mathbf{k},-1} \\ b^*_{-\mathbf{k},+1} \end{pmatrix} \tag{8.15}$$

and

$$\beta_k = c\sqrt{\frac{\hbar}{2\varepsilon_k}} \ , \ \varepsilon_{\boldsymbol{k}} = c\sqrt{k^2 + k_0^2} \ , \ k_0 = mc/\hbar. \tag{8.16}$$

The time dependence of the operators is $c_{\boldsymbol{k}\sigma}$, $b_{\boldsymbol{k}\sigma} \sim e^{-i\varepsilon_k t}$, where $\sigma = \pm 1$ is the spin label. The c's and the b's operators satisfy usual bosonic commutation relations for four distinct types of bosons. The c-operators correspond to particles (electrons), while the b-operators correspond to antiparticles (positrons). For a free field we have

$$\psi = \frac{1}{\sqrt{V}} \sum_{\boldsymbol{k}} c\sqrt{\frac{\hbar}{2\varepsilon_{\boldsymbol{k}}}} \begin{pmatrix} c_{\boldsymbol{k}} e^{-i\varepsilon_{\boldsymbol{k}} t + i\boldsymbol{kr}} \\ b^*_{\boldsymbol{k}} e^{i\varepsilon_{\boldsymbol{k}} t - i\boldsymbol{kr}} \end{pmatrix} \ , \tag{8.17}$$

where $c_{\boldsymbol{k}} = \begin{pmatrix} c_{\boldsymbol{k},+1} \\ c_{\boldsymbol{k},-1} \end{pmatrix}$ and $b^*_{\boldsymbol{k}} = \begin{pmatrix} b^*_{\boldsymbol{k},+1} \\ b^*_{\boldsymbol{k},-1} \end{pmatrix}$. According to equation (8.2), the Fourier transform is

$$\psi(k) = 2\pi c\sqrt{\frac{\hbar}{2\varepsilon_{\boldsymbol{k}}}} \begin{pmatrix} c_{\boldsymbol{k}} \delta(\omega - \varepsilon_{\boldsymbol{k}}) \\ b^*_{-\boldsymbol{k}} \delta(\omega + \varepsilon_{\boldsymbol{k}}) \end{pmatrix} \ , \tag{8.18}$$

where $b^*_{-\boldsymbol{k}} = \begin{pmatrix} b^*_{\boldsymbol{k},-1} \\ b^*_{\boldsymbol{k},+1} \end{pmatrix}$.

In the presence of the electromagnetic field we need to use the covariant derivative $D_\mu = \partial_\mu - \frac{e}{ic\hbar} A_\mu$ in equation (8.12), according to $\psi \longrightarrow \psi + u^\mu D_\mu \psi$; we get the Klein-Gordon equation

$$\left(p_\mu - \frac{e}{c} A_\mu\right)\left(p^\mu - \frac{e}{c} A^\mu\right)\Psi - \frac{ie\hbar}{2c}\sigma^{\mu\nu} F_{\mu\nu}\Psi = m^2 c^2 \Psi \tag{8.19}$$

with electromagnetic field, where $\sigma^{\mu\nu} = \frac{1}{2}[\gamma^\mu, \gamma^\nu]$ is the spin matrix and $F_{\mu\nu} = \partial_\mu A_\nu - \partial_\nu A_\mu$ is the electromagnetic field. This equation can be written as

$$\begin{gathered}\frac{1}{c^2}\frac{\partial^2\psi}{\partial t^2} - \Delta\psi + \frac{m^2c^2}{\hbar^2}\psi + \frac{e}{c\hbar^2}p_\mu A^\mu\psi + \frac{e}{c\hbar^2}A_\mu p^\mu\psi - \\ -\frac{e^2}{c^2\hbar^2}A_\mu A^\mu\psi - \frac{e}{c\hbar}\sum \mathbf{H}\psi + \frac{ie}{c\hbar}\boldsymbol{\alpha}\mathbf{E}\psi = 0 \ ,\end{gathered} \tag{8.20}$$

where $\boldsymbol{H}$ is the magnetic field, $\boldsymbol{E}$ is the electric field,

$$\boldsymbol{\Sigma} = \begin{pmatrix} \boldsymbol{\sigma} & 0 \\ 0 & \boldsymbol{\sigma} \end{pmatrix} \ , \ \boldsymbol{\alpha} = \begin{pmatrix} 0 & \boldsymbol{\sigma} \\ \boldsymbol{\sigma} & 0 \end{pmatrix} \ , \tag{8.21}$$

and $\boldsymbol{\sigma}$ are the Pauli matrices,

$$\begin{gathered}\sigma_x = \begin{pmatrix} 0 & 1 \\ 1 & 0 \end{pmatrix} \ , \ \sigma_y = \begin{pmatrix} 0 & -i \\ i & 0 \end{pmatrix} \ , \\ \sigma_z = \begin{pmatrix} 1 & 0 \\ 0 & -1 \end{pmatrix} \ .\end{gathered} \tag{8.22}$$

8.4 Free bosons

The free equation of motion of the boson field (the Klein-Gordon equation (8.13)) reads

$$\frac{1}{c^2}\frac{\partial^2\psi}{\partial t^2} - \Delta\psi + \frac{m^2c^2}{\hbar^2}\psi = 0 \ . \tag{8.23}$$

If we multiply this equation by $\dot{\psi}^*$, the equation for ψ^* by $\dot{\psi}$, where ψ^* is the adjoint of ψ (transposed conjugate), and add the two equations, we get

$$\begin{gathered}\frac{\partial}{\partial t}\left(\frac{1}{c^2}\dot{\psi}^*\dot{\psi} + \partial_i\psi^*\partial_i\psi + \frac{m^2c^2}{\hbar^2}\psi^*\psi\right) - \\ -\partial_i\left(\dot{\psi}^*\partial_i\psi + \partial_i\psi^*\dot{\psi}\right) = 0 \ ,\end{gathered} \tag{8.24}$$

where $i = 1, 2, 3$; hence, we can see that

$$w_e = \frac{1}{c^2}\dot{\psi}^*\dot{\psi} + \partial_i\psi^*\partial_i\psi + \frac{m^2c^2}{\hbar^2}\psi^*\psi \tag{8.25}$$

is the energy density and

$$\mathbf{g}_e = -\frac{1}{c^2}\left(\dot{\psi}^* grad\psi + grad\psi^* \cdot \dot{\psi}\right) \tag{8.26}$$

is the momentum density. The total energy is

$$\begin{aligned} W_e = \int d\mathbf{r} w_e = \sum_{\mathbf{k}\sigma} \beta_k^2 \tfrac{2\varepsilon_k^2}{c^2} \left(c^*_{\mathbf{k}\sigma} c_{\mathbf{k}\sigma} + b_{\mathbf{k}\sigma} b^*_{\mathbf{k}\sigma}\right) = \\ = \sum_{\mathbf{k}\sigma} \hbar\varepsilon_k \left(c^*_{\mathbf{k}\sigma} c_{\mathbf{k}\sigma} + b_{\mathbf{k}\sigma} b^*_{\mathbf{k}\sigma}\right) = \\ = \sum_{\mathbf{k}\sigma} \hbar\varepsilon_k \left(c^*_{\mathbf{k}\sigma} c_{\mathbf{k}\sigma} + b^*_{\mathbf{k}\sigma} b_{\mathbf{k}\sigma} + 1\right) \ . \end{aligned} \tag{8.27}$$

The time dependence $c_{\mathbf{k}\sigma}$, $b_{\mathbf{k}\sigma} \sim e^{-i\varepsilon_k t}$ is established by the canonical equations of motion of these operators with the hamiltonian W_e. Similarly, the total momentum is

$$\begin{aligned} \mathbf{G}_e = \int d\mathbf{r}\mathbf{g}_e = \sum_{\mathbf{k}\sigma} \hbar\mathbf{k} \left(c^*_{\mathbf{k}\sigma} c_{\mathbf{k}\sigma} + b_{\mathbf{k}\sigma} b^*_{\mathbf{k}\sigma}\right) = \\ = \sum_{\mathbf{k}\sigma} \hbar\mathbf{k} \left(c^*_{\mathbf{k}\sigma} c_{\mathbf{k}\sigma} + b^*_{\mathbf{k}\sigma} b_{\mathbf{k}\sigma} + 1\right) \ . \end{aligned} \tag{8.28}$$

If we multiply the equations for ψ and ψ^* by ψ^* and ψ, respectively, and subtract the two equations from one another, we get another law of conservation, which reads

$$\frac{1}{c^2}\frac{\partial}{\partial t}\left(\psi^*\dot{\psi} - \dot{\psi}^*\psi\right) - \partial_i\left(\psi^*\partial_i\psi - \partial_i\psi^* \cdot \psi\right) = 0 \ ; \tag{8.29}$$

hence,

$$\begin{aligned} Q = \tfrac{ie}{\hbar c^2} \int d\mathbf{r} \left(\psi^*\dot{\psi} - \dot{\psi}^*\psi\right) = e\sum_{\mathbf{k}\sigma} \left(c^*_{\mathbf{k}\sigma} c_{\mathbf{k}\sigma} - b_{\mathbf{k}\sigma} b^*_{\mathbf{k}\sigma}\right) = \\ = e\sum_{\mathbf{k}\sigma} \left(c^*_{\mathbf{k}\sigma} c_{\mathbf{k}\sigma} - b^*_{\mathbf{k}\sigma} b_{\mathbf{k}\sigma} - 1\right) \end{aligned} \tag{8.30}$$

is the electric charge (where e is the particle charge) and

$$\begin{aligned} \mathbf{J} = -\tfrac{ie}{\hbar} \int d\mathbf{r} \left(\psi^* grad\psi - grad\psi^* \cdot \psi\right) = \\ = e\sum_{\mathbf{k}\sigma} \tfrac{c^2\mathbf{k}}{\varepsilon_k} \left(c^*_{\mathbf{k}\sigma} c_{\mathbf{k}\sigma} - b_{\mathbf{k}\sigma} b^*_{\mathbf{k}\sigma}\right) = \\ = e\sum_{\mathbf{k}\sigma} \tfrac{c^2\mathbf{k}}{\varepsilon_k} \left(c^*_{\mathbf{k}\sigma} c_{\mathbf{k}\sigma} - b^*_{\mathbf{k}\sigma} b_{\mathbf{k}\sigma} - 1\right) \end{aligned} \tag{8.31}$$

is the electric current. With the notation

$$\rho = \frac{ie}{\hbar c^2}\left(\psi^* \dot{\psi} - \dot{\psi}^* \psi\right) \ , \ j_i = -\frac{ie}{\hbar}\left(\psi^* \partial_i \psi - \partial_i \psi^* \cdot \psi\right) \ , \qquad (8.32)$$

$$j^\mu = (c\rho, \mathbf{j}) = \frac{ie}{\hbar}\left(\psi^* \partial^\mu \psi - (\partial^\mu \psi^*)\psi\right) \ , \qquad (8.33)$$

equation (8.29) is the continuity equation $\partial_\mu j^\mu = 0$.

8.5 Quasi-classical electron

The equation of motion (8.20) for the bosonic field with electromagnetic interaction can be obtained from a lagrangian. A hamiltonian formalism can be set up, with the free boson energy and the free photon energy as free hamiltonians. In the quantum-field theory the states are labelled by the occupation numbers of free bosons and free photons. Their time evolution is described by a Schroedinger equation with the interaction hamiltonian (which is not relativistically invariant, but its solution in the limit of infinite times - the scattering matrix - is a constant, irrelevant for the relativistic invariance). In the electromagnetic-interaction hamiltonian only the transverse photons are active, *i.e.* the longitudinal part of the vector potential is eliminated by the scalar potential, according to the Lorenz gauge, and the scalar potential is determined by the charge density (if it exists). This way, we get transition amplitudes (or scattering matrix) between field-quantum states, which define electromagnetic quantum effects. In these transitions the motion of the relativistic electrons is a field-quantum motion, defined by the emission or absorption of boson quanta. During these processes the quantum states of the electrons change. This quantum-field motion is lost in the quasi-classical description of the electron.

First, we note that the free Klein-Gordon equation (8.23) in Fourier transforms reads

$$\left(\omega^2/c^2 - k^2 - k_0^2\right)\psi(k) = 0 \ , \qquad (8.34)$$

or

$$\left(\omega^2 - \varepsilon_{\boldsymbol{k}}^2\right)\psi(k) = 0 \ . \qquad (8.35)$$

The solution of this equation is

$$\psi(k) = 2\pi C(\boldsymbol{k})\delta(\omega - \varepsilon_{\boldsymbol{k}}) + 2\pi B^*(-\boldsymbol{k})\delta(\omega + \varepsilon_{\boldsymbol{k}}) \ , \tag{8.36}$$

where the bispinors $C(\boldsymbol{k})$ and $B^*(-\boldsymbol{k})$ are established according to equation (8.18).

The Fourier transform of the current density given by equation (8.33) is

$$j^\mu(k) = (c\rho(k), \boldsymbol{j}(k)) = \frac{e}{\hbar}\sum_{k_1}(2k_1 - k)^\mu \psi^*(k_1 - k)\psi(k_1) \ . \tag{8.37}$$

Let us compute the inverse Fourier transform of the density. It is given by

$$\rho(t, \boldsymbol{r}) = \tfrac{e}{2V}\textstyle\sum_{\boldsymbol{k},\boldsymbol{k}_1} \frac{\varepsilon_{\boldsymbol{k}_1}+\varepsilon_{\boldsymbol{k}_1-\boldsymbol{k}}}{\sqrt{\varepsilon_{\boldsymbol{k}_1}\varepsilon_{\boldsymbol{k}_1-\boldsymbol{k}}}} \cdot$$

$$\cdot\{c^*_{\boldsymbol{k}_1-\boldsymbol{k}}c_{\boldsymbol{k}_1}e^{-i\left(\varepsilon_{\boldsymbol{k}_1}-\varepsilon_{\boldsymbol{k}_1-\boldsymbol{k}}\right)t+i\boldsymbol{k}\boldsymbol{r}} - \tag{8.38}$$

$$-b_{-\boldsymbol{k}_1+\boldsymbol{k}}b^*_{-\boldsymbol{k}_1}e^{i\left(\varepsilon_{\boldsymbol{k}_1}-\varepsilon_{\boldsymbol{k}_1-\boldsymbol{k}}\right)t+i\boldsymbol{k}\boldsymbol{r}}\} \ .$$

Let us fix a $\boldsymbol{k}_1^0$ value for $\boldsymbol{k}_1$. For $k_1^0 \gg k_0$ the frequency $\varepsilon_{\boldsymbol{k}_1^0-\boldsymbol{k}}$ can be approximated by

$$\varepsilon_{\boldsymbol{k}_1^0-\boldsymbol{k}} = \varepsilon_{\boldsymbol{k}_1^0} - \boldsymbol{v}_1\boldsymbol{k} + \dots \ , \tag{8.39}$$

where

$$\boldsymbol{v}_1 = \frac{c\boldsymbol{k}_1^0}{\sqrt{k_1^{02} + k_0^2}} = \frac{c^2\boldsymbol{k}_1^0}{\varepsilon_{\boldsymbol{k}_1^0}} \tag{8.40}$$

is the group velocity. Within this approximation

$$\frac{\varepsilon_{\boldsymbol{k}_1^0} + \varepsilon_{\boldsymbol{k}_1^0-\boldsymbol{k}}}{\sqrt{\varepsilon_{\boldsymbol{k}_1^0}\varepsilon_{\boldsymbol{k}_1^0-\boldsymbol{k}}}} = 2 + O(k^2) \ . \tag{8.41}$$

For small values of $\boldsymbol{k}$ in comparison with $\boldsymbol{k}_1^0$ and small variations of $\boldsymbol{k}_1$ around the fixed $\boldsymbol{k}_1^0$-value the summation over $\boldsymbol{k}_1$ of the product $c^*_{\boldsymbol{k}_1-\boldsymbol{k}}c_{\boldsymbol{k}_1} \simeq c^*_{\boldsymbol{k}_1}c_{\boldsymbol{k}_1}$ (the mean value over a fixed number of electrons) may be taken equal to unity (equal weights $c^*_{\boldsymbol{k}_1}c_{\boldsymbol{k}_1}$). A similar approximation holds for the b-term in equation (8.38). Consequently, we get

$$\rho(t, \boldsymbol{r}) = e\delta(\boldsymbol{r} - \boldsymbol{v}_1 t) - e\delta(\boldsymbol{r} + \boldsymbol{v}_1 t) \ . \tag{8.42}$$

We can see that this is the density of a pointlike classical charge e moving with velocity $\boldsymbol{v}_1$ and a density of a pointlike classical charge $-e$ moving with velocity $-\boldsymbol{v}_1$. This is the charge density of a classical electron and a classical positron. If we have more electrons (positrons) we replace the weights $c^*_{\boldsymbol{k}_1}c_{\boldsymbol{k}_1}$ $(b_{-\boldsymbol{k}_1}b^*_{-\boldsymbol{k}_1})$ by their corresponding number. This is the quasi-classical approximation to the electron (positron). It is worth noting that in this approximation the quantum-field effects are removed, due to the mean value of the operators $c^*_{\boldsymbol{k}_1-\boldsymbol{k}}c_{\boldsymbol{k}_1}(b_{-\boldsymbol{k}_1+\boldsymbol{k}}b^*_{-\boldsymbol{k}_1})$ over states with a fixed number of electrons (positrons). Similarly, the current density is given by

$$\boldsymbol{j}(t,\boldsymbol{r}) = e\boldsymbol{v}_1\delta(\boldsymbol{r}-\boldsymbol{v}_1 t) + e\boldsymbol{v}_1\delta(\boldsymbol{r}+\boldsymbol{v}_1 t) \tag{8.43}$$

(we can verify the continuity equation $\partial_\mu j^\mu = 0$).

The meaning of the quasi-classical approximation is as follows. We localize an electron (positron) in a small spatial region, as tightly as possible. Then, we leave it free. The electron will move with a certain velocity $\boldsymbol{v}_1$, sufficiently large, such that this uniform motion will be preserved over a large distance and a long time. The distance covered by this electron is $k_1^0/k_0 \gg 1$ times the Compton wavelength λ_c. For an electron energy $1GeV$ this distance is $\simeq 8$Å, which shows that the Zitterbewegung is still very influential at pretty high energies. The errors in the electron localization along this trajectory are of the order of $\lambda_c = 1/k_0$. Over such a long distance the contributions of the wavevectors smaller than k_0 (neglected in our approximation) become important, the quantum-mechanical delocalization effects appear and the trajectory is lost. We may localize again the electron, somewhere, and the procedure will be repeated. We will have a collection of classical trajectories, over long but finite distances, randomly distributed, both as origin and direction.

8.6 Quasi-classical electromagnetic field

Let us compute the electromagnetic potentials generated by a quasi-classical electron which moves with velocity $\boldsymbol{v}$. They are given by

$$\begin{aligned} &\tfrac{1}{c^2}\tfrac{\partial^2\Phi}{\partial t^2} - \Delta\Phi = 4\pi e\delta(\boldsymbol{r}-\boldsymbol{v}t) \ , \\ &\tfrac{1}{c^2}\tfrac{\partial^2\boldsymbol{A}}{\partial t^2} - \Delta\boldsymbol{A} = \tfrac{4\pi e\boldsymbol{v}}{c}\delta(\boldsymbol{r}-\boldsymbol{v}t) \ , \end{aligned} \tag{8.44}$$

or, in Fourier transforms,

$$\begin{aligned} &k^2\Phi(k) = -4\pi e\tfrac{2\pi}{\sqrt{V}}\delta(\omega-\boldsymbol{vk}) \ , \\ &k^2\boldsymbol{A}(k) = -\tfrac{4\pi e\boldsymbol{v}}{c}\tfrac{2\pi}{\sqrt{V}}\delta(\omega-\boldsymbol{vk}) \ . \end{aligned} \tag{8.45}$$

From the first equation (8.45) we get

$$\Phi(t,\boldsymbol{r}) = \frac{4\pi e}{V}\sum_{\boldsymbol{k}}\frac{e^{-i\boldsymbol{vk}t+i\boldsymbol{kr}}}{k^2-(\boldsymbol{vk})^2/c^2} \ ; \tag{8.46}$$

it is convenient to introduce the parallel and the perpendicular variables with respect to $\boldsymbol{v}$ by $\boldsymbol{k}=(\boldsymbol{k}_\perp,k_\parallel)$, $\boldsymbol{r}=(\boldsymbol{r}_\perp,r_\parallel)$, which lead to

$$\begin{aligned} &\Phi(t,\boldsymbol{r}) = \tfrac{e}{\pi^2(1-v^2/c^2)}\cdot \\ &\cdot\int d\boldsymbol{k}_\perp\int_0^\infty dk_\parallel\tfrac{\cos k_\parallel(r_\parallel-vt)}{k_\parallel^2+k_\perp^2/(1-v^2/c^2)}e^{i\boldsymbol{k}_\perp\boldsymbol{r}_\perp} \ . \end{aligned} \tag{8.47}$$

This integral can be effected by using

$$\int_0^\infty dx\frac{\cos ax}{b^2+x^2} = \frac{\pi}{2b}e^{-|a|b} \ , \ \int_0^\infty dxJ_0(x)e^{-\lambda x} = \frac{1}{\sqrt{1+\lambda^2}} \ , \tag{8.48}$$

where J_0 is the Bessel function of the first rank and the zeroth order $(\lambda\geq 0)$. The final result is

$$\begin{aligned} &\Phi(t,\boldsymbol{r}) = \tfrac{e}{\sqrt{(r_\parallel-vt)^2+(1-v^2/c^2)r_\perp^2}} \ , \\ &\boldsymbol{A}(t,\boldsymbol{r}) = \tfrac{e\boldsymbol{v}/c}{\sqrt{(r_\parallel-vt)^2+(1-v^2/c^2)r_\perp^2}} \ . \end{aligned} \tag{8.49}$$

We can check the Lorenz gauge $\frac{1}{c}\frac{\partial\Phi}{\partial t}+div\boldsymbol{A}=0$. The fields are given by

$$\begin{aligned}\boldsymbol{E}&=e\frac{-\boldsymbol{v}v(r_{\parallel}-vt)/c^2+\left(r_{\parallel}-vt,\left(1-v^2/c^2\right)\boldsymbol{r}_{\perp}\right)}{\left[(r_{\parallel}-vt)^2+(1-v^2/c^2)r_{\perp}^2\right]^{3/2}}\ ,\\ \boldsymbol{H}&=e\left(1-v^2/c^2\right)\frac{\boldsymbol{v}\times\boldsymbol{r}}{\left[(r_{\parallel}-vt)^2+(1-v^2/c^2)r_{\perp}^2\right]^{3/2}}\ .\end{aligned}\tag{8.50}$$

This is the Lienard-Wiechert field of a uniformly moving pointlike charge.

8.7 Quasi-classical Klein-Gordon electron

We turn now to the Klein-Gordon equation (8.20), where we use the electromagnetic potentials produced by a quasi-classical electron. We wish to estimate the effects of the self-interaction. With Fourier transforms and making use of the Lorenz gauge this equation reads

$$\begin{aligned}&-(k^2-k_0^2)\psi(k)+\tfrac{2e}{c\hbar}\textstyle\sum_{k_1}k_1^{\mu}A_{\mu}(k-k_1)\psi(k_1)-\\ &-\tfrac{e^2}{c^2\hbar^2}\textstyle\sum_{k_1k_2}A_{\mu}(k-k_1-k_2)A^{\mu}(k_1)\psi(k_2)-\\ &-\tfrac{e}{c\hbar}\textstyle\sum_{k_1}\boldsymbol{H}(k-k_1)\boldsymbol{\Sigma}\psi(k_1)+\\ &+\tfrac{ie}{c\hbar}\textstyle\sum_{k_1}\boldsymbol{E}(k-k_1\boldsymbol{\alpha}\psi(k_1)=0\ ,\end{aligned}\tag{8.51}$$

where

$$\begin{aligned}&-k^2A_{\mu}(k)=\tfrac{4\pi}{c}j_{\mu}(k)=\\ &=\tfrac{4\pi e}{c\hbar}\textstyle\sum_{k_1}(2k_1-k)_{\mu}\psi^*(k_1-k)\psi(k_1)\ .\end{aligned}\tag{8.52}$$

We solve this equation by a perturbation-theoretical approach, the perturbation parameter being e. The zeroth order approximation is

$$(k^2-k_0^2)\psi_0(k)=0\ ,\tag{8.53}$$

which gives

$$\psi_0(k)=2\pi c\sqrt{\frac{\hbar}{2\varepsilon_{\boldsymbol{k}}}}\begin{pmatrix}c_{\boldsymbol{k}}\delta(\omega-\varepsilon_{\boldsymbol{k}})\\ 0\end{pmatrix}\tag{8.54}$$

(equation (8.18)). We write $\psi = \psi_0 + \psi_1$, $\psi_1 \sim e^2$, where ψ_1 is given by

$$-(k^2 - k_0^2)\psi_1(k) + \tfrac{2e}{c\hbar}\sum_{k_1} k_1^\mu A_\mu(k-k_1)\psi_0(k_1) -$$
$$-\tfrac{e}{c\hbar}\sum_{k_1} \boldsymbol{H}(k-k_1)\boldsymbol{\Sigma}\psi_0(k_1) + \tag{8.55}$$
$$+\tfrac{ie}{c\hbar}\sum_{k_1} \boldsymbol{E}(k-k_1)\boldsymbol{\alpha}\psi_0(k_1) = 0 \ ,$$

the field being

$$A_\mu(k) = -\frac{4\pi e}{c\hbar k^2}\sum_{k_1}(2k_1 - k)_\mu \psi_0^*(k_1 - k)\psi_0(k_1) \tag{8.56}$$

(and $\boldsymbol{H}$ and $\boldsymbol{E}$ derived from this potential). The first interacting term in equation (8.55) reads

$$V(k) = -\tfrac{8\pi e^2}{c^2\hbar^2}\sum_{k_1 k_2} \tfrac{k_1^\mu (2k_2 + k_1 - k)_\mu}{(k-k_1)^2} \cdot$$
$$\cdot \psi_0^*(k_2 + k_1 - k)\psi_0(k_2)\psi_0(k_1) \ . \tag{8.57}$$

This term brings two contributions: a direct one,

$$V_d(k) = -\tfrac{8\pi e^2}{c^2\hbar^2}\sum_{k_1 k_2} \tfrac{k_1^\mu (2k_2 + k_1 - k)_\mu}{(k-k_1)^2} \cdot$$
$$\cdot \left[\psi_0^*(k_2 + k_1 - k)\psi_0(k_2)\right]\psi_0(k_1) \tag{8.58}$$

and an exchange one,

$$V_{ex}(k) \simeq -\tfrac{8\pi e^2}{c^2\hbar^2}\sum_{k_1 k_2} \tfrac{k_1^\mu (2k_2 + k_1 - k)_\mu}{(k-k_1)^2} \cdot$$
$$\cdot [\psi_0^*(k_2 + k_1 - k)\psi_0(k_1)]\psi_0(k_2) \ . \tag{8.59}$$

The exchange contribution implies $\boldsymbol{k}_1 = \boldsymbol{k}_2 = \boldsymbol{k}$, which brings a factor $1/V$ for the summation over $\boldsymbol{k}_{1,2}$; therefore, the exchange contribution is zero. The direct contribution implies $A_\mu(k - k_1)$, which, according to equations (8.45), is proportional to $\delta(\omega - \omega_1 - \boldsymbol{v}(\boldsymbol{k} - \boldsymbol{k}_1))$; the integration over ω_1 in equation (8.58) leads to

$$V_d(k) = -\tfrac{16\pi^2 e^2}{c\sqrt{2\hbar}V}\sum_{\boldsymbol{k}_1} \tfrac{\varepsilon_{\boldsymbol{k}_1} - \boldsymbol{v}\boldsymbol{k}_1}{\sqrt{\varepsilon_{\boldsymbol{k}_1}}\left[(\omega - \varepsilon_{\boldsymbol{k}_1})^2/c^2 - (\boldsymbol{k} - \boldsymbol{k}_1)^2\right]} \cdot$$
$$\cdot \delta(\omega - \varepsilon_{\boldsymbol{k}_1} - v(\boldsymbol{k} - \boldsymbol{k}_1))c_{\boldsymbol{k}} \ . \tag{8.60}$$

for $\boldsymbol{k}_1$ close to $\boldsymbol{k}$ we have $\varepsilon_{\boldsymbol{k}_1} = \varepsilon_{\boldsymbol{k}} + v(\boldsymbol{k}_1 - \boldsymbol{k})$, which implies $\delta(\omega - \varepsilon_{\boldsymbol{k}})$, the summation over $\boldsymbol{k}_1$ yields a factor $\frac{4\pi}{3}(\Delta k_1)^3$, where $\Delta k_1 = \mid \boldsymbol{k} - \boldsymbol{k}_1 \mid$, and the factor $k^2 - k_0^2$ in equation (8.55) can be approximated by

$$\omega^2 - \varepsilon_{\boldsymbol{k}}^2 = \Delta\varepsilon_{\boldsymbol{k}} \cdot 2\varepsilon_{\boldsymbol{k}} = \frac{c^2 k \Delta k_1}{\varepsilon_{\boldsymbol{k}}} \cdot 2\varepsilon_{\boldsymbol{k}} \ , \tag{8.61}$$

such that we get finally

$$c_{\boldsymbol{k}}^{(1)d} = \frac{2e^2}{3\pi c\hbar} \frac{k_0^2}{k\sqrt{k^2 + k_0^2(v^2/3c^2 - 1)}} c_{\boldsymbol{k}} \ , \tag{8.62}$$

where

$$\psi_1^d(k) = 2\pi c \sqrt{\frac{\hbar}{2\varepsilon_{\boldsymbol{k}}}} \begin{pmatrix} c_{\boldsymbol{k}}^{(1)d} \delta(\omega - \varepsilon_{\boldsymbol{k}}) \\ 0 \end{pmatrix} \ ; \tag{8.63}$$

in equation (8.60) we averaged $(\boldsymbol{v}\boldsymbol{k}_1)^2$ over the orientations of $\boldsymbol{k}_1$. We can see that the direct interaction brings a correction proportional to the fine-structure constant $\alpha = e^2/c\hbar$.

The $\boldsymbol{H}$- and $\boldsymbol{E}$-interaction terms in equation (8.55) bring no contribution to the quasi-classical electron (as expected, because they are quantum-mechanical contributions). Indeed, from equations (8.45) we get

$$\Phi(k) = \frac{c\boldsymbol{v}\boldsymbol{A}(k)}{v^2} \tag{8.64}$$

and the fields

$$\begin{aligned} \boldsymbol{E}(k) &= i\tfrac{8\pi^2 e}{\sqrt{V}}(-\omega\boldsymbol{v}/c^2 + \boldsymbol{k})\tfrac{\delta(\omega - \boldsymbol{v}\boldsymbol{k})}{k^2} = \\ &= -i\left(\tfrac{\omega}{c}\boldsymbol{A} + \tfrac{c\boldsymbol{v}\boldsymbol{A}(k)}{v^2}\boldsymbol{k}\right) \ , \end{aligned} \tag{8.65}$$

$$\boldsymbol{H}(k) = i\tfrac{8\pi^2 e}{\sqrt{V}}\tfrac{\boldsymbol{v}\times\boldsymbol{k}}{c}\tfrac{\delta(\omega - \boldsymbol{v}\boldsymbol{k})}{k^2} = i\boldsymbol{A}(k) \times \boldsymbol{k} \ .$$

In equation (8.55) $\boldsymbol{E}(k - k_1)$ and $\boldsymbol{H}(k - k_1)$ yield the factors $\boldsymbol{k} - \boldsymbol{k}_1$ and $\omega - \omega_1 = \boldsymbol{v}(\boldsymbol{k} - \boldsymbol{k}_1)$ apart from the contribution of the potential $A_\mu(k - k_1)$. Consequently, an additional factor of the order $\mid \boldsymbol{k} - \boldsymbol{k}_1 \mid = \Delta k_1$ occurs in these terms, in comparison with the interaction V_d, which leads to vanishing $\boldsymbol{E}$- and $\boldsymbol{H}$-interactions.

The correction $c_{\boldsymbol{k}}^{(1)d}$ given by equation (8.62) can be written as

$$c_{\boldsymbol{k}}^{(1)d} = \frac{1}{2}\alpha A c_{\boldsymbol{k}} \ , \ A = \frac{4}{3\pi}\frac{k_0^2}{k\sqrt{k^2+k_0^2(v^2/3c^2-1)}} \ . \tag{8.66}$$

According to equations (8.27) and (8.30) the energy and the charge of the quasi-classical electron are renormalized by

$$\varepsilon_{\boldsymbol{k}} \rightarrow (1+\alpha A)\varepsilon_{\boldsymbol{k}} \ , \ e \rightarrow (1+\alpha A)e \ . \tag{8.67}$$

Let us assume an external monochromatic electromagnetic radiation with the vector potential

$$\boldsymbol{A}(t,\boldsymbol{r}) = \boldsymbol{A}_0 \cos(\Omega t - \boldsymbol{K}\boldsymbol{r}) \tag{8.68}$$

($\boldsymbol{K}\boldsymbol{A}_0 = 0$); its Fourier transform is

$$A(k) = \pi\sqrt{V}A_0\left[\delta(\omega+\Omega)\delta_{\boldsymbol{k},-\boldsymbol{K}} + \delta(\omega-\Omega)\delta_{\boldsymbol{k},\boldsymbol{K}}\right] \ . \tag{8.69}$$

The interaction V in equation (8.55) is

$$V = \tfrac{e}{c\hbar}\boldsymbol{k}\boldsymbol{A}_0\left[\psi(k+K)+\psi(k-K)\right] \ ; \tag{8.70}$$

since K is much smaller than k, we may write approximately

$$V \simeq \frac{2e}{c\hbar}\boldsymbol{k}\boldsymbol{A}_0\psi(k) \ . \tag{8.71}$$

This interaction modifies the frequency $\varepsilon_{\boldsymbol{k}}/c$ of the quasi-classical electron by a factor $1+ec\boldsymbol{k}\boldsymbol{A}_0/\hbar\varepsilon_{\boldsymbol{k}}^2$, which implies a correction of the order of the fine-structure constant. The Fourier transforms of the electric field and the magnetic field are

$$\begin{aligned}\boldsymbol{E}(k) &= \tfrac{\pi\sqrt{V}}{2i}\tfrac{\Omega}{c}\boldsymbol{A}_0\left[\delta(\omega+\Omega)\delta_{\boldsymbol{k},-\boldsymbol{K}} - \delta(\omega-\Omega)\delta_{\boldsymbol{k},\boldsymbol{K}}\right] \ , \\ \boldsymbol{H}(k) &= \tfrac{\pi\sqrt{V}}{2i}\boldsymbol{K}\times\boldsymbol{A}_0\left[\delta(\omega+\Omega)\delta_{\boldsymbol{k},-\boldsymbol{K}} - \delta(\omega-\Omega)\delta_{\boldsymbol{k},\boldsymbol{K}}\right] \ .\end{aligned} \tag{8.72}$$

The $\boldsymbol{E}$- and $\boldsymbol{H}$-interactions include factors $\psi(k+K)-\psi(k-K)$, which, for small values of K in comparison to k, are vanishing.

The quasi-classical relativistic electron is a classical particle, slightly affected by the electromagnetic interaction. It is worth noting that its self-interaction implies the cutoff k_0 (which enters automatically in equations, from its definition), which amounts to an electron radius of the order of the Compton wavelength λ_c; indeed, the self-energy given by equations (8.66) and (8.67) goes formally to ∞ for $k_0 \to \infty$. Obviously, the existence of k_0 arises from a non-vanishing mass m.

We get the classical relativistic electron by using the Klein-Gordon equation, not because this equation would have a quasi-classical limit, but because it is relativistically invariant. This invariance ensures the compatibility of this equation with the classical limit. Moreover, the self-energy governed by the Compton wavelength obtained in this approach indicates the consistency of the cutoff treatment. The relativistic character of the Klein-Gordon equation does not preclude, of course, its using for quantum-mechanical effects, when the electron states are changed, and the electron is not classical anymore. The interaction with an external electromagnetic field described above is vanishing for a classical field, but it leads to the Compton effect for quantum-field photons.

9 Electromagnetism in Matter

9.1 Maxwell equations in matter

Let us assume a charge density ρ and a current density $\boldsymbol{j}$ in matter. This charge and current distribution generates an electric field and a magnetic field. These fields polarize and magnetize the matter, *i.e.* they induce a charge density ρ_i and a current density $\boldsymbol{j}_i$. External fields do the same thing. The induced charge density arises from a distribution of dipoles, *i.e.* a local displacement of charges. It can be written as $\rho_i = -div\boldsymbol{P}$, where $\boldsymbol{P}$ is called polarization; obviously, it is a dipole density. The corresponding total charge is distributed over the surface. The corresponding current density is $\boldsymbol{j}_i^{(1)} = \dot{\boldsymbol{P}}$. In addition, a magnetization current density may appear, which can be written as $\boldsymbol{j}_i^{(2)} = c \cdot curl\boldsymbol{M}$, where $\boldsymbol{M}$ is called magnetization; it is a density of magnetic moments (c denotes the speed of light in vacuum). The total electric field $\boldsymbol{E}$ and the total magnetic field $\boldsymbol{H}$ satisfy the Maxwell equations

$$div\boldsymbol{E} = 4\pi\rho - 4\pi div\boldsymbol{P} \ ,$$

$$curl\boldsymbol{H} = \frac{1}{c}\frac{\partial \boldsymbol{E}}{\partial t} + \frac{4\pi}{c}\boldsymbol{j} + \frac{4\pi}{c}\dot{\boldsymbol{P}} + 4\pi curl\boldsymbol{M} \ , \tag{9.1}$$

$$curl\boldsymbol{E} = -\frac{1}{c}\frac{\partial \boldsymbol{H}}{\partial t} \ , \ div\boldsymbol{H} = 0 \ .$$

We can see that these equations can be written as

$$div\boldsymbol{D} = 4\pi\rho \ ,$$

$$curl\,(\boldsymbol{H} - 4\pi\boldsymbol{M}) = \tfrac{1}{c}\tfrac{\partial\boldsymbol{D}}{\partial t} + \tfrac{4\pi}{c}\boldsymbol{j} \ , \tag{9.2}$$

$$curl\boldsymbol{E} = -\tfrac{1}{c}\tfrac{\partial\boldsymbol{H}}{\partial t} \ , \ div\boldsymbol{H} = 0 \ ,$$

where $\boldsymbol{D} = \boldsymbol{E} + 4\pi\boldsymbol{P}$ is called electric displacement. By a strange historical tradition $\boldsymbol{H}$ is denoted *by* $\boldsymbol{B}$; it is called magnetic induction. The formation $\boldsymbol{H} - 4\pi\boldsymbol{M}$ becomes $\boldsymbol{B} - 4\pi\boldsymbol{M}$ and is called magnetic field, denoted by $\boldsymbol{H}$. The Maxwell equations become

$$div\boldsymbol{D} = 4\pi\rho \ ,$$

$$curl\boldsymbol{H} = \tfrac{1}{c}\tfrac{\partial\boldsymbol{D}}{\partial t} + \tfrac{4\pi}{c}\boldsymbol{j} \ , \tag{9.3}$$

$$curl\boldsymbol{E} = -\tfrac{1}{c}\tfrac{\partial\boldsymbol{B}}{\partial t} \ , \ div\boldsymbol{B} = 0 \ .$$

We must keep in mind that the physical nature is the same for $\boldsymbol{E}$ and $\boldsymbol{B}$, and, similarly, $\boldsymbol{D}$ and $\boldsymbol{H}$ have the same physical nature. $\boldsymbol{E}$ and $\boldsymbol{B}$ are generated by all charges and currents, $\boldsymbol{D}$ and $\boldsymbol{H}$ are generated only by external charges and currents.

From equations (9.3) we get

$$\boldsymbol{E}\tfrac{\partial\boldsymbol{D}}{\partial t} + \boldsymbol{H}\tfrac{\partial\boldsymbol{B}}{\partial t} = c\,(\boldsymbol{E}curl\boldsymbol{H} - \boldsymbol{H}curl\boldsymbol{E}) - 4\pi\boldsymbol{j}\boldsymbol{E} \ ; \tag{9.4}$$

since $\boldsymbol{E}curl\boldsymbol{H} - \boldsymbol{H}curl\boldsymbol{E} = -div\,(\boldsymbol{E}\times\boldsymbol{H})$, we get immediately

$$\frac{1}{4\pi}\left(\boldsymbol{E}\frac{\partial\boldsymbol{D}}{\partial t} + \boldsymbol{H}\frac{\partial\boldsymbol{B}}{\partial t}\right) + \frac{c}{4\pi}div\,(\boldsymbol{E}\times\boldsymbol{H}) + \boldsymbol{j}\boldsymbol{E} = 0 \ , \tag{9.5}$$

or

$$\frac{\partial w_{em}}{\partial t} + div\boldsymbol{S} + \boldsymbol{j}\boldsymbol{E} = 0 \ , \tag{9.6}$$

where w_{em} is the energy density of the field $\boldsymbol{S} = \frac{c}{4\pi}\boldsymbol{E}\times\boldsymbol{H}$ is the density of the energy flux (Poynting vector) and $\boldsymbol{j}\boldsymbol{E}$ is the work per unit time and unit volume done by the field upon charges ($-\boldsymbol{j}\boldsymbol{E}$ is the work done by charges upon the field). Equation (9.6) is the energy conservation. The density of interaction energy is $\int dt\boldsymbol{jE}$. We note that the time variation of the energy density is

$$\frac{\partial w_{em}}{\partial t} = \frac{1}{4\pi}\left(\boldsymbol{E}\frac{\partial \boldsymbol{D}}{\partial t} + \boldsymbol{H}\frac{\partial \boldsymbol{B}}{\partial t}\right) . \tag{9.7}$$

Also, these equations may include external fields.

Similarly,

$$\boldsymbol{p} = \frac{1}{c^2}\boldsymbol{S} = \frac{1}{4\pi c}\boldsymbol{E}\times\boldsymbol{H} \tag{9.8}$$

is a momentum density; its time variation is the force density acting upon the field in the medium, *i.e.* upon the medium by the field ouside the medium. It is called the Abraham momentum.

The Lorentz force acting upon charges and currents is given, from equations (9.3), by

$$\begin{aligned} \boldsymbol{f} &= \rho\boldsymbol{E} + \tfrac{1}{c}\boldsymbol{j}\times\boldsymbol{B} = \\ &= \tfrac{1}{4\pi}\left(\boldsymbol{E}div\boldsymbol{D} + curl\boldsymbol{H}\times\boldsymbol{B} - \tfrac{1}{c}\tfrac{\partial \boldsymbol{D}}{\partial t}\times\boldsymbol{B}\right) = \\ &= \tfrac{1}{4\pi}[\boldsymbol{E}div\boldsymbol{D} + \boldsymbol{H}div\boldsymbol{B} + \\ &+ curl\boldsymbol{H}\times\boldsymbol{B} + curl\boldsymbol{E}\times\boldsymbol{D} - \tfrac{1}{c}\tfrac{\partial}{\partial t}(\boldsymbol{D}\times\boldsymbol{B})] \end{aligned} \tag{9.9}$$

this equation can be written as

$$f_i = t_i - \frac{1}{4\pi c}\frac{\partial}{\partial t}(\boldsymbol{D}\times\boldsymbol{B})_i \ , \tag{9.10}$$

where

$$t_i = \partial_j\left(E_iD_j + H_iB_j\right) - D_j\partial_iE_j - B_j\partial_iH_j \tag{9.11}$$

is the force arising from the Maxwell stress. We can see that

$$\boldsymbol{g} = \frac{1}{4\pi c}\boldsymbol{D}\times\boldsymbol{B} \tag{9.12}$$

is the momentum density of the field in the medium acting upon the charges embedded in the medium; it is called the Minkowski momentum.

9.2 Electric permittivity and magnetic permeability

In linear media there should exist a linear relation between $\boldsymbol{P}$ and $\boldsymbol{D}$, on one hand, and $\boldsymbol{E}$, on the other; as well as between $\boldsymbol{M}$ and $\boldsymbol{B}$, on one hand, and $\boldsymbol{H}$, on the other. In isotropic media we may write $\boldsymbol{P} = \chi_e \boldsymbol{E}$, $\boldsymbol{D} = \varepsilon \boldsymbol{E}$, $\varepsilon = 1 + 4\pi\chi_e$ and, similarly, $\boldsymbol{M} = \chi_m \boldsymbol{H}$, $\boldsymbol{B} = \mu \boldsymbol{H}$, $\mu = 1 + 4\pi\chi_m$; $\chi_{e,m}$ are the electric and magnetic susceptibilities, ε is the electric permittivity (dielectric function), μ is the magnetic permeability. (A similar relation between $\boldsymbol{P}$ and $\boldsymbol{D}$, or between $\boldsymbol{M}$ and $\boldsymbol{B}$, defines the electric and magnetic polarizability, respectively). By thermodynamic stability we can show that $\varepsilon > 1$ and $\mu > 0$. In general, the media are dispersive, *i.e.* ε and μ depend on the frequency (or, even on the wavevector, for spatial dispersion); an energy loss is always present in dispersive media. The transparent media have a real constant ε and μ in a frequency range.

Let us introduce the real constants ε and μ in the Maxwell equations (9.3) together with the potentials Φ and $\boldsymbol{A}$, defined by

$$\boldsymbol{E} = -\frac{1}{c}\frac{\partial \boldsymbol{A}}{\partial t} - grad\Phi \ , \ \ \boldsymbol{B} = curl\boldsymbol{A} \ ; \tag{9.13}$$

we get the wave equations

$$\begin{aligned} \frac{\varepsilon\mu}{c^2}\frac{\partial^2\Phi}{\partial t^2} - \Delta\Phi = \frac{4\pi\rho}{\varepsilon} \ , \\ \frac{\varepsilon\mu}{c^2}\frac{\partial^2\boldsymbol{A}}{\partial t^2} - \Delta\boldsymbol{A} = \frac{4\pi\mu\boldsymbol{j}}{c} \ , \end{aligned} \tag{9.14}$$

providing the Lorenz gauge

$$\frac{\varepsilon\mu}{c}\frac{\partial\Phi}{\partial t} + div\boldsymbol{A} = 0 \tag{9.15}$$

is satisfied. According to these equations the electromagnetic waves in media propagate with velocity $v = c/\sqrt{\varepsilon\mu}$ and dispersion relation given by $\omega^2 = v^2 k^2$, where ω is the frequency and $\boldsymbol{k}$ is the wavevector. The product $\varepsilon\mu$ is the squared refraction index n^2. The Lorenz gauge eliminates one component from the four components of the potentials, in accordance with the three unknowns of the Maxwell equations. This elimination is performed by a relationship between the scalar potential and the longitudinal vector potential. The remaining Coulomb energy is cancelled out by the corresponding interaction energy, such that we are left with transverse modes, interacting with charges (the density of interaction energy is $\rho\Phi - \frac{1}{c}\boldsymbol{jA}$). A monochromatic free mode is described by

$$\begin{aligned} \boldsymbol{E} &= -\tfrac{\omega}{c}\boldsymbol{A}_0 \sin(\boldsymbol{kr} - \omega t + \alpha) \ , \\ \boldsymbol{B} &= -\boldsymbol{k} \times \boldsymbol{A}_0 \sin(\boldsymbol{kr} - \omega t + \alpha) \ , \end{aligned} \tag{9.16}$$

where $\boldsymbol{A}_0$ is the amplitude of the vector potential and α is a constant phase. Making use of equations (9.5), (9.6) and (9.12), these fields lead to

$$\begin{aligned} w_{em} &= \tfrac{\varepsilon\omega^2}{4\pi c^2} A_0^2 \sin(\boldsymbol{kr} - \omega t + \alpha) \ , \\ \boldsymbol{g} &= \tfrac{\varepsilon\omega^2}{4\pi c^2 v} A_0^2 \tfrac{\boldsymbol{k}}{k} \sin(\boldsymbol{kr} - \omega t + \alpha) \end{aligned} \tag{9.17}$$

and $w_{em} = vg = cg/n$.

9.2.1 Longitudinal field

The Fourier component of the Lorenz gauge leads to

$$\boldsymbol{A}_{l\boldsymbol{k}} = \frac{i\boldsymbol{k}}{k^2}\frac{\varepsilon\mu}{xc}\dot{\Phi}_{\boldsymbol{k}} \ , \ \boldsymbol{E}_{l\boldsymbol{k}} = -\frac{i\boldsymbol{k}}{k^2}\left(\frac{\varepsilon\mu}{c^2}\ddot{\Phi}_{\boldsymbol{k}} + k^2\Phi_{\boldsymbol{k}}\right) = -\frac{4\pi i\boldsymbol{k}}{\varepsilon k^2}\rho_{\boldsymbol{k}} \ , \tag{9.18}$$

according to equation (9.14). The longitudinal field energy (equation (9.7)) is

$$W_{eml} = \frac{\varepsilon}{8\pi}\int d\boldsymbol{r}\boldsymbol{E}_l^2 = \sum_{\boldsymbol{k}} \frac{2\pi}{\varepsilon k^2}\rho_{\boldsymbol{k}}\rho_{-\boldsymbol{k}} \ ; \tag{9.19}$$

this is the Coulomb energy V_c. The interaction energy corresponding to the longitudinal degrees of freedom is

$$\begin{aligned} &\int dt d\boldsymbol{r}\boldsymbol{j}\boldsymbol{E}_l = \int dt \sum\nolimits_{\boldsymbol{k}} \tfrac{4\pi i\boldsymbol{k}\boldsymbol{j}_{\boldsymbol{k}}}{\varepsilon k^2}\rho_{-\boldsymbol{k}} = \\ &= -\int dt \sum\nolimits_{\boldsymbol{k}} \tfrac{4\pi\dot{\rho}_{\boldsymbol{k}}}{\varepsilon k^2}\rho_{-\boldsymbol{k}} = -\sum\nolimits_{\boldsymbol{k}} \tfrac{2\pi}{\varepsilon k^2}\rho_{\boldsymbol{k}}\rho_{-\boldsymbol{k}} \ , \end{aligned} \tag{9.20}$$

i.e. $-V_c$. We see that the longitudinal degrees of freedom do not take part in dynamics.

9.3 Quantization

For transverse free modes we may write $\boldsymbol{E} = -\frac{1}{c}\frac{\partial \boldsymbol{A}}{\partial t}$, $\boldsymbol{B} = curl\boldsymbol{A}$, where $div\boldsymbol{A} = 0$. Making use of Fourier transforms, we get the total energy and the total momentum

$$W_{em} = \int dV w_{em} = \sum_{\boldsymbol{k}} \frac{\varepsilon}{8\pi c^2}\left(\dot{\boldsymbol{A}}_{\boldsymbol{k}}\dot{\boldsymbol{A}}_{-\boldsymbol{k}} + \omega_k^2 \boldsymbol{A}_{\boldsymbol{k}}\boldsymbol{A}_{-\boldsymbol{k}}\right) ,$$
$$\boldsymbol{G} = \int dV \boldsymbol{g} = \sum_{\boldsymbol{k}} \frac{i\varepsilon}{4\pi c^2}\boldsymbol{k}\dot{\boldsymbol{A}}_{\boldsymbol{k}}\boldsymbol{A}_{-\boldsymbol{k}} , \tag{9.21}$$

where $\omega_k = vk$. Introducing the photon operators $a_{\boldsymbol{k}}$

$$\boldsymbol{A}_{\boldsymbol{k}} = c\sqrt{\frac{2\pi\hbar}{\varepsilon\omega_k}}\boldsymbol{e}_{\boldsymbol{k}}\left(a_{\boldsymbol{k}} + a^*_{-\boldsymbol{k}}\right) ,$$
$$\dot{\boldsymbol{A}}_{\boldsymbol{k}} = -ic\omega_k\sqrt{\frac{2\pi\hbar}{\varepsilon\omega_k}}\boldsymbol{e}_{\boldsymbol{k}}\left(a_{\boldsymbol{k}} - a^*_{-\boldsymbol{k}}\right) , \tag{9.22}$$

where $\boldsymbol{e}_{\boldsymbol{k}}$ are unit vectors perpendicular to $\boldsymbol{k}$ (polarization vectors) and $\hbar$ is Planck's constant, we get

$$W_{em} = \sum_{\boldsymbol{k}} \frac{1}{2}\hbar\omega_k\left(a_{\boldsymbol{k}}a^*_{\boldsymbol{k}} + a^*_{\boldsymbol{k}}a_{\boldsymbol{k}}\right) ,$$
$$\boldsymbol{G} = \sum_{\boldsymbol{k}} \frac{1}{2}\hbar\boldsymbol{k}\left(a_{\boldsymbol{k}}a^*_{\boldsymbol{k}} + a^*_{\boldsymbol{k}}a_{\boldsymbol{k}}\right) , \tag{9.23}$$

which is the photon representation of the electromagnetic field in matter. We note that $(w_{em}/c, \boldsymbol{g}) = (g/n, \boldsymbol{g}) = (\hbar k/n, \hbar\boldsymbol{k}) = (\hbar\omega/c, \hbar\boldsymbol{k})$, which is a relativistic vector.

9.4 Dual potentials

Let an external electric field $\boldsymbol{E}_0$ and an external magnetic field $\boldsymbol{H}_0$ act upon a sample of matter. The electric field produces a polarization characterized by a local displacement $\boldsymbol{u}$ of the charges q with concentration n.[1] The magnetic field induces a magnetization, which we

[1] P. Drude, "Zur Elektronentheorie der Metalle", Ann. Phys. **306** 566 (1900); *ibid*, "Zur Elektronentheorie der Metalle, 2. Teile. Galvanomagnetische und thermomagnetische Effecte", **308** 369 (1900); H. A. Lorentz, *The Theory of Electrons*, Teubner, Leipzig (1916).

neglect. The displacement $\boldsymbol{u}$ generates a charge density $\rho = -nqdiv\boldsymbol{u}$ and the corresponding current density $\boldsymbol{j} = nq\dot{\boldsymbol{u}}$. Apart from the fulfilment of the continuity equation, it is worth noting that $nq(-cdiv\boldsymbol{u}, \dot{\boldsymbol{u}})$ is a relativistic vector (where c is the speed of light in vacuum). Indeed, $nqdV$ is a constant, where dV in $dV/V = div\boldsymbol{u}$ is the element of the volume V; therefore, $nqV div\boldsymbol{u} \cdot dt\frac{dx^i}{dt}$, $i = 0, 1, 2, 3$, is a relativistic vector, where $dx^i = (cdt, d\boldsymbol{r})$. Leaving aside the irrelevant factors, we get $cdiv\boldsymbol{u}$ for the time component and

$$\begin{aligned} & div\boldsymbol{u}\frac{dx^\alpha}{dt} = \frac{\partial u^\beta}{\partial x^\beta}\frac{dx^\alpha}{dt} = \frac{du^\beta}{dx^\beta}\frac{dx^\alpha}{dt} = \\ & = \frac{du^\alpha}{dt} = (\dot{\boldsymbol{u}})^\alpha \ , \ \alpha, \beta = 1, 2, 3 \end{aligned} \tag{9.24}$$

for the spatial components.

The Gauss equation reads

$$div\boldsymbol{E} = -4\pi nqdiv\boldsymbol{u} \ , \tag{9.25}$$

which shows that the displacement $\boldsymbol{u}$ generates an internal electric field $\boldsymbol{E}$. We can see that $\boldsymbol{u}$ behaves as an electric field; $\boldsymbol{P} = nq\boldsymbol{u}$ is the polarization (density of electrical dipoles). The most general solution of this equation is

$$\boldsymbol{E} = -4\pi nq\boldsymbol{u} + curl\boldsymbol{A} \ , \tag{9.26}$$

where $\boldsymbol{A}$ is an undetermined vector potential. Similarly, the Maxwell-Ampere equation is

$$\begin{aligned} & curl\boldsymbol{H} = \frac{1}{c}\frac{\partial \boldsymbol{E}}{\partial t} + \frac{4\pi}{c}\boldsymbol{j} = \\ & = -\frac{4\pi nq}{c}\dot{\boldsymbol{u}} + \frac{1}{c}curl\dot{\boldsymbol{A}} + \frac{4\pi nq}{c}\dot{\boldsymbol{u}} = \frac{1}{c}curl\dot{\boldsymbol{A}} \ . \end{aligned} \tag{9.27}$$

The most general solution of this equation is

$$\boldsymbol{H} = \frac{1}{c}\dot{\boldsymbol{A}} + grad\Phi \ , \tag{9.28}$$

where Φ is an undetermined scalar potential. We have already four unknowns $\boldsymbol{A}$, Φ for the three unknowns of the Maxwell equations. But the Faraday equation

$$curl\boldsymbol{E} = -4\pi nq \cdot curl\boldsymbol{u} - \Delta\boldsymbol{A} + grad\, div\boldsymbol{A} = -\frac{1}{c^2}\ddot{\boldsymbol{A}} - \frac{1}{c}grad\dot{\Phi} \tag{9.29}$$

leads to the wave equation

$$\frac{1}{c^2}\ddot{\boldsymbol{A}} - \Delta \boldsymbol{A} = 4\pi nq \cdot curl\boldsymbol{u} \tag{9.30}$$

and the condition

$$\frac{1}{c}\dot{\Phi} + div\boldsymbol{A} = 0 \ , \tag{9.31}$$

which is a Lorenz-type gauge. The Gauss equation for the magnetic field is

$$div\boldsymbol{H} = \frac{1}{c}div\dot{\boldsymbol{A}} + \Delta\Phi = -\frac{1}{c^2}\ddot{\Phi} + \Delta\Phi = 0 \ , \tag{9.32}$$

which indicates that Φ may be taken zero.

It is convenient to write down the wave equations for the fields. Making use of the definitions (9.26) and (9.28), equation (9.30) and the condition (9.31), we get

$$\begin{gathered} \tfrac{1}{c^2}\ddot{\boldsymbol{E}} - \Delta\boldsymbol{E} = -\tfrac{4\pi nq}{c^2}\ddot{\boldsymbol{u}} + 4\pi nq \cdot grad\, div\boldsymbol{u} \ , \\ \tfrac{1}{c^2}\ddot{\boldsymbol{H}} - \Delta\boldsymbol{H} = \tfrac{4\pi nq}{c}curl\dot{\boldsymbol{u}} \ . \end{gathered} \tag{9.33}$$

We can see from the definitions (9.26) and (9.28) that the representations of the electric field and the magnetic field in terms of potentials are inter-changed; this motivates the denomination of dual potentials. Equations (9.33) can also be obtained by using the regular potentials ($\boldsymbol{E} = -\frac{1}{c}\dot{\boldsymbol{A}} - grad\Phi$, $\boldsymbol{H} = curl\boldsymbol{A}$).

From the Maxwell equations

$$\begin{gathered} div\boldsymbol{E} = -4\pi nq div\boldsymbol{u} \ , \ div\boldsymbol{H} = 0 \ , \\ curl\boldsymbol{E} = -\tfrac{1}{c}\tfrac{\partial \boldsymbol{H}}{\partial t} \ , \ curl\boldsymbol{H} = \tfrac{1}{c}\tfrac{\partial \boldsymbol{E}}{\partial t} + \tfrac{4\pi nq}{c}\dot{\boldsymbol{u}} \end{gathered} \tag{9.34}$$

we get immediately

$$\frac{1}{8\pi}\frac{\partial}{\partial t}\left(E^2 + H^2\right) + \frac{c}{4\pi}div\left(\boldsymbol{E} \times \boldsymbol{H}\right) + nq\dot{\boldsymbol{u}}\boldsymbol{E} = 0 \ . \tag{9.35}$$

The equation of motion of the charges q with mass m is non-relativistic, since the external fields are weak and the charge density and the current density are small; $\boldsymbol{u}$ is small and varies slightly and slowly in space and time; its variation reads[2]

$$m\ddot{\boldsymbol{u}} + m\omega_c^2\boldsymbol{u} + m\gamma\dot{\boldsymbol{u}} = q(\boldsymbol{E} + \boldsymbol{E}_0) \ , \tag{9.36}$$

where ω_c is a characteristic frequency of the charges and γ is a (small) dissipation coefficient. We neglect the charge interaction in this equation. From this equation we can express $\dot{\boldsymbol{u}}\boldsymbol{E}$ in equation (9.35) in terms of $\boldsymbol{u}$ and $\boldsymbol{E}_0$; we get

$$\begin{gathered}\tfrac{1}{8\pi}\tfrac{\partial}{\partial t}\left(E^2 + H^2\right) + \tfrac{\partial}{\partial t}\left(\tfrac{1}{2}nm\dot{u}^2 + \tfrac{1}{2}nm\omega_c^2u^2\right) + \\ +nm\gamma\dot{u}^2 + \tfrac{c}{4\pi}div\left(\boldsymbol{E}\times\boldsymbol{H}\right) = nq\dot{\boldsymbol{u}}\boldsymbol{E}_0 \ .\end{gathered} \tag{9.37}$$

According to this equation the work done per unit time and unit volume by the external field is transferred to the field energy, the motion energy of the charges, the radiation and the dissipation, as expected. Equation (9.37) is the energy conservation.

9.5 Dielectric function. Eigenmodes

The external field $\boldsymbol{E}_0$ determines the displacement $\boldsymbol{u}$ and the internal field $\boldsymbol{E}$ by the coupled equation (9.36) and the first equation (9.33); the internal magnetic field is determined by the second equation (9.33). Let us assume that the external field is monochromatic, with frequency ω, written as $\boldsymbol{E}_0e^{-i\omega t}$.

Let us assume first that the displacement is longitudinal, *i.e.* $curl\boldsymbol{u} = 0$. Then, by the second equation (9.33), $\boldsymbol{H} = 0$, by equation (9.30) $\boldsymbol{A} = 0$, by equation (9.32) $\Phi = 0$ and, by the first equation (9.33), $\boldsymbol{E} = -4\pi nq\boldsymbol{u}$; inserted in equation (9.36) it leads to

$$\boldsymbol{u} = -\frac{q\boldsymbol{E}_0}{m}\frac{1}{\omega^2 - \omega_c^2 - \omega_0^2 + i\omega\gamma} \ , \tag{9.38}$$

where $\omega_0 = (4\pi nq^2/m)^{1/2}$ is the plasma frequency. The total field is

$$\boldsymbol{E}_t = \boldsymbol{E}_0 + \boldsymbol{E} = \boldsymbol{E}_0\frac{\omega^2 - \omega_c^2 + i\omega\gamma}{\omega^2 - \omega_c^2 - \omega_0^2 + i\gamma\omega} \tag{9.39}$$

[2] M. Apostol, *Essays in Electromagnetism and Matter*, Lambert, Saarbruecken (2013).

and the dielectric function, defined by $\boldsymbol{E}_0 = \varepsilon \boldsymbol{E}_t$, is[3]

$$\varepsilon = 1 - \frac{\omega_0^2}{\omega^2 - \omega_c^2 + i\omega\gamma} \; ; \tag{9.40}$$

the electric susceptibility defined by $\boldsymbol{P} = nq\boldsymbol{u} = \chi_e \boldsymbol{E}_t$ is

$$\chi_e = -\frac{\omega_0^2}{4\pi\left(\omega^2 - \omega_c^2 + i\omega\gamma\right)} \; , \tag{9.41}$$

and $\varepsilon = 1 + 4\pi\chi_e$ (the polarizability is defined by $P = \alpha E_0 = \alpha\varepsilon E_t = \chi_e E_t$). Similarly, the current is

$$\begin{aligned} \boldsymbol{j} = nq\dot{\boldsymbol{u}} = & \tfrac{\omega_0^2 \boldsymbol{E}_0}{4\pi} \tfrac{i\omega}{\omega^2 - \omega_c^2 - \omega_0^2 + i\gamma\omega} = \\ = & \tfrac{\omega_0^2 \boldsymbol{E}_t}{4\pi} \tfrac{i\omega}{\omega^2 - \omega_c^2 + i\omega\gamma} \; , \end{aligned} \tag{9.42}$$

whence the conductivity

$$\sigma = \frac{\omega_0^2}{4\pi} \frac{i\omega}{\omega^2 - \omega_c^2 + i\omega\gamma} \; . \tag{9.43}$$

We note that the eigenmodes $\omega_L^2 = \omega_c^2 + \omega_0^2$ are given by the vanishing of the denominator in equation (9.38), *i.e.* by $\varepsilon = 0$. Let us compute the energy density given by equation (9.37). Making use of $\boldsymbol{E} = -4\pi nq\boldsymbol{u}$, we get

$$\begin{aligned} \mathcal{E} = & \tfrac{1}{8\pi}E^2 + \tfrac{1}{2}nm\dot{u}^2 + \tfrac{1}{2}nm\omega_c^2 u^2 = \\ = & \tfrac{1}{2}nm\dot{\boldsymbol{u}}^2 + \tfrac{1}{2}nm\left(\omega_c^2 + \omega_0^2\right)\boldsymbol{u}^2 \; , \end{aligned} \tag{9.44}$$

which indicates harmonic oscillators with eigenfrequency $\omega_L = \left(\omega_c^2 + \omega_0^2\right)^{1/2}$. The displacement $\boldsymbol{u}$ is proportional to the external field $\boldsymbol{E}_0$, according to equation (9.38), and it oscillates with frequency ω; the energy reads, for instance,

$$\mathcal{E} = \frac{1}{2}nm\omega^2 u_0^2 \cos^2\omega t + \frac{1}{2}nm\omega_L^2 u_0^2 \sin^2\omega t \; , \tag{9.45}$$

[3] R. H. Lyddane, R. G. Sachs and E. Teller, "On the polar vibrations of alkali halides", Phys. Rev. **59** 673 (1941); C. Kittel, *Introduction to Solid State Physics*, Wiley, NJ (2005).

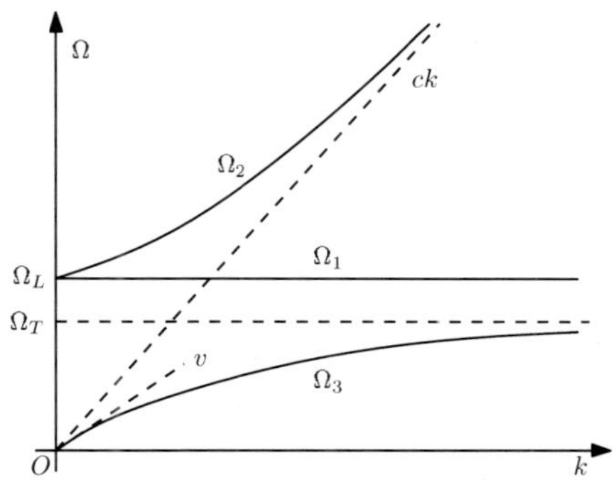

Figure 9.1: The polaritonic eigenmodes given by $\varepsilon\omega^2 - c^2k^2 = 0$, where $\omega_L^2 = \omega_0^2 + \omega_c^2$ and $v = c\omega_c/\omega_L$.

where u_0 is the displacement amplitude (the longitudinal field is uniform, as between the plates of a capacitor). We can see that the energy oscillates; from u_0 given by equation (9.38) we can compute the energy transferred to the sample.

Let us assume now a transverse displacement, *i.e.* $div\boldsymbol{u} = 0$. Equations (9.30), (9.33) and (9.36) read

$$\begin{gathered}\ddot{\boldsymbol{u}} + \omega_c^2\boldsymbol{u} + \gamma\dot{\boldsymbol{u}} = \frac{q}{m}\left(\boldsymbol{E} + \boldsymbol{E}_0\right) \ , \\ \ddot{\boldsymbol{E}} - c^2\Delta\boldsymbol{E} = -4\pi nq\ddot{\boldsymbol{u}} \ , \\ \ddot{\boldsymbol{H}} - c^2\Delta\boldsymbol{H} = 4\pi nqc \cdot curl\dot{\boldsymbol{u}} \ , \\ \ddot{\boldsymbol{A}} - c^2\Delta\boldsymbol{A} = 4\pi nqc^2 \cdot curl\boldsymbol{u} \ . \end{gathered} \tag{9.46}$$

Obviously, for $\boldsymbol{E}_0 \sim e^{i(\boldsymbol{kr}-\omega t)}$, where $\omega^2 = c^2k^2$, all the quantities in equations (9.46) have the same spatial and temporal dependence and, according to these equations, $\boldsymbol{u} = 0$ and $\boldsymbol{E} = -\boldsymbol{E}_0$. The fields $\boldsymbol{E}$ and $\boldsymbol{H}$ have not a physical cause. It follows that a transverse external field $\boldsymbol{E}_0$ with $\omega^2 = c^2k^2$ cannot be propagated in matter. This is the Ewald-Oseen extinction theorem.[4] However, an external field can excite electromagnetic eigenmodes in matter. These eigenmodes should

[4] P. P. Ewald, "Ueber die Grundlagen der Kristalloptik", Thesis, Munich (1912),

be worked out first, because the equation of motion (9.36) is not valid anymore. Let us assume $\boldsymbol{E}_0 = 0$ in the above equations. We get

$$\begin{aligned} \left(\omega^2 - c^2k^2\right)\boldsymbol{E} &= -4\pi nq\omega^2\boldsymbol{u} \ , \\ \left(\omega^2 - \omega_c^2 + i\omega\gamma\right)\boldsymbol{u} &= -\tfrac{q}{m}\boldsymbol{E} \ , \end{aligned} \tag{9.47}$$

whence

$$\left(\omega^2 - \omega_c^2 + i\omega\gamma\right)\boldsymbol{u} = \frac{\omega_0^2\omega^2}{\omega^2 - c^2k^2}\boldsymbol{u} \ , \tag{9.48}$$

or

$$\left(\omega^2 - c^2k^2\right)\boldsymbol{u} = \frac{\omega_0^2\omega^2}{\omega^2 - \omega_c^2 + i\omega\gamma}\boldsymbol{u} \ , \tag{9.49}$$

which leads to

$$\left(\varepsilon\omega^2 - c^2k^2\right)\boldsymbol{u} = 0 \ . \tag{9.50}$$

We can see that eigenmodes with the dispersion relation given by

$$\varepsilon\frac{\omega^2}{c^2} - k^2 = 0 \tag{9.51}$$

can be propagated in matter. The dispersion relation given by this equation defines the polaritons. The frequency ω is the frequency of the external field, but the magnitude k of the wavevector is changed according to equation (9.51). There are two branches of polariton frequencies, one starting at $\left(\omega_0^2 + \omega_c^2\right)^{1/2}$ and going to ck for $k \longrightarrow \infty$, another starting at 0 with the slope $v = c\omega_c/\omega_L$ and going to ω_c for $k \longrightarrow \infty$, as shown in Fig. 9.1.

We note that $K = \omega/c$ in equation (9.51) is the magnitude of the wavevector of the external field, and ω/k is the phase velocity of the polaritons; it follows that $v = c/\sqrt{\varepsilon}$, where $\sqrt{\varepsilon} = n$ is the refractive index; therefore, the polaritons are in the refracted ray, and, comparing the velocities, we get the Snell law of refraction $\sin i/\sin r = n$, where i is the incidence angle and r is the refraction angle. It is worth noting that equation (9.50) determines both the magnitude and the direction of the wavevector $\boldsymbol{k}$ (the latter by Huygens' principle).

Ann. Phys. **49 (354)** 1 (1916); C. W. Oseen, "Ueber die Wechselwirkung zwischen zwei elektrischen Dipolen und ueber die Drehung der Polarisationsebene in Kristallen und Fluessigkeiten", Ann. Phys. **48 (353)** 1 (1915); M. Born and E. Wolf, *Principles of Optics*, Pergamon (London, 1959) and references therein.

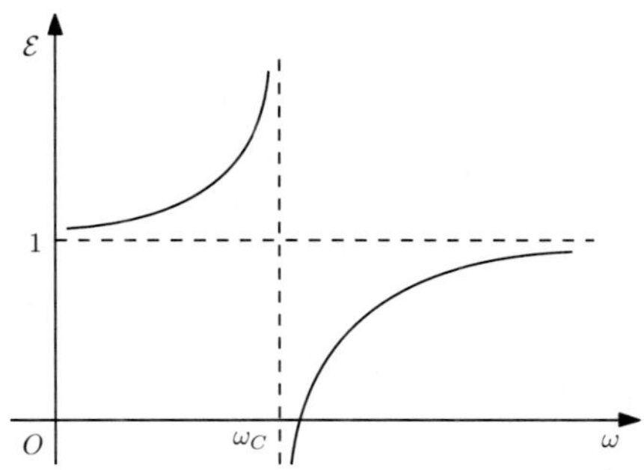

Figure 9.2: Dielectric function $\varepsilon = 1 - \omega_0^2/(\omega^2 - \omega_c^2)$.

According to equation (9.50), the free displacement $\boldsymbol{u}$ goes like $e^{i(\boldsymbol{kr}-\omega(k)t)}$, where $\omega(k)$ are the polariton eigenfrequencies given by equation (9.51). Since $\boldsymbol{k}$ is usually spread over a finite range, it follows that the group velocity $\boldsymbol{v}_g = \partial\omega(k)/d\boldsymbol{k}$ is relevant in the free polariton propagation.

From equations (9.46) and (9.51) we get the fields

$$\boldsymbol{E} = \tfrac{4\pi nq}{\varepsilon-1}\boldsymbol{u} \ , \ \boldsymbol{H} = \tfrac{4\pi nqc}{\omega(\varepsilon-1)}\boldsymbol{k}\times\boldsymbol{u} \ , \tag{9.52}$$

where ω is given by equation (9.51) and $\boldsymbol{uk} = 0$. It is worth noting that ω is the external frequency and k is changed in comparison with K: it is slightly greater than K and rotated by the angle $i - r$. Also, it is worth noting that the spatial dependence of the internal fields and the displacement $\boldsymbol{u}$ is governed by $\boldsymbol{k}$. The energy density given by equation (9.37) becomes

$$\mathcal{E} = \frac{1}{2}nm\dot{\boldsymbol{u}}^2 + \frac{1}{2}nm\left[\omega_c^2 + \frac{(\varepsilon+1)}{(\varepsilon-1)^2}\omega_0^2\right]\boldsymbol{u}^2 \ . \tag{9.53}$$

The dielectric constant is shown in Fig. 9.2. For usual dielectrics ω_c is high and ω_0 is low. For usual frequencies ω the dielectric function ε is almost a flat curve, slightly above unity. We say that the medium is dispersionless (*i.e.* it is transparent). We may approximate the above equation by

$$\mathcal{E} = \frac{1}{2}nm\dot{\boldsymbol{u}}^2 + \frac{1}{2}nm(\omega_c^4/\omega_0^2)\boldsymbol{u}^2 \ . \tag{9.54}$$

In this energy the displacement $\boldsymbol{u}$ goes like $\cos(\omega t - \boldsymbol{kr})$, where ω is the frequency of the external field and $\boldsymbol{k}$ is given by equation (9.50). The energy given by equation (9.53) has both a time and a spatial dependence. The amplitude of the displacement is obtained from the boundary conditions at the surface, which require the continuity of the tangential electric components and the normal magnetic components. The energy transferred to the medium can be obtained from the Poynting vectors.

9.6 Polarization

The situation is completely different for small bodies. Let us assume the transverse external fields $\boldsymbol{E}_0$ and $\boldsymbol{H}_0$ and a body with dimension much smaller than the wavelength of the external fields. We wish to solve equations (9.46) in this case. Since the dimension of the body is much smaller than the wavelength we may leave aside the spatial dependence over the body. It follows, form equations (9.46), that $\boldsymbol{H} = 0$ and

$$\boldsymbol{u} = -\frac{q\boldsymbol{E}_0}{m}\frac{1}{\omega^2 - \omega_L^2 + i\omega\gamma} \tag{9.55}$$

(which defines the polarizability). We shold include the surface effects, which amounts to writing the displacement as $\boldsymbol{u}\theta(r_s - r)$, where $\boldsymbol{r}_s$ is the vector which defines the surface of the body. In this case we get an internal magnetic field, proportional to the suface of the body (if the body is not very symmetrical). We leave aside this magnetic field. The situation is similar with a longitudinal electric field, except for the presence of the external magnetic field $\boldsymbol{H}_0$. This magnetic field acts upon the induced polarization current with a force density

$$\boldsymbol{f} = \frac{nq}{c}\dot{\boldsymbol{u}} \times \boldsymbol{H}_0 = \frac{\omega_0^2}{4\pi c}\frac{i\omega}{\omega^2 - \omega_L^2 + i\omega\gamma}\boldsymbol{E}_0 \times \boldsymbol{H}_0 \; ; \tag{9.56}$$

we can see that this force is the rate of change per unit time of the Abraham momentum density $\boldsymbol{E}_0 \times \boldsymbol{H}_0/4\pi c$ (Poynting vector divided by c^2). It is directed along the propagation wavevector. It differs from the radiation pressure, which is proportional to $\boldsymbol{E}_0 \times \boldsymbol{H}_0/4\pi$, by the presence of the polarizability. The force given by equation (9.56) can be written as

$$f = \frac{\omega_0^2 E_0^2}{8\pi c}\left(\frac{i}{\omega - \omega_L + i\omega\gamma} + \frac{i}{\omega + \omega_L + i\omega\gamma}\right) \ , \tag{9.57}$$

which, in the limit $\gamma \rightarrow 0$, becomes

$$f = \frac{\omega_0^2 E_0^2}{8c}\left[\delta(\omega - \omega_L) + \delta(\omega + \omega_L)\right] \ . \tag{9.58}$$

We can see that it is a resonance force.[5] For a numerical estimation we can take $f \simeq \omega_0^2 E_0^2/8c\gamma$, where $E_0 = 10^{-8} esu$ (corresponding approximately to $10^{-6} w/cm^2$) and $\omega_0 = 10^6 s^{-1}$ (a typical piezoelectric eigenfrequency; $\omega_c = 0$); a tentative value for γ could be $\gamma = 10^8 s^{-1}$. We get $f = 10^{-23} dyn/cm^3$, which is an extremely small force. Of course, the body can be polarized by another, independent source, whose frequency may match the frequency of the external field.

The polarization dynamics (Drude model) leads to the Ewald-Oseen extinction theorem, in conjunction with the refracted ray and the equation of motion of the polarization for the polaritonic eigenmodes. An external electromagnetic wave acts upon small bodies with a polarization force which has a resonant character. In metals, the characteristic frequency ω_c is zero, ε acquires large negative values and electromagnetic waves cannot be propagated. The magnetization can be included by a current density given by $c \cdot curl\boldsymbol{M}$, where $\boldsymbol{H}$ is denoted by $\boldsymbol{B}$ and $\boldsymbol{H} = \boldsymbol{B} - 4\pi\boldsymbol{M}$.

9.7 Polarizable matter

The electromagnetic field is generated by electric charges, usually in motion. The motion of the electric charges in usual matter is non-relativistic, since their velocity is much smaller than the light velocity c. For a sufficiently dense condensed matter the motion of the electric charges is also in the quasi-classical limit of the quantum-mechanical motion, so a quantum-statistical average is appropriate. Therefore, we get a classical (both non-quantum and non-relativistic)

[5] M. Apostol, S. Ilie, A. Petrut, M. Savu and S. Toba, "A generalization of the dipolar force", J. Appl. Phys. **112** 024905 (2012).

motion for the electric charges in matter and, consequently, a classical electromagnetic field. The electromagnetic field in matter admits a mechanical representation in terms of the matter polarization, which can be decomposed in harmonic oscillators. The polarization is identified by a displacement field in the position of the mobile charges. The collective motion of the polarization is governed, in general, by three distinct, infinite sets of harmonic oscillators, corresponding to longitudinal, plasmon (one set) and transverse, polariton (two sets) frequencies. This collective motion can be quantized, oferring an example of emergent dynamics. The electromagnetic field generated by the polarization motion has a corresponding mechanical representation in terms of the displacement field. This representation is given here for the electromagnetic field energy, Poynting vector, Lorentz force and charge-field interaction energy.

The electromagnetic coupling between two polarizable bodies can be examined in this framework, and the corresponding plasmonic and polaritonic eigenfrequencies are identified. The Casimir and van der Waals-London forces can be derived, with coefficients which depend on the nature of the two bodies. These forces arise from the vacuum fluctuations of the polarization. The electromagnetic field (generated by these fluctuations) brings its own conribution to the forces acting between two polarizable bodies. When a dielectric, at least, is present, the electromagnetic-field force goes like $-1/d^2$, where d is the separation distance between the two bodies. When a conductor, at least, is present, the electromagnetic-field force is of Casimir type ($\sim -1/d^4$).

The important surface (or shape) effects cannot be accounted for within such an approach, as a consequence of the use of Fourier decompositions, which do not represent correctly the sharp surfaces (*e.g.*, the well-known Gibbs phenomenon). In order to include such surface effects, we should keep the dependence on the direct-space coordinate transverse to the surface, which is a much more difficult task. In this context, the particular case of pointlike, δ-particles can be analyzed.

We adopt a generic model of matter polarization consisting of N identical mobile charges q, with mass m and density $n = N/V$, moving in a rigid, neutralizing background of volume V. A small displacement field $\mathbf{u}(\mathbf{r}, t)$ in the position $\mathbf{r}$ of these charges gives, at time t, a local density imbalance $\delta n = -n div \mathbf{u}$ and a polarization charge density $\rho = -nq div\mathbf{u}$. We can see that $\mathbf{P} = nq\mathbf{u}$ is the polarization. Therefore, the displacement field $\mathbf{u}(\mathbf{r}, t)$ is a representation for the polarization field $\mathbf{P}(\mathbf{r}, t)$. The displacement field obeys the Newton law of motion

$$m\ddot{\mathbf{u}} = q\mathbf{E} - m\omega_c^2\mathbf{u} - m\gamma\dot{\mathbf{u}} + q\mathbf{E}_0 \ , \tag{9.59}$$

where $\mathbf{E}$ is the polarization electric field generated by the polarization charges (and currents), ω_c is a characteristic frequency, γ is a (small) damping factor and $\mathbf{E}_0$ is an external electric field. This is the well-known Lorentz-Drude (plasma) model of polarizable matter,[6] which assumes a homogeneous, isotropic matter, without spatial dispersion, represented by a field of harmonic oscillators of frequency ω_c. Taking the temporal Fourier transform of equation (9.59), with $\mathbf{E}_t = \mathbf{E} + \mathbf{E}_0$ the total electric field, we get the electric susceptibility $\chi(\omega) = P/E_t$ and the dielectric function

$$\varepsilon(\omega) = 1 + 4\pi\chi(\omega) = \frac{\omega^2 - \omega_c^2 - \omega_p^2}{\omega^2 - \omega_c^2 + i\omega\gamma} = \frac{\omega^2 - \omega_L^2}{\omega^2 - \omega_T^2 + i\omega\gamma} \ , \tag{9.60}$$

where $\omega_p = \sqrt{4\pi nq^2/m}$ is the plasma frequency. This is also well known as the Lydane-Sachs-Teller dielectric function,[7] with the longitudinal frequency $\omega_L = \sqrt{\omega_c^2 + \omega_p^2}$ and the transverse frequency $\omega_T = \omega_c$. The model can be generalized by including the spatial dispersion, several characteristic frequencies ω_c, or by adding an external magnetic field, etc. It is worth noting the absence of the magnetic part of the Lorentz force in equation (9.59), according to the non-relativistic motion of the slight displacement $\mathbf{u}$. It is easy to see that, apart from relativistic contributions, it would introduce non-linearities in equation (9.59), which are beyond our assumption of a small displacement $\boldsymbol{u}$. Using spatial Fourier transforms, this approximation can be formulated as $\mathbf{k}\mathbf{u}(\mathbf{k}) \ll 1$, where $\mathbf{k}$ is the wavevector.

In general, an additional displacement $\mathbf{u}_0$ can be introduced in such a model, originating in external causes, subjected to collisions and

[6] P. Drude, "Zur Elektronentheorie der Metalle", Ann. Phys. **306** 566 (1900); "2. Teile. Galvanomagnetische und thermomagnetische Effekte", **308** 369 (1900); H. A. Lorentz, *The Theory of Electrons*, Teubner, Leipzig (1916); N. W. Ashcroft and N. D. Mermin, *Solid State Physics*, Saunders College, Philadelphia (1976); G. D. Mahan, *Many-Particle Physics*, Plenum, NY (1980).

[7] C. Kittel, *Introduction to Solid State Physics*, Wiley, NJ (2005).

obeying a different, averaged equation of motion, $m\dot{\mathbf{u}}_0 = q\mathbf{E}_t\tau$, where τ is a relaxation time; as it is well known, it gives rise to a density of "conduction" current $\mathbf{j}_0 = nq\dot{\mathbf{u}}_0 = (nq^2\tau/m)\mathbf{E}_t$ and the conductivity $\sigma = nq^2\tau/m$. We can see that it implies $\omega_c = 0$ in equation (9.59), a condition which defines the conductors; for dielectrics, $\omega_c \neq 0$. We leave aside the conduction current $\mathbf{j}_0$.

The displacement $\mathbf{u}$ generates also a density of polarization current $\mathbf{j} = nq\dot{\mathbf{u}}$, which satisfies the continuity equation $\dot{\rho} + div\mathbf{j} = 0$. This equation allows an additional current, written as $\mathbf{j}_m = c \cdot curl\mathbf{M}$ (since $div\mathbf{j}_m = 0$); it is easy to see, from Maxwell's equations, that $\mathbf{M}$ is the magnetization. Therefore, the electromagnetic sources are represented both by the displacement $\mathbf{u}$ and magnetization $\mathbf{M}$. These fields ($\mathbf{u}$ and $\mathbf{M}$) are determined, $\mathbf{u}$ by equation (9.59) and $\mathbf{M}$ by the well-known equation of motion $\dot{\mathbf{M}} = \gamma\mathbf{B} \times \mathbf{M}$ for the magnetization, where $\mathbf{B}$ is the magnetic field and γ is the gyromagnetic factor. With such a representation for the sources, the Maxwell equations in matter are completely soluble. Here, we leave aside the magnetization, and consider only non-magnetic matter. With the polarization charge density ρ and current density $\mathbf{j}$ established above, and with usual notations, the Maxwell equations read

$$\begin{gathered} div\mathbf{E} = 4\pi\rho = -4\pi div\mathbf{P} = \\ = -4\pi nq div\mathbf{u} \ , \ div\mathbf{H} = 0 \ , \\ curl\mathbf{E} = -\tfrac{1}{c}\tfrac{\partial\mathbf{H}}{\partial t} \ , \ curl\mathbf{H} = \tfrac{1}{c}\tfrac{\partial\mathbf{E}}{\partial t} + \tfrac{4\pi}{c}\mathbf{j} = \\ = \tfrac{1}{c}\tfrac{\partial\mathbf{E}}{\partial t} + \tfrac{4\pi}{c}\dot{\mathbf{P}} = \tfrac{1}{c}\tfrac{\partial\mathbf{E}}{\partial t} + \tfrac{4\pi}{c}nq\dot{\mathbf{u}} \ . \end{gathered} \tag{9.61}$$

We solve these equations, together with equation (9.59), for the electric field $\mathbf{E}$, the magnetic field $\mathbf{H}$ and the displacement field $\mathbf{u}$ (polarization $\mathbf{P} = nq\mathbf{u}$).

To this end, as usually, we introduce the electromagnetic potentials $\mathbf{A}$ and Φ, through $\mathbf{E} = -(1/c)\partial\mathbf{A}/\partial t - grad\Phi$ and $\mathbf{H} = curl\mathbf{A}$, subjected to the Lorenz gauge $div\mathbf{A} + (1/c)\partial\Phi/\partial t = 0$, and use Fourier transforms of the type

$$\mathbf{u}(\mathbf{r}, t) = \frac{1}{2\pi\sqrt{N}}\sum_{\mathbf{k}}\int d\omega\mathbf{u}(\mathbf{k}, \omega)e^{-i\omega t + i\mathbf{kr}} \tag{9.62}$$

(and similar transforms for all the other functions of position and time). We note the well-known symmetry property $\mathbf{u}^*(-k,-\omega)=\mathbf{u}(\mathbf{k},\omega)$, corresponding to the real-valued field $\mathbf{u}(\mathbf{r},t)$. As it is well known, such a Fourier representation does not account for surface effects, *i.e.* effects associated with sharp surfaces. In order to account for finite-size (surface) effects an expansion in eigenfunctions of the laplacian (with outgoing-waves boundary conditions at infinity) is appropiate, preserving the dependence on the direct-space coordinate perpendicular to the surface. It is easy to see that, in general, such a decomposition implies curvilinear coordinates, and leads to serious technical difficulties. The density of these eigenmodes (which appears in summations over states) differs, in general, from the density of the plane waves which appear in the Fourier transformations. Therefore, additional contributions, which depend on the extension and the shape of the bodies, may appear. In addition, the sharpness of the surfaces generates the well-known "depolarizing" factors, especially in the long-wavelength limit, which cannot be taken into account by Fourier transforms.

As it is well known, the Maxwell equations (9.61) lead to the wave equations with sources

$$\frac{1}{c^2}\frac{\partial^2\Phi}{\partial t^2}-\Delta\Phi=4\pi\rho\ ,\ \frac{1}{c^2}\frac{\partial^2\mathbf{A}}{\partial t^2}-\Delta\mathbf{A}=\frac{4\pi}{c}\mathbf{j} \tag{9.63}$$

for the electromagnetic potentials. Taking the Fourier transforms, we get

$$\mathbf{A}(\mathbf{k},\omega)=4\pi i\lambda F(\mathbf{k},\lambda)\mathbf{u}(\mathbf{k},\omega) \tag{9.64}$$

and $\Phi(\mathbf{k},\omega)=\mathbf{k}\mathbf{A}(\mathbf{k},\omega)/\lambda$, where $\lambda=\omega/c$, $F(k,\lambda)=(\lambda^2-k^2)^{-1}$ is the Green function (for the Helmholtz equation) and the factor nq is left aside (it is restored in the final formulae). The latter relationship is the Lorenz gauge, which expresses the charge conservation, *i.e.* the continuity equation. In order to account for the retardation, the function F must actually be written as $F(k,\lambda)=(\lambda^2-k^2+i\lambda 0^+)^{-1}$, as it can be seen by taking the Fourier transforms of the retarded

Kirchhoff's potentials

$$\begin{aligned}\Phi(\mathbf{r},t) &= \int d\mathbf{r}' \frac{\rho(\mathbf{r}',t-\left|\mathbf{r}-\mathbf{r}'\right|/c)}{\left|\mathbf{r}-\mathbf{r}'\right|} \ , \\ \mathbf{A}(\mathbf{r},t) &= \frac{1}{c}\int d\mathbf{r}' \frac{\mathbf{j}(\mathbf{r}',t-\left|\mathbf{r}-\mathbf{r}'\right|/c)}{\left|\mathbf{r}-\mathbf{r}'\right|} \ ,.\end{aligned} \tag{9.65}$$

which are solutions of equations (9.63). For simplicity, we often leave aside the arguments $\mathbf{k}$, ω in the Fourier transforms. Making use of equation (9.64), we get straightforwardly

$$\mathbf{E} = -4\pi F\left[\lambda^2\mathbf{u} - \mathbf{k}(\mathbf{ku})\right] \ , \ \mathbf{H} = -4\pi\lambda F\mathbf{k}\times\mathbf{u} \ . \tag{9.66}$$

It is convenient to introduce the longitudinal displacement $u_1 = \mathbf{ku}/k$ and the transverse displacement $u_2 = \mathbf{k}_\perp\mathbf{u}/k$, where $\mathbf{k}_\perp$ is a vector perpendicular to $\mathbf{k}$ and of the same magnitude as $\mathbf{k}$, and to write $\mathbf{u} = u_1\mathbf{k}/k + u_2\mathbf{k}_\perp/k$. We note the symmetry property $-u^*_{1,2}(-k,-\omega) = u_{1,2}(k,\omega)$. In summations over states we take into account that there are in fact two transverse components u_2 of the displacement, corresponding to two polarizations. Making use of these notations we get

$$E_1 = -4\pi u_1 \ , \ E_2 = -4\pi\lambda^2 F u_2 \ . \tag{9.67}$$

We can see that the longitudinal component of the internal (polarization) field compensates the longitudinal component of the polarization, as expected, while the transverse polarization introduces a spatial dispersion. Introducing these fields in the equation of motion (9.59) we get

$$\begin{aligned}(\omega^2-\omega_c^2-\omega_p^2+i\omega\gamma)u_1 &= -\frac{q}{m}E_{01} \ , \\ (\omega^2-\omega_c^2-\omega_p^2\lambda^2F+i\omega\gamma)u_2 &= -\frac{q}{m}E_{02} \ ,\end{aligned} \tag{9.68}$$

whence the longitudinal and transverse polarizabilities

$$\begin{aligned}\alpha_1 = E_{01}/P_1 &= -\frac{1}{4\pi}\frac{\omega_p^2}{\omega^2-\omega_c^2-\omega_p^2+i\omega\gamma} \ , \\ \alpha_2 = E_{02}/P_2 &= -\frac{1}{4\pi}\frac{\omega_p^2}{\omega^2-\omega_c^2-\omega_p^2\lambda^2F+i\omega\gamma} \ .\end{aligned} \tag{9.69}$$

It is worth noting that for finite-size bodies the plasma frequency in equations (9.68) is modified by the "depolarizing" factors. For instance, for a half-space it acquires the well-known form $\omega_p/\sqrt{2}$ of the surface plasmon,[8] for a sphere in the dipole approximation it becomes the spherical plasmon $\omega_p/\sqrt{3}$.[9] For uniform fields (in the long-wavelength limit $\mathbf{k} \to 0$) the two polarizabilities coincide, as expected. Making use of equations (9.68) and (9.69) we get straightforwardly the electric susceptibility

$$\chi = \frac{\alpha_1}{1 - 4\pi\alpha_1} = \frac{\alpha_2}{1 - 4\pi\lambda^2 F\alpha_2} = -\frac{1}{4\pi}\frac{\omega_p^2}{\omega^2 - \omega_c^2} , \tag{9.70}$$

the same for both longitudinal and transverse field components, as given by equation (9.60). We can see that the electric susceptibility does not exhibit spatial dispersion, in contrast to the transverse polarizability.

9.7.1 Magnetization

The magnetization $\mathbf{M}$ (wich gives the current $\mathbf{j}_m = c \cdot curl\mathbf{M}$) can be written as

$$\mathbf{M} = \frac{1}{2c}\mathbf{R} \times \mathbf{j}_m = \frac{nq}{2c}(\mathbf{r}_a - \mathbf{r}) \times \mathbf{v} = \frac{q}{2mc}\mathbf{L} , \tag{9.71}$$

where $\mathbf{R} = \mathbf{r}_a - \mathbf{r}$, $\mathbf{v}$ is the velocity of the mobile charges q placed at $\mathbf{r}_a$ and $\mathbf{L}$ is the angular momentum (per unit volume) (the electromagnetic momentum $nq\mathbf{A}/c$, where $\mathbf{A}$ is the vector potential, is neglected). The equation of motion for the above quantity can be written as

$$\frac{d\mathbf{L}}{dt} = \frac{1}{c}\mathbf{R} \times (\mathbf{j}_m \times \mathbf{B}) = \frac{nq}{c}\mathbf{R} \times (\mathbf{v} \times \mathbf{B}) , \tag{9.72}$$

where we include only the torque of the magnetic part of the Lorentz force corresponding to the (total) induction $\mathbf{B}$. Using $\mathbf{v} = d\mathbf{R}/dt$ and

[8] M. Apostol and G. Vaman, "Electromagnetic field interacting with a semi-infinite plasma", J. Opt. Soc. Am. **A26** 1747 (2009).

[9] M. Apostol and G. Vaman, "Plasmons and difraction of an electromagnetic plane wave by a metallic sphere", Progr. Electrom. Res. PIER **98** 97 (2009).

averaging equation (9.72) over a finite motion, it is easy to get

$$\frac{d\mathbf{M}}{dt} = \frac{q}{2mc}\mathbf{M} \times \mathbf{B} \tag{9.73}$$

(since the average of dR^2/dt and $d(\mathbf{R}(\mathbf{R}B)/dt$ are vanishing; accordingly, $\mathbf{R}\times(\mathbf{R}\times\mathbf{B}) = -\mathbf{R}\times(\mathbf{R}\times\mathbf{B})$ by such an averaging). This is the well-known Larmor's equation of motion.[10] If the velocity $\mathbf{v}$ in equation (9.71) is $\dot{\mathbf{u}}$, then the current $\mathbf{j}_m$ is the polarization current $\partial\mathbf{P}/\partial t$, and $\mathbf{M}$ would be superfluous: its role would already be played by $\mathbf{P}$ and the equation of motion (9.73) would already be included in the equation of motion for $\mathbf{P}$. A macroscopic orbital magnetism, particularly induced by an electric field, is already described by $\mathbf{P}$, and even a "magnetization" $\mathbf{M} = (1/2c)\mathbf{j}\times\mathbf{r}$ can be introduced ($\mathbf{j} = c\cdot curl\mathbf{M}$), where $\mathbf{j} = \partial\mathbf{P}/\partial t$ (negative of a magnetic moment of the unit volume). However, it is not particularly useful, and it is already included in the solution for the fields in matter.

As it is well known, for the usual magnetic matter the magnetization $\mathbf{M}$ arises from magnetic moments $\boldsymbol{\mu}$ at the atomic scale, both orbital and spin-like, whose origin is quantal and relativistic. They give rise to paramagnetism and diamagnetism, usually under the action of an external magnetic field. For instance, the electron has a magnetic moment $\boldsymbol{\mu} = -(|e|\,\hbar/mc)\,\mathbf{s}$, where $\hbar$ is Planck's constant, m is the electron mass and $\mathbf{s}$ is the electron spin ($s = 1/2$, Bohr's magneton $\mu_B = |e|\,\hbar/2mc$). The leading contribution to the quantum-mechanical equation of motion of a magnetic moment (spin) is given by,[11]

$$\dot{\mathbf{M}} = \gamma\mathbf{B}\times\mathbf{M}\ , \tag{9.74}$$

where $\gamma = \mu/\hbar$ is the gyromagnetic factor (compare with equation (9.73)). Such an equation of motion could be used in conjunction with Maxwell equations and equation of motion (9.59) for the polarization $\mathbf{P}$. We have now four equations for the four unknowns, and their

[10]See, for instance, L. Landau and E. Lifshitz, *Course of Theoretical Physics*, vol. 2, *The Classical Theory of Fields*, vol. 9, *Statistical Physics*, part 2, Elsevier (Oxford, 2004).

[11]See, for instance, L. Landau and E. Lifshitz, *Course of Theoretical Physics*, vol. 2, *The Classical Theory of Fields*, vol. 9, *Statistical Physics*, part 2, Elsevier (Oxford, 2004).

solution can be determined. It is worth noting that, in general, these equations are coupled, so that we have a dependence of the form $\mathbf{P}(\mathbf{E}, \mathbf{H})$ and $\mathbf{M}(\mathbf{E}, \mathbf{H})$, *i.e.* we have electro-magnetic effects, and not purely an electric or a magnetic susceptibility.

Usually, this is not the case. Due to its relativistic origin, the magnetization can be neglected in usual matter. This amounts to put the magnetization zero ($\mathbf{M} = 0$) (and magnetic permeability equal to unity, $\mu = 1$) and to make not a distinction between magnetic induction $\mathbf{B}$ and the magnetic field $\mathbf{H}$. All these considerations are in accordance with the real situation where most of the substances are, practically, non-magnetic indeed.

There is, however, one important exception: ferromagnetic matter (with all its great varities). Therein, there exists a large magnetization (usually along an axis, of "easy" magnetization, which makes the ferromagnets anisotropic). In principle, equation (9.74) can be used in this case, which would bring in discussion magnons and, in general, magnetostatic modes, with their own contribution to the dispersion of the magnetic permeability (though for rather low frequency). In addition, equation (9.74) shows that the relation between $\mathbf{H}$ and $\mathbf{B}$ may not be linear. However, we may assume that the ferromagnetic magnetization acts mainly as an external (magnetic) field in Maxwell equations, so it does not produce an appreciable trouble.

9.8 Polarization eigenmodes

The longitudinal polarizability does not depend on the wavevector $\mathbf{k}$. Its singularity (leaving aside the damping parameter γ) gives the longitudinal ("plasmon") mode $\Omega_1 = \omega_L = \sqrt{\omega_c^2 + \omega_p^2}$, as resulting from two oscillatory motions, one with the local, characteristic frequency ω_c and another collective, driven by the polarization field, with the frequency ω_p. It can also be obtained from the zero of the dielectric function, $\varepsilon = 0$. The singularities of the transverse polarizability correspond to propagating polaritons. They are given by the well-known dispersion relation $\varepsilon\omega^2 = c^2k^2$. For $\omega_c = 0$ (conductors), they acquire the well-known form $\Omega_2(k) = \sqrt{\omega_p^2 + c^2k^2}$, which again, exhibits the characteristic pattern of two oscillatory motions (plasmons with frequency ω_p and photons with frequency ck). For $\omega_c \neq 0$, we have two polaritonic branches, $\Omega_{2,3}(k)$, corresponding to the occurrence of a third force, with frequency ω_c. In the long-wavelenght limit ($k \to 0$) they behave like

$$\Omega_2 \simeq \sqrt{\omega_L^2 + \omega_p^2 c^2 k^2/\omega_L^2} \ , \ \Omega_3 \simeq \frac{\omega_T}{\omega_L} ck \ , \tag{9.75}$$

while in the short-wavelenght limit ($k \to \infty$) they go like

$$\Omega_2 \simeq \sqrt{c^2 k^2 + \omega_p^2} \ , \ \Omega_3 \simeq \omega_T = \omega_c \ . \tag{9.76}$$

We can see that the frequency Ω_3 in the long-wavelength limit corresponds to "photons" with a renormalized phase velocity $v = c\omega_T/\omega_L$ ($\Omega_3 = vk$). These two polaritonic branches arise from the transverse mode ω_c and the photonic mode ck, split in the region $\omega_c \simeq ck$ by the plasmonic coupling ω_p. If ω_p is a small parameter, we may take approximately ω_c and ck for the two transverse frequencies, as for non-polarizable matter.

For the sake of the generality we preserve three distinct frequencies, Ω_1 for the longitudinal component and $\Omega_{2,3}(k)$ for the transverse components. They are shown schematically in Fig. 9.3

According to equations (9.68) the frequencies $\Omega_{1,2,3}$ are the eigenfrequencies of three types of harmonic oscillators, given by

$$\begin{gathered} u_1(\mathbf{k},\omega) = 2\pi\delta(\omega-\Omega_1)u_1(\mathbf{k}) \ , \\ u_2(\mathbf{k},\omega) = 2\pi\delta(\omega-\Omega_2)u_2^{(1)}(\mathbf{k}) + 2\pi\delta(\omega-\Omega_3)u_2^{(2)}(\mathbf{k}) \ , \end{gathered} \tag{9.77}$$

or

$$\begin{gathered} u_1(\mathbf{k},t) = u_1(\mathbf{k})e^{-i\Omega_1 t} \ , \\ u_2(\mathbf{k},t) = u_2^{(1)}(\mathbf{k})e^{-i\Omega_2 t} + u_2^{(2)}(\mathbf{k})e^{-i\Omega_3 t} \ . \end{gathered} \tag{9.78}$$

Equations (9.77) and (9.78) include only the positive frequencies. The negative frequencies $-\Omega_{1,2,3}$ are included whenever the case. We emphasize that the functions $u_1(\mathbf{k},\omega)$ and $u_2^{(1,2)}(\mathbf{k},\omega)$ are solutions of

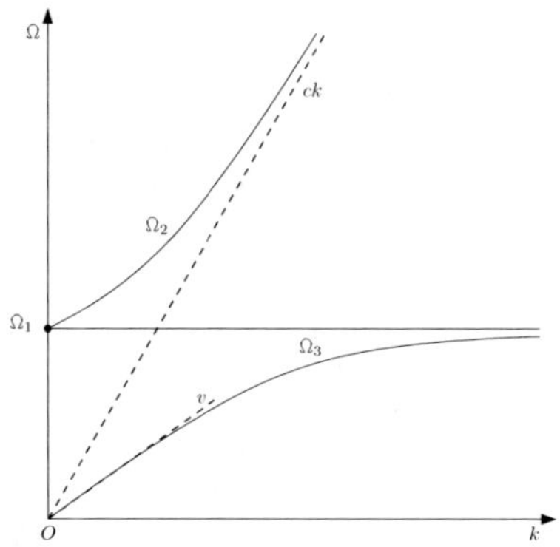

Figure 9.3: The polarization longitudinal eigenmode Ω_1 and the two transverse eigenmodes $\Omega_{2,3}$, according to equations (9.68). We note the velocity v of the renormalized "photons" (ck is the photon frequency).

equations (9.68) for a vanishing external field. Making use of equations (9.67), we get the field generated by the polarization

$$E_1(\mathbf{k},t) = -4\pi u_1(\mathbf{k})e^{-i\Omega_1 t} ,$$

$$E_2(\mathbf{k},t) = -4\pi\frac{\Omega_2^2-\omega_c^2}{\omega_p^2}u_2^{(1)}(\mathbf{k})e^{-i\Omega_2 t}- \tag{9.79}$$

$$-4\pi\frac{\Omega_3^2-\omega_c^2}{\omega_p^2}u_2^{(2)}(\mathbf{k})e^{-i\Omega_3 t} .$$

According to equations (9.78), the dynamics of the coordinates $u_1(\mathbf{k},t)$, $u_2^{(2,3)}(\mathbf{k},t)$ is governed by a harmonic-oscillator hamiltonian

$$H = \sum_{\mathbf{k}}\left[\frac{1}{2m}\left|p_1\right|^2 + \frac{1}{2}m\Omega_1^2\left|u_1\right|^2\right] +$$

$$+\sum_{\mathbf{k}}\left[\frac{1}{2m}\left|p_2^{(1)}\right|^2 + \frac{1}{2}m\Omega_2^2\left|u_2^{(1)}\right|^2\right] + \tag{9.80}$$

$$+\sum_{\mathbf{k}}\left[\frac{1}{2m}\left|p_2^{(2)}\right|^2 + \frac{1}{2}m\Omega_3^2\left|u_2^{(2)}\right|^2\right] ,$$

where p_1 and $p_2^{(1,2)}$ are the corresponding momenta (here the negative frequencies are automatically included). By taking the inverse (spatial) Fourier transform we may get a very complicated interaction, involving the displacement field $\mathbf{u}(\mathbf{r}, t)$, generated by the polarization field in the direct space. More interesting, we can quantize the motion of the harmonic oscillators described by the hamiltonian given above. For instance, the mean square displacements are given by equations like $\overline{|u_1|^2} = (\hbar/m\Omega_1)(n_1 + 1/2)$, where n_1 is the quantum number (and $\hbar$ is Planck's constant). These oscillators have also an energy of the form $\hbar\Omega_1(n_1 + 1/2)$ (including the zero-point energy $\hbar\Omega_1/2$), etc. We can also take a statistical average corresponding to a given temperature, or we can account for the quantum-statistical fluctuations. According to the results given above, the internal, polarization field can be represented entirely in terms of the coordinates u_1 and $u_2^{(1,2)}$, thus acquiring a mechanical representation. It is worth noting that the quantum numbers $n_{1,2,3}$ depend on the wavevector $\mathbf{k}$ (as well as the frequencies $\Omega_{2,3}(k)$, which depend only on the magnitude k of the wavevector). The energy quanta associated with the frequencies $\Omega_{1,2,3}$ are much higher than the usual temperatures, so we may neglect the temperature effects, and limit ourselves to the zero-point contributions (vacuum fluctuations), except for the long-wavelength limit of the frequency $\Omega_3 = vk$, whose contribution, usually, is comparatively small.

By usual procedure, from Maxwell's equations (9.34) we get the well-known energy conservation

$$\frac{1}{8\pi}\frac{\partial}{\partial t}(E^2 + H^2) + \frac{c}{4\pi}div(\mathbf{E} \times \mathbf{H}) = -nq\mathbf{E}\dot{\mathbf{u}} \ , \tag{9.81}$$

where the *rhs* is the rate of change of the density of the kinetic energy of the mobile charges. We may write equation (9.81) as

$$\frac{\partial}{\partial t}(u + t) = -div\mathbf{s} \ , \tag{9.82}$$

where $u = (E^2 + H^2)/8\pi$ is the density of the electromagnetic field energy, $t = nm\dot{\mathbf{u}}^2/2$ is the density of kinetic energy and $\mathbf{s} = c(\mathbf{E} \times \mathbf{H})/4\pi$ is the density of the Poynting vector.

We estimate below the time averages of total quantities, like the energy

$$U = \frac{1}{8\pi T}\int dtd\mathbf{r}(E^2 + H^2) \ , \qquad (9.83)$$

where T is a time sufficiently long in comparison with any relevant frequency. Making use of the Fourier transforms and of equations (9.66) and (9.67) we get straightforwardly

$$U = \frac{nq^2}{T}\sum_{\mathbf{k}}\int d\omega\left[|u_1|^2 + 2\frac{\omega^2}{\omega^2 - c^2k^2}|u_2|^2\right] \ , \qquad (9.84)$$

where we have introduced the factor 2 for the two transverse polarizations. We insert here the solutions $u_{1,2}(\mathbf{k},\omega)$ given by equations (9.77) and take into account that $\delta^2(\omega) = (T/2\pi)\delta(\omega)$. In addition, we note that the negative frequencies give the same contribution as the positive ones, so we account for them by a pre-factor 2. We get

$$U = 4\pi nq^2\sum_{\mathbf{k}}\left[|u_1(\mathbf{k})|^2 + \frac{2\Omega_2^2}{\Omega_2^2 - c^2k^2}\left|u_2^{(1)}(\mathbf{k})\right|^2\right] + \qquad (9.85)$$
$$+8\pi nq^2\sum_{\mathbf{k}}\frac{\Omega_3^2}{\Omega_3^2 - c^2k^2}\left|u_2^{(2)}(\mathbf{k})\right|^2 \ .$$

The mean square fluctuations in equations (9.85) can be replaced by their zero-point values of the form $|u_1(\mathbf{k})|^2 = \hbar/2m\Omega_1$, etc; we get

$$U = \frac{2\pi\hbar nq^2}{m}\sum_{\mathbf{k}}\left[\frac{1}{\Omega_1} + \frac{2\Omega_2}{\Omega_2^2 - c^2k^2} + \frac{2\Omega_3}{\Omega_3^2 - c^2k^2}\right] \ . \qquad (9.86)$$

The summation (integration) over $\mathbf{k}$ in equation (9.86) is divergent. It is reasonable to introduce an ultraviolet cutoff $k_c \simeq 1/a$, where a is the mean separation distance between the mobile charges (density $n = 1/a^3$). Under these circumstances we may neglect ck in comparison with the relevant frequencies occurring in equation (9.86), and replace approximately the integrand by $1/\overline{\Omega} \simeq 1/\Omega_1 + 2/\omega_L + 2/\omega_T$. We get an estimate

$$U \simeq N\frac{q^2}{a}\frac{\hbar^2/ma^2}{\hbar\overline{\Omega}} \qquad (9.87)$$

for the energy. In equation (9.87) we can identify the Coulomb energy q^2/a per particle, the particle localization energy $\hbar^2/ma^2$ and the energy quanta $\hbar\overline{\Omega}$. We can see that matter stores a large amount of electromagnetic energy, arising from the zero-point motion of its polarization.

Similarly, we can estimate the Poynting vector

$$\mathbf{S} = \frac{c}{4\pi T}\int dtd\mathbf{r}\mathbf{E}\times\mathbf{H} =$$

$$= 16\pi nq^2c^2\sum_{\mathbf{k}}\{\frac{\Omega_2^3}{(\Omega_2^2-c^2k^2)^2}\left|u_2^{(1)}(\mathbf{k})\right|^2 + \tag{9.88}$$

$$+\frac{\Omega_3^3}{(\Omega_3^2-c^2k^2)^2}\left|u_2^{(2)}(\mathbf{k})\right|^2\}\mathbf{k}\ .$$

We can see that it is given only by the transverse components of the displacement field $\mathbf{u}$ and it is directed along the propagation wavevectors $\mathbf{k}$, as expected. The total kinetic energy of the mobile charges is given by

$$K = \frac{n}{T}\int dtd\mathbf{r}\frac{1}{2}m\dot{\mathbf{u}}^2 =$$

$$= 2\sum_{\mathbf{k}}\{\frac{1}{2}m\Omega_1^2\left|u_1(\mathbf{k})\right|^2 + m\Omega_2^2\left|u_2^{(1)}(\mathbf{k})\right|^2 + \tag{9.89}$$

$$+m\Omega_3^2\left|u_2^{(2)}(\mathbf{k})\right|^2\}\ .$$

It can be estimated as $K\simeq N\cdot\hbar\overline{\Omega}$, and we can see that it is comparable with the electromagnetic energy U. By comparing with equation (9.80), we can see that K includes also the potential energy (therefore it is the total mechanical energy) of the harmonic oscillators, by the pre-factor 2.

It is also well-known that we get the Lorentz force

$$f_i = \rho E_i + \frac{1}{c}(\mathbf{j}\times\mathbf{H})_i = \partial_j\sigma_{ij} - \frac{\partial}{\partial t}g_i \tag{9.90}$$

from the Maxwell equations (9.34), arising from the stress tensor

$$\sigma_{ij} = \frac{1}{4\pi}\left[E_iE_j + H_iH_j - \frac{1}{2}\delta_{ij}(E^2+H^2)\right] \tag{9.91}$$

and the electromagnetic momentum $\mathbf{g} = \mathbf{s}/c^2$, where $\rho = -nq div\mathbf{u}$ and $\mathbf{j} = nq\dot{\mathbf{u}}$ are the polarization charge and current densities. We give here the representation of the Lorentz force in terms of the displacement (polarization) field

$$\mathbf{f} = \frac{1}{T}\int dtd\mathbf{r}\left(\rho\mathbf{E} + \frac{1}{c}\mathbf{j}\times\mathbf{H}\right) =$$

$$= -8\pi i n q^2 \sum_{\mathbf{k}}\{|u_1(\mathbf{k})|^2 + \frac{2\Omega_2^2}{\Omega_2^2 - c^2k^2}\left|u_2^{(1)}(\mathbf{k})\right|^2 + \tag{9.92}$$

$$+\frac{2\Omega_3^2}{\Omega_3^2 - c^2k^2}\left|u_2^{(2)}(\mathbf{k})\right|^2\}\mathbf{k}\ .$$

Since, usually, u_1 and $u_2^{(1,2)}$ depend only on the magnitude of the wavevector (and not of its direction), the force given by equation (9.92) is zero. Similarly, we get the charge-field interaction energy

$$E_{int} = \frac{1}{T}\int dtd\mathbf{r}\left(\rho\Phi - \frac{1}{c}\mathbf{j}\mathbf{A}\right) =$$

$$= 8\pi n q^2 \sum_{\mathbf{k}}\{|u_1(\mathbf{k})|^2 + \frac{2\Omega_2^2}{\Omega_2^2 - c^2k^2}\left|u_2^{(1)}(\mathbf{k})\right|^2 + \tag{9.93}$$

$$+\frac{2\Omega_3^2}{\Omega_3^2 - c^2k^2}\left|u_2^{(2)}(\mathbf{k})\right|^2\}\ .$$

Writing $\mathbf{f} = \sum_{\mathbf{k}}\mathbf{f}(\mathbf{k})$ and $E_{int} = \sum_{\mathbf{k}}E_{int}(\mathbf{k})$, we can see that the Fourier components of the force are given by $\mathbf{f}(\mathbf{k}) = -i\mathbf{k}E_{int}(\mathbf{k})$, as expected.

9.9 Two electromagnetically coupled bodies

We consider two polarizable bodies, denoted by a and b, one (a) placed at the origin $\mathbf{r} = 0$ and another (b) placed at $\mathbf{r}_0$. The displacement field for the body a is $\mathbf{u}_a(\mathbf{r}, t) = \mathbf{u}(\mathbf{r}, t)$ and the displacement field for the body b is denoted by $\mathbf{u}_b(\mathbf{r}, t) = \mathbf{v}(\mathbf{r} - \mathbf{r}_0, t)$. We use Fourier transforms of the type given by equation (9.62), with the pre-factor $N^{-1/2} = (N_a + N_b)^{-1/2}$, where $N_{a,b}$ are the number of mobile charges in body a and, respectively, b, both enclosed in the same volume V. The densities are given by $n_{a,b} = N_{a,b}/V$. The Fourier transform of $\mathbf{u}_b(\mathbf{r}, t)$

acquires an exponential factor $e^{-i\mathbf{kr}_0}$, $\mathbf{u}_b(\mathbf{k},\omega)=\mathbf{v}(\mathbf{k},\omega)e^{-i\mathbf{kr}_0}$, which appears in the field $\mathbf{E}_b$ generated by the body b, according to equations (9.67). Consequently, we can write the equations of motion for $\mathbf{u}_a=\mathbf{u}$ as

$$\begin{gathered}(\omega^2-\omega_{ca}^2-\omega_{pa}^2)u_1=\omega_{pb}^2v_1e^{-i\mathbf{kr_0}}\ ,\\(\omega^2-\omega_{ca}^2-\omega_{pa}^2\lambda^2F)u_2=\omega_{pb}^2\lambda^2Fv_2e^{-i\mathbf{kr_0}}\ .\end{gathered}\tag{9.94}$$

We can see that the b-field corresponds to the propagating wavevector $-\mathbf{k}$ and the source placed at $\mathbf{r}_0$, *i.e.* it contains the factor $e^{i(-\mathbf{k})\mathbf{r}_0}=e^{-i\mathbf{kr_0}}$, in accordance with the retardation requirements (for positive frequencies). In order to fulfil the retardation requirements, we must limit ouselves only to fields outgoing from their sources. Similar equations of motion can be written for $\mathbf{u}_b$, where the field originates in the source a. In the b-frame this field corresponds to the propagating wavevector $\mathbf{k}$ and source placed at $-\mathbf{r}_0$, *i.e.* it must contains the factor $e^{i\mathbf{k}(-\mathbf{r}_0)}=e^{-i\mathbf{kr_0}}$. In addition we must restrict to $\mathbf{kr}_0>0$ for positive frequencies and $\mathbf{kr}_0<0$ for negative frequencies. We get

$$\begin{gathered}(\omega^2-\omega_{cb}^2-\omega_{pb}^2)v_1=\omega_{pa}^2u_1e^{-i\mathbf{kr_0}}\ ,\\(\omega^2-\omega_{cb}^2-\omega_{pb}^2\lambda^2F)v_2=\omega_{pa}^2\lambda^2Fu_2e^{-i\mathbf{kr_0}}\ .\end{gathered}\tag{9.95}$$

We can see that the polarizations of the two bodies are coupled, through their interaction, which depends on their mutual position $\mathbf{r}_0$. The two homogeneous systems of equations (9.94) and (9.95) have eigenfrequencies which depend on this interaction, given by the dispersion equations

$$\begin{gathered}\{(\omega^2-\omega_{ca}^2-\omega_{pa}^2)(\omega^2-\omega_{cb}^2-\omega_{pb}^2)-\\-\omega_{pa}^2\omega_{pb}^2e^{-2i\mathbf{kr_0}}\}u_1=0\ ,\\\{(\omega^2-\omega_{ca}^2-\omega_{pa}^2\lambda^2F)(\omega^2-\omega_{cb}^2-\omega_{pb}^2\lambda^2F)-\\-\omega_{pa}^2\omega_{pb}^2\lambda^4F^2e^{-2i\mathbf{kr_0}}\}u_2=0\ .\end{gathered}\tag{9.96}$$

They correspond to the equations

$$(4\pi)^2\alpha_{a1,2}\alpha_{b1,2}e^{-2i\mathbf{kr}_0}=1\tag{9.97}$$

for the polarizabilities of the two bodies.

Equations (9.96), or (9.97), require $2\mathbf{kr}_0 = \pi n$, where n is a positive integer, $n = 0, 1, 2....$. For realistic values of the parameters $\omega_{ca,b}$ and $\omega_{pa,b}$ we get real solutions of these equations for n an even integer, so we set $\mathbf{kr}_0 = \pi n$, $n = 0, 1, 2.....$ The interaction shifts the longitudinal modes $\omega_{La,b}$ of the two bodies, the new longitudinal frequencies being given by

$$\Omega^2_{L1,2} = \frac{1}{2}\left[\omega^2_{La} + \omega^2_{Lb} \pm \sqrt{(\omega^2_{La} - \omega^2_{Lb})^2 + 4\omega^2_{pa}\omega^2_{pb}}\right] \tag{9.98}$$

(we note that they do not depend on the wavevector $\mathbf{k}$); for conductors we have only one longitudinal frequency $\Omega_{L1} = \sqrt{\omega^2_{pa} + \omega^2_{pb}}$, but we keep both frequencies $\Omega_{L1,2}$ for the sake of the generality. The solution of the first equation (9.96) is given by

$$u_1(\mathbf{k}, \omega) = 2\pi\delta(\omega - \Omega_{L1})u_1^{(1)}(\mathbf{k}) + 2\pi\delta(\omega - \Omega_{L2})u_1^{(2)}(\mathbf{k}) \tag{9.99}$$

for $\mathbf{kr}_0 > 0$ and a similar contribution for negative frequencies and $\mathbf{kr}_0 < 0$. The coordinate v_1 is obtained directly from equations (9.96).

Similarly, from the second equation (9.96) we get four frequencies, in general, for the polaritonic transverse modes, denoted by Ω_{Ti}, $i = 1, 2, 3, 4$. In the long-wavelengh limit they behave like

$$\begin{gathered}\Omega^2_{T1,2} \simeq \tfrac{1}{2}\{\omega^2_{La} + \omega^2_{Lb}\pm \\ \pm\sqrt{(\omega^2_{La} + \omega^2_{Lb})^2 - 4\omega^2_{ca}\omega^2_{cb} - 4\omega^2_{ca}\omega^2_{pb} - 4\omega^2_{cb}\omega^2_{pa}}\}\end{gathered} \tag{9.100}$$

(plus a small contribution of the form $const \cdot k^2$) and

$$\Omega_{T3,4} \simeq vk \ , \ v = c\frac{\omega_{ca}\omega_{cb}}{\sqrt{\omega^2_{ca}\omega^2_{cb} + \omega^2_{ca}\omega^2_{pb} + \omega^2_{cb}\omega^2_{pa}}} \ ; \tag{9.101}$$

for $k \to \infty$,

$$\Omega^2_{T1,2} \simeq c^2k^2 + \omega^2_{pa} + \omega^2_{pb} \ , \ \Omega_{T3,4} \simeq \omega_{ca,b} \ . \tag{9.102}$$

For small couplings ($\omega_{pa,b} \ll \omega_{ca,b}$) these frequencies reduce approximately to the two transverse modes $\omega_{ca,b}$ and two photon mode

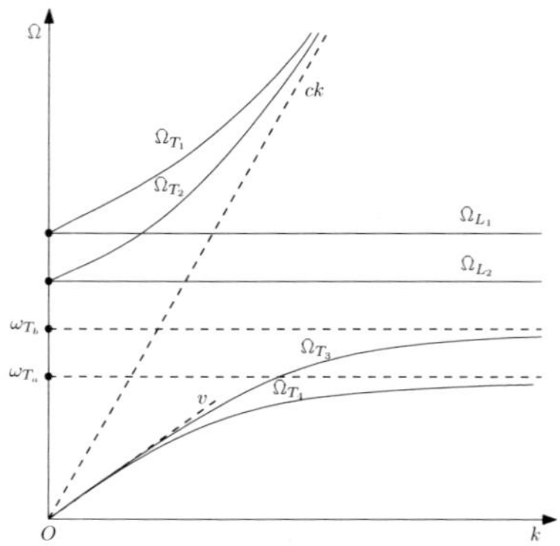

Figure 9.4: The longitudinal ($\Omega_{L1,2}$) and transverse ($\Omega_{T1,2,3,4}$) eigenmodes for two electromagnetically-coupled bodies a and b, as given by the roots of equations (9.96). The transverse modes $\omega_{Ta,b} = \omega_{ca,b}$ are also shown, together with the velocity v of the two modes $\Omega_{T3,4}$ (the slope of the curves $\Omega_{T3,4}$ at the origin). ck is the photon frequency.

$\omega \simeq ck$. For conductors ($\omega_{ca} = \omega_{cb} = 0$) we get only one polaritonic branch $\Omega_T = \sqrt{\omega_{pa}^2 + \omega_{pb}^2 + c^2k^2}$. The eigenfrequencies $\Omega_{L1,2}$ and $\Omega_{T1,\dots4}$ of two electromagnetically coupled bodies a and b are shown schematically in Fig. 9.4.

The solution of the second equation (9.96) can be written as

$$u_2(\mathbf{k},\omega) = 2\pi \sum_{i=1}^{4} \delta(\omega - \Omega_{Ti}) u_2^{(i)}(\mathbf{k}) \tag{9.103}$$

for $\mathbf{kr}_0 > 0$ and a corresponding decomposition for negative frequencies and $\mathbf{kr}_0 < 0$. The coordinate v_2 is obtained directly from the second equation (9.96) as a function of u_2.

It is worth computing the Lorentz force acting on one body on behalf of the other; for instance, the force by which body a acts upon body b. We follow the same procedure as that which led to equation (9.92) and use the coupled equations (9.94) and (9.95) in order to eliminate the coordinates $v_{1,2}$ in favour of the coordinates $u_{1,2}$. We get

$$\mathbf{f}_{ab} = \frac{1}{T}\int dt d\mathbf{r}\left(\rho_b \mathbf{E}_a + \frac{1}{c}\mathbf{j}_b \times \mathbf{H}_a\right) =$$

$$= -\frac{2in_a n_b q^2 V}{T\omega_{pb}^2 N}\sum_{\mathbf{k}}\int d\omega(\omega^2 - \omega_{ca}^2 - \omega_{pb}^2)\left|u_1(\mathbf{k},\omega)\right|^2 \mathbf{k} -$$

$$-\frac{4in_a n_b q^2 V}{T\omega_{pb}^2 N}\sum_{\mathbf{k}}\int d\omega(\omega^2 - \omega_{ca}^2 - \omega_{pa}\lambda^2 F)\left|u_2(\mathbf{k},\omega)\right|^2 \mathbf{k} = \tag{9.104}$$

$$= \frac{in_a n_b q^2 \omega_{pa}^2 V}{2\pi T\omega_{pb}^2 N}\sum_{\mathbf{k}}\int d\omega$$

$$\left[\frac{1}{\alpha_{a1}}\left|u_1(\mathbf{k},\omega)\right|^2 + \frac{2}{\alpha_{a2}}\left|u_2(\mathbf{k},\omega)\right|^2\right]\mathbf{k} \ ,$$

where $\alpha_{a1,2}$ are the polarizabilities of the body a, as given by equations (9.69). If we eliminate the coordinates $u_{1,2}$ in favour of the coordinates $v_{1,2}$ (by using equations (9.94) and (9.95), we get a similar expression for the force $\mathbf{f}_{ab}$, with a and b interchanged; this corresponds to the force $\mathbf{f}_{ba}$ and, since the sign does not change, we conclude that $\mathbf{f}_{ab} = \mathbf{f}_{ba} = 0$. Indeed, when introducing the decompositions given by equations (9.99) and (9.103), we note that for positive frequencies $\mathbf{kr}_0 > 0$, while $\mathbf{kr}_0 < 0$ for negative frequencies. Since both contributions are equal, we can write

$$\begin{gathered}\mathbf{f}_{ab} = -\frac{4\pi in_a n_b q^2 V}{\omega_{pb}^2 N}\sum_{\mathbf{k}}\sum_{i=1}^{2} \\ (\Omega_{Li}^2 - \omega_{La}^2)\left|u_1^{(i)}(\mathbf{k})\right|^2 \mathbf{k} - \\ -\frac{8\pi in_a n_b q^2 V}{\omega_{pb}^2 N}\sum_{\mathbf{k}}\sum_{i=1}^{4} \\ (\Omega_{Ti}^2 - \omega_{ca}^2 - \frac{\omega_{pa}^2\Omega_{Ti}^2}{\Omega_{Ti}^2 - c^2k^2})\left|u_2^{(i)}(\mathbf{k})\right|^2 \mathbf{k} \ ,\end{gathered} \tag{9.105}$$

where the summation extends over the whole $\mathbf{k}$-space. We can see, indeed, that $\mathbf{f}_{ab} = 0$, as long as the amplitudes $u_{1,2}^{(i)}$ do not depend on the direction of the wavevector $\mathbf{k}$.

9.10 Casimir force

Equations (9.94) and (9.95) describe two pairs of coupled harmonic oscillators. One pair, say $v_{1,2}$ is completely determined by the motion of the pair $u_{1,2}$; its energy is taken up in the modified frequencies given by the dispersion equations (9.96). Therefore, we have, in fact, only one pair of harmonic oscillators for the two coupled bodies a and b, of coordinates $u_{1,2}$ and eigenfrequencies Ω_{Li}, $i = 1, 2$ and Ω_{Ti}, $i = 1, 2, 3, 4$, governed by a harmonic-oscillator hamiltonian of the type given in equation (9.80). Its ground-state (zero-point) energy reads

$$E = \sum_{\mathbf{k}} \left(\sum_{i=1}^{2} \frac{1}{2} \hbar \Omega_{Li} + \sum_{i=1}^{4} \hbar \Omega_{Ti}(k) \right) , \qquad (9.106)$$

where we have introduced a factor 2 in order to account for the two transverse polarizations. We may choose to eliminate $u_{1,2}$ in favour of the coordinates $v_{1,2}$. The energy given by equation (9.106) remains unchanged. Both situations are equally valid, though they introduce an asymmetry, in the sense that in each case one pair of coordinates are completely determined by the other pair.

These frequencies are obtained from equations (9.96) by requiring $\mathbf{k}\mathbf{r}_0 = \pi n$. It is convenient to introduce the component κ of the wavevector $\mathbf{k}$ along the position vector $\mathbf{r}_0$, whose magnitude is denoted by d ($r_0 = d$). Therefore, we have $\kappa_n = \pi n/d$ and $k_n^2 = k_\perp^2 + \kappa_n^2$, where $\mathbf{k}_\perp$ is the transverse wavevector, *i.e.* the component of the wavevector $\mathbf{k}$ perpendicular to $\mathbf{r}_0$. The energy given by equation (9.106) can then be written as

$$E = \frac{S}{2\pi} \sum_{n=0} \int dk_\perp k_\perp \left(\sum_{i=1}^{2} \frac{1}{2} \hbar \Omega_{Li} + \sum_{i=1}^{4} \hbar \Omega_{Ti}(k_n) \right) , \qquad (9.107)$$

where S is the transverse area. We estimate the change brought about by the finite distance d in the energy E by using the Euler-Maclaurin formula:[12]

$$\Delta E = \sum_{m=1} \frac{(-1)^m B_m (\pi/d)^{2m-1}}{(2m)!} f^{(2m-1)}(0) , \qquad (9.108)$$

[12] E. T. Whittaker and G. N. Watson, *A Course of Modern Analysis*, Cambridge Mathematical Library, Cambridge (2004), p. 128.

where B_m are the Bernoulli's numbers and

$$f(\kappa) = \frac{S}{2\pi} \int dk_\perp k_\perp$$
$$\left(\sum_{i=1}^{2} \frac{1}{2}\hbar\Omega_{Li} + \sum_{i=1}^{4} \hbar\Omega_{Ti}(\sqrt{k_\perp^2 + \kappa^2})\right) . \tag{9.109}$$

It is easy to see that the longitudinal frequencies Ω_{Li} do not contribute to equation (9.108). Similarly, the transverse frequencies $\Omega_{T1,2}$, which, according to equation (9.100), go like $\sim\sqrt{const + \kappa^2}$ in the long-wavelength limit, do not contribute to equation (9.108), since all their odd-order derivatives are vanishing for $\kappa = 0$. We are left with the contributions arising from $\Omega_{T3,4}$, which behave like vk in the long-wavelength limit, according to equation (9.101). Equation (9.108) becomes

$$\Delta E = \frac{\hbar v S}{2\pi} \sum_{m=1} \frac{(-1)^m B_m (\pi/d)^{2m-1}}{(2m)!} \left(\int_{\kappa^2} du \sqrt{u} \right)_0^{(2m-1)} , \tag{9.110}$$

where we have introduced $u = k_\perp^2 + \kappa^2$ The only contribution to equation (9.110) comes from the third-order derivative. We get ($B_2 = 1/30$)

$$\Delta E = -\frac{\pi^2 \hbar v S}{360} \cdot \frac{1}{d^3} \tag{9.111}$$

and an attractive force

$$F = -\frac{\pi^2 \hbar v S}{120} \cdot \frac{1}{d^4} \tag{9.112}$$

acting between the two bodies. This is the Casimir force, arising between two polarizable bodies from the polarization vacuum fluctuations. It differs from the classical formula $F = -\pi^2 \hbar c S / 240 d^4$, derived for two conducting half-spaces,[13] by a factor 2, arising from the two branches $\Omega_{T3,4}$ (which behave identically in the long-wavelength

[13] H. B. G. Casimir and D. Polder, "The influence of retardation on the London-van der Waals forces," Phys. Rev. **73** 360 (1948); H. Casimir, "On the attraction between two perfectly conducting plates," Proc. Kon. Ned. Ak. Wet. **51** 793 (1948); L. Landau and E. Lifshitz, *Course of Theoretical Physics*, vol. 5 (*Statistical Physics*), part. 2, Butterworth-Heinemann, Oxford (2003).

limit), as well as by the presence of the renormalized velocity v instead of c. The nature of the bodies is incorporated in the polariton velocity

$$v = c\frac{\omega_{ca}\omega_{cb}}{\sqrt{\omega_{ca}^2\omega_{cb}^2 + \omega_{ca}^2\omega_{pb}^2 + \omega_{cb}^2\omega_{pa}^2}} \tag{9.113}$$

as given by equation (9.101). It is worth noting that this velocity v differs from the corresponding pre-factor in the formula given for two dielectrics. For identical bodies there remains only one vk-branch, with velocity v given by equation (9.113) for $\omega_{c,pa} = \omega_{c,pb}$. The formula (9.112) for the force should then be modified accordingly.

The effect of the temperature $T = 1/\beta$ can be incorporated in equation (9.110) by the change

$$\int_{\kappa^2} du\sqrt{u} \rightarrow \int_{\kappa^2} du\sqrt{u}\coth\left[\frac{1}{2}\beta\hbar v\sqrt{u}\right] \ . \tag{9.114}$$

For realistic values of the parameters we have $\beta\hbar v/d \gg 1$, so we get a temperature correction factor $\simeq \coth(\beta\hbar v/d)$ in the expression of the force. In the opposite limit (very high temperature, $\beta\hbar v/d \ll 1$) the force is vanishing.

The situation described above corresponds to two dielectrics ($\omega_{ca,b} \neq 0$). Let us assume that we have a conductor a ($\omega_{ca} = 0$) and a dielectric b ($\omega_{cb} \neq 0$). The two transverse modes $\Omega_{T3,4}$ disappear in this case, and the two longitudinal modes $\Omega_{L1,2}$ do not contribute to the energy. We are left with the two transverse modes $\Omega_{T1,2}$, which, in the long-wavelength limit go like

$$\Omega_{T1,2}^2 \simeq \Omega_{L1,2}^2 + v_{1,2}^2 k^2 \ , \tag{9.115}$$

where

$$v_{1,2}^2 = \frac{1}{2}c^2\left[1 \pm \frac{\omega_{pa}^2 + \omega_{pb}^2 - \omega_{cb}^2}{\sqrt{(\omega_{pa}^2 + \omega_{pb}^2 + \omega_{cb}^2)^2 - 4\omega_{pa}^2\omega_{cb}^2}}\right] \ . \tag{9.116}$$

It is worth noting that, while Ω_{T1} goes like $\sim ck$ for $k \rightarrow \infty$, the mode Ω_{T2} approaches ω_{cb} in the same limit. The frequencies $\Omega_{T1,2}$ given by equation (9.115) can be written as

$$\Omega_{T1,2} = v_{1,2}\sqrt{k_\perp^2 + \Omega_{L1,2}^2/v_{1,2}^2 + \kappa^2} \tag{9.117}$$

and we can see that $k_\perp$ may acquire imaginary values, such that $k_\perp^2 + \Omega_{L1,2}^2/v_{1,2}^2 = k_\perp^{'2}$ behaves as the square of a new transverse wavevector $k_\perp^{'}$ ranging from 0 to $\Omega_{L1,2}/v_{1,2}$ (a similar situation for two dielectrics would imply imaginary frequencies $\Omega_{3,4}$ which are not physically acceptable). The transverse **k**_-waves are damped waves in this case. It is easy to see that the Casimir force will have the same expression as given by equation (9.112) above, with the proper replacement of the velocity v by $(v_1 + v_2)/2$. The same situation holds for two conductors, where we have only one polaritonic branch $\Omega_{T1} = \sqrt{\omega_{pa}^2 + \omega_{pb}^2 + c^2k^2}$. In this case, we get the classical result $F = -\pi^2\hbar cS/240d^4$ (corresponding to two half-spaces).

9.11 van der Waals-London force

In the non-retarded regime $\lambda = 0$, which corresponds to $k = 0$ for free waves. It follows that $\kappa = -ik_\perp$ acquires purely imaginary values (the waves are damped along the $\mathbf{r}_0$). The second equations (9.96) give the transverse modes $\omega_{ca,b}$, which do not depend on $k_\perp$ and, consequently do not contribute to the change in energy. We are left with the longitudinal modes, given by the first equation (9.96) which reads now

$$(\omega^2 - \omega_{La}^2)(\omega^2 - \omega_{Lb}^2) - \omega_{pa}^2\omega_{pb}^2 e^{-2k_\perp d} = 0 \ . \tag{9.118}$$

For realistic situations we may take $\omega_{La} \simeq \omega_{Lb}$ and neglect $(\omega_{La}^2 - \omega_{Lb}^2)^2$ in comparison with $\omega_{pa}^2\omega_{pb}^2$. Further, we may expand the solutions of equation (9.118) in powers of $\omega_{pa}\omega_{pb}e^{-k_\perp d}$, and subtract the energy corresponding to $d \to \infty$. We get the approximate change in the ground-state (zero-point) energy

$$\Delta E \simeq -\frac{\hbar S}{4\pi\sqrt{2}} \frac{\omega_{pa}^2\omega_{pb}^2}{(\omega_{La}^2 + \omega_{Lb}^2)^{3/2}} \int_0 dk_\perp k_\perp e^{-2k_\perp d} \tag{9.119}$$

and the van der Waals-London force[14]

$$F \simeq -\frac{\hbar S}{8\pi\sqrt{2}} \frac{\omega_{pa}^2\omega_{pb}^2}{(\omega_{La}^2 + \omega_{Lb}^2)^{3/2}} \cdot \frac{1}{d^3} \ . \tag{9.120}$$

[14] R. Eisenschitz and F. London, "Ueber das Verhaeltnis der van der Waalsschen Krafte zu den homoopolaren Bindungskraften", Z. Physik **60** 491 (1930); F.

We note that the pre-factor in equation (9.120) differs from a known result.[15].

Comparing equations (9.112) and (9.120), we get a crossover distance of the order $d_1 \simeq c/\omega_p$, which separates the ranges of the Casimir and van der Waals-London forces (ω_p being a frequency of the order of the plasma frequencies of the two bodies).

It is worth noting the solutions of equation (9.118) for two identical conductors: $\Omega^2 = \omega_p^2(1 \pm e^{-kd})$. They differ from the well-known solutions $\Omega^2 = \omega_p^2(1 \pm e^{-kd})/2$ corresponding to two half-spaces separated by distance d, which exhibit the surface plasmons (for instance, for $d \to \infty$ we get the well-known surface plasmon $\omega_p/\sqrt{2}$). This is a particular example that our approach does not account for surface effects, as we discussed before. The surface effects may bring significant changes both in numerical coefficients, as well as in the d-dependence of the forces, especially through the depolarizing fields the surfaces generate, with contributions that may differ for identical, or distinct bodies. Such effects are more significant for the non-retarded regime.

In order to make this situation more clear, we resort here to another well-known decomposition of the spherical wave. Indeed, we preserve the coordinate z along the direction between the two bodies, and use

London, "Zur Theorie und Systematik der Molekularkrafte", Z. Physik **63** 245 (1930); "The general theory of molecular forces", Trans. Faraday Soc. **33** 8 (1937); E. Lifshitz, ZhETF **29** 94 (1956) (Sov. Phys. JETP **2** 73 (1956)); I. E. Dzyaloshinskii, E. M. Lifshitz and L. P. Pitaevskii, "The general theory of van der Waals forces," Adv. Phys. **10** 165 (1961); J. Schwinger, "Casimir effect in source theory", Lett. Math. Phys. **1** 43 (1975); J. Schwinger, L. L. deRaad, Jr., and K. A. Milton, "Casimir effect in dielectrics", Ann. Phys. **115** 1 (1978); N. G. Van Kampen, B. R. A. Nijboer and K. Schram, "On the macroscopic theory of van der Waals forces," Phys. Lett. **A26** 307 (1968); T. H. Boyer, "Quantum electromagnetic zero-point energy of a conducting spherical shell and the Casimir model for a charged particle", Phys. Rev. **174** 1764 (1968); "Recalculations of long-range van der Waals potentials", Phys. Rev. **180** 19 (1969); P. W. Milloni, *The Quantum Vacuum*, Academic Press, San Diego (1994).

[15]L. Landau and E. Lifshitz, *Course of Theoretical Physics*, vol. 5 (*Statistical Physics*), part. 2, Butterworth-Heinemann, Oxford (2003).

only transverse Fourier transforms. The decomposition reads[16]

$$\frac{e^{i\lambda R}}{R}=\frac{i}{2\pi}\int d\mathbf{k}\frac{1}{\kappa}e^{i\mathbf{kr}}e^{i\kappa|z|} \ . \tag{9.121}$$

If we further take the Fourier transform with respect to the z-coordinate we get the function $F(k,\lambda)$ used here. Now, the derivatives of the function $e^{i\kappa|z|}$ are given by

$$\frac{\partial}{\partial z}e^{i\kappa|z|}=i\kappa sgn(z)e^{i\kappa|z|} \ , \ \frac{\partial^2}{\partial z^2}e^{i\kappa|z|}=2i\kappa\delta(z)-\kappa^2 e^{i\kappa|z|} \ . \tag{9.122}$$

We can see easily that, while the Fourier transform reproduces the first derivative of the function $e^{i\kappa|z|}$, it does not reproduce the second derivative (as a consequence of the Gibbs phenomenon in the Fourier transform of the function $sgn(z)$). Such second-derivative-terms are responsible for the surface charges.

9.12 Electromagnetic field energy

It is worth examining the electromagnetic field energy, as given by equation (9.83), for the total field generated by the two bodies. The fields are given by $\mathbf{E}=\mathbf{E}_a+\mathbf{E}_b$ and $\mathbf{H}=\mathbf{H}_a+\mathbf{H}_b$, where $\mathbf{E}_{a,b}$ and $\mathbf{H}_{a,b}$ are given by equation (9.66) and (9.67). We get straightforwardly

$$U=\frac{nq^2}{T}\sum_{\mathbf{k}}\int d\omega\left(|w_1|^2+2\lambda^2 F|w_2|^2\right) \ , \tag{9.123}$$

where $\mathbf{w}=(n_a/n)\mathbf{u}+(n_b/n)\mathbf{v}$. We choose first to eliminate the coordinates $v_{1,2}$ in favour of the coordinates $u_{1,2}$, by using equations (9.94). Introducing the zero-point mean values for $|u_{1,2}|^2$, we get

$$\begin{gathered}U=\frac{\pi\hbar n_a^2q^2}{nm}\sum_{\mathbf{k}}\sum_{i=1}^2\left(\frac{\Omega_{Li}^2-\omega_{ca}^2}{\omega_{pa}^2}\right)^2\frac{1}{\Omega_{Li}}+\\+\frac{2\pi\hbar n_a^2q^2}{nm}\sum_{\mathbf{k}}\sum_{i=1}^4\left(\frac{\Omega_{Ti}^2-\omega_{ca}^2}{\omega_{pa}^2}\right)^2\frac{\Omega_{Ti}^2-c^2k^2}{\Omega_{Ti}^3} \ .\end{gathered} \tag{9.124}$$

[16] M. Apostol and G. Vaman, "Electromagnetic field interacting with a semi-infinite plasma", J. Opt. Soc. Am. **A26** 1747 (2009); I. S. Gradshteyn and I. M. Ryzhik, *Table of Integrals, Series and Products*, eds. A. Jeffrey and D. Zwillinger, 6th edition, Academic Press, San Diego (2000), pp. 714-715, 6.677.1,2.

If we eliminate $u_{1,2}$ in favour of the coordinates $v_{1,2}$, and use the zero-point mean square fluctuations for $|v_{1,2}|^2$, we get a similar equation for U, with a and b interchanged. Since both situations are equally valid we take the mean value, corresponding to a symmetrized contribution arising from the two bodies.

Let us assume that we have two dielectrics. It is easy to see that the change in energy brought by the finite distance d comes from the two branches $\Omega_{T3,4} = vk$ of "renormalized" photons. The corresponding energy can be written as

$$U_{3,4} \simeq -\frac{2\pi\hbar q^2}{nm}\left[\left(\frac{n_a\omega_{ca}^2}{\omega_{pa}^2}\right)^2 + \left(\frac{n_b\omega_{cb}^2}{\omega_{pb}^2}\right)^2\right] \cdot \frac{c^2-v^2}{v^3}\sum_{n=0}\sum_{\mathbf{k}_\perp}\frac{1}{\sqrt{k_\perp^2+\kappa_n^2}} . \tag{9.125}$$

We apply the usual procedure of extracting the finite change in energy given by the Euler-Maclaurin formula (9.108). The contribution comes from the first-order derivative in equation (9.108). We get

$$\Delta U = -\frac{\pi\hbar q^2 S}{12nm}\left[\left(\frac{n_a\omega_{ca}^2}{\omega_{pa}^2}\right)^2 + \left(\frac{n_b\omega_{cb}^2}{\omega_{pb}^2}\right)^2\right]\frac{c^2-v^2}{v^3}\cdot\frac{1}{d} . \tag{9.126}$$

We can see that the electromagnetic field generates an attractive force ($v < c$) which goes like $-1/d^2$; it acts at long distance, beyond $d_2 \simeq a\sqrt{\frac{mc^2}{q^2/a}}$, where mc^2 is the rest energy of the mobile charges and q^2/a is their Coulomb energy (a being the mean separation distance between the charges). It is easy to see that $d_2 = c\sqrt{a^3/(q^2/m)} \simeq c/\omega_p \simeq d_1$, *i.e.* the $-1/d^4$-Casimir force and the $-1/d^2$-electromagnetic field force are competitive for moderate distances. According to equation (9.114), the effect of the temperature on this d^2-force could be more visible, as the force acts at a longer distance.

If body a is a conductor ($\omega_{ca} = 0$), then its contribution to the change in the electromagnetic field energy goes like $-1/d^3$ (as for a Casimir force). The ratio of this force to the Casimir force is of the order of $(\hbar^2/ma^2)(q^2/a)/(\hbar\omega_p)^2$. The contribution of the dielectric ($\omega_{cb} \neq 0$) to the energy goes like $-1/d$; it is given by the corresponding term (ω_{cb}) in equation (9.126). Both contributions arise from the damped $\Omega_{T1,2}$-modes. Finally, for two conductors, the electromagnetic field energy contributes a Casimir force, given by $F=-\hbar cS/480\pi d^4$. It is much smaller (by factor $1/2\pi^3$) than the Casimir force $F=-\pi^2\hbar cS/240d^4$ arising from polarization.

9.13 Pointlike particles

We specialize the above calculations to pointlike, δ-particles, since such localized particles exhibit certain important particularities. Let us consider a displacement field given by

$$\mathbf{u}(\mathbf{r},t)=\mathbf{u}(t)a^3\delta(\mathbf{r}) \tag{9.127}$$

for a δ-particle of volume a^3 placed at $\mathbf{r}=0$. According to equation (9.62), its Fourier transform is

$$\mathbf{u}(\mathbf{k},\omega)=\frac{a^3\sqrt{N}}{V}\mathbf{u}(\omega)\ , \tag{9.128}$$

where N is the number of mobile charges in the particle (density $n=N/a^3$). The charge and current densities are given by

$$\rho=-i\frac{nqa^3\sqrt{N}}{V}\mathbf{ku}\ ,\ \boldsymbol{j}=-i\omega\frac{nqa^3\sqrt{N}}{V}\mathbf{u}\ . \tag{9.129}$$

We get the electric and magnetic fields

$$\begin{gathered}\mathbf{E}=-4\pi F\tfrac{nqa^3\sqrt{N}}{V}\left[\lambda^2\mathbf{u}-\mathbf{k}(\mathbf{ku})\right]\ ,\\ \mathbf{H}=-4\pi\lambda F\tfrac{nqa^3\sqrt{N}}{V}\mathbf{k}\times\mathbf{u}\end{gathered} \tag{9.130}$$

and the components

$$E_1=-4\pi\frac{nqa^3\sqrt{N}}{V}u_1\ ,\ E_2=-4\pi\lambda^2F\frac{nqa^3\sqrt{N}}{V}u_2\ . \tag{9.131}$$

The internal, polarization field $\mathbf{E}$ in the equation of motion (9.59) does not appear anymore in this case, in view of the lack of spatial extension of the δ-particle. We can use the external field $\mathbf{E}_0$ at the location $\mathbf{r}=0$ of the particle, and the equation of motion reads (with $\gamma=0$)

$$(\omega^2-\omega_c^2)\mathbf{u}(\omega)=-\frac{q}{m}\mathbf{E}_0(\mathbf{r}=0,\omega)\ . \tag{9.132}$$

We can see that there is only one polarizability, which coincides with the electric susceptibility

$$\alpha(\omega)=\chi(\omega)=-\frac{1}{4\pi}\frac{\omega_p^2}{\omega^2-\omega_c^2}\ . \tag{9.133}$$

In the absence of an external field equation (9.132) has harmonic-oscillator solutions

$$\mathbf{u}(\omega)=2\pi\delta(\omega-\omega_c)\mathbf{u}\ , \tag{9.134}$$

where $\mathbf{u}$ are now constant amplitudes, corresponding to $\mathbf{u}(t)=\mathbf{u}e^{-i\omega_c t}$. Their dynamics is governed by the hamiltonian

$$H=\frac{1}{2m}\mathbf{p}^2+\frac{1}{2}m\omega_c^2\mathbf{u}^2\ , \tag{9.135}$$

where $\mathbf{p}$ is the momentum. The zero-point mean square fluctuations are $\hbar/2m\omega_c$ for each component.

Following the calculations which led to equation (9.85), the energy of the electromagnetic field can be written as

$$U=4\pi\frac{(nqa^3)^2}{V}\sum_{\mathbf{k}}\left(|u_1|^2+\frac{2\omega_c^2}{\omega_c^2-c^2k^2}|u_2|^2\right)\ , \tag{9.136}$$

where, for zero-point fluctuations $|u_{1,2}|^2$ are replaced by $\hbar/2m\omega_c$. Similarly, we can have a representation in terms of the $\mathbf{u}$-coordinates for the Poynting vector, the "self-Lorentz force" and the charge-field "self-interaction", though, the latter two quantities are in fact meaningless, since the field acts only outside the particle.

We consider another pointlike particle of volume b^3 placed at $\mathbf{r}_0$, with the displacement field $\mathbf{v}(t)b^3\delta(\mathbf{r}-\mathbf{r}_0)$, and follow the treatment given in the preceding sections. The coupled equations (9.94) and (9.95) become

$$\begin{aligned}(\omega^2-\omega_{ca}^2)a^3u_1&=\omega_{pb}^2b^3v_1e^{-i\mathbf{k}\mathbf{r}_0}\ ,\\(\omega^2-\omega_{ca}^2)a^3u_2&=\omega_{pb}^2b^3\lambda^2Fv_2e^{-i\mathbf{k}\mathbf{r}_0}\end{aligned} \tag{9.137}$$

and

$$(\omega^2 - \omega_{cb}^2)b^3 v_1 = \omega_{pa}^2 a^3 u_1 e^{-i\mathbf{k}\mathbf{r}_0} \ ,$$
$$(\omega^2 - \omega_{cb}^2)b^3 v_2 = \omega_{pa}^2 a^3 \lambda^2 F u_2 e^{-i\mathbf{k}\mathbf{r}_0} \ . \tag{9.138}$$

From the dispersion equations we get two longitudinal modes

$$\Omega_{L1,2}^2 = \frac{1}{2}\left[\omega_{ca}^2 + \omega_{cb}^2 \pm \sqrt{(\omega_{ca}^2 - \omega_{ct}^2)^2 + 4\omega_{pa}^2\omega_{pb}^2}\right] \tag{9.139}$$

corresponding to the oscillator coordinates

$$u_1(\omega) = 2\pi\delta(\omega - \Omega_{L1})u_1^{(1)} + 2\pi\delta(\omega - \Omega_{L2})u_1^{(2)} \tag{9.140}$$

and, in general, four branches of transverse modes Ω_{Ti}, $i = 1, ...4$. In the long-wavelength limit they go like

$$\Omega_{T1,2}^2 \simeq \Omega_{L1,2}^2 + const \cdot k^2 \ , \ \Omega_{T3,4} \simeq vk \ , \tag{9.141}$$

where the velocity v is given by

$$v = c\sqrt{\frac{\omega_{ca}\omega_{cb}}{\omega_{ca}\omega_{cb} + \omega_{pa}\omega_{pb}}} \ . \tag{9.142}$$

We can see that all these transverse modes depend on the magnitude of the wavevector **k**, while our model, given by equation (9.127), for δ-particles does not allow for a **k**-dependence of the displacement field. Therefore, all these "transverse" solutions of the dispersion equations are not acceptable, and we are left with the longitudinal modes only. We conclude that there is not a Casimir force (nor an electromagnetic field force) for two δ-particles (the longitudinal modes do not give such forces). This is an expected result, because the transverse modes are spatially-dispersive and the δ-particles, in view of their lack of spatial extension, cannot accomodate them. The same conclusion holds for the interaction of a δ-particle and a spatially-extended body.

Similarly, in the non-retarded regime, the dispersion equation (9.118) (with $\omega_{La,b} \to \omega_{ca,b}$) gives damped modes which cannot be sustained by the δ-localized particles. Therefore, we conclude that our approach gives not a van der Waals-London force for two δ-particles.

The δ-particles offer the opportunity of a direct-space approach. Indeed, the displacement field given by equation (9.127) gives a current density $\mathbf{j}(\mathbf{r},\omega) = -i\omega nqa^3\mathbf{u}(\omega)\delta(\mathbf{r})$ (and a charge density $\rho(\mathbf{r},\omega) = -nqa^3[\mathbf{u}(\omega)grad]\delta(\mathbf{r}))$. By equations (9.65) we get immediately the vector potential

$$\mathbf{A} = -i\lambda nqa^3\mathbf{u}f(r) \ , \tag{9.143}$$

where $f(r) = e^{i\lambda r}/r$ is the (outgoing) spherical wave. From the Lorenz gauge we get the scalar potential

$$\Phi = -nqa^3\frac{\mathbf{ur}}{r}f^{'}(r) \ , \tag{9.144}$$

so we have straightforwardly the electric field

$$\mathbf{E} = nqa^3\left[\lambda^2 f\mathbf{u} + \frac{f^{'}}{r}\mathbf{u} - \frac{(\mathbf{ru})f^{'}}{r^3}\mathbf{r} + \frac{(\mathbf{ru})f^{''}}{r^2}\mathbf{r}\right] \ . \tag{9.145}$$

We estimate this field, produced by particle a, at the location $\mathbf{r} = \mathbf{r}_0 = (0,0,d)$ of the particle b. It is convenient to use the longitudinal projection $\parallel$ on the z-direction and the transverse projection $\perp$ on the $(x,\ y)$-plane. We get

$$E_{a\parallel} = n_a qa^3(\lambda^2 f + f^{''})u_\parallel \ , \ \mathbf{E}_{a\perp} = n_a qa^3(\lambda^2 f + \frac{1}{d}f^{'})\mathbf{u}_\perp \ , \tag{9.146}$$

where $f = e^{i\lambda d}/d$. The equations of motion for the displacement field $\mathbf{v}(t)b^3\delta(\mathbf{r}-\mathbf{r}_0)$ of the particle b read

$$\begin{aligned} (\omega^2 - \omega_{cb}^2)v_\parallel &= -\tfrac{\omega_{pa}^2}{4\pi}a^3(\lambda^2 f + f^{''})u_\parallel \ , \\ (\omega^2 - \omega_{cb}^2)v_\perp &= -\tfrac{\omega_{pa}^2}{4\pi}a^3(\lambda^2 f + \tfrac{1}{d}f^{'})u_\perp \ . \end{aligned} \tag{9.147}$$

Similarly, we compute the field produced by particle b at the location $\mathbf{r} = 0$ of the particle a (in equation (9.145) $\mathbf{r}$ is replaced by $\mathbf{r} - \mathbf{r}_0$), and get the equations of motion

$$\begin{aligned} (\omega^2 - \omega_{ca}^2)u_\parallel &= -\tfrac{\omega_{pb}^2}{4\pi}b^3(\lambda^2 f + f^{''})v_\parallel \ , \\ (\omega^2 - \omega_{ca}^2)u_\perp &= -\tfrac{\omega_{pb}^2}{4\pi}b^3(\lambda^2 f + \tfrac{1}{d}f^{'})v_\perp \ . \end{aligned} \tag{9.148}$$

From equations (9.147) and (9.148) we get the dispersion equations

$$
\begin{aligned}
(\omega^2-\omega_{ca}^2)(\omega^2-\omega_{cb}^2) &= \tfrac{\omega_{pa}^2\omega_{pb}^2}{(4\pi)^2}a^3b^3(\lambda^2 f+f'')^2 \ , \\
(\omega^2-\omega_{ca}^2)(\omega^2-\omega_{cb}^2) &= \tfrac{\omega_{pa}^2\omega_{pb}^2}{(4\pi)^2}a^3b^3(\lambda^2 f+\tfrac{1}{d}f')^2 \ .
\end{aligned}
\tag{9.149}
$$

It is easy to see that the two equations (9.149) are not compatible with one another, so we set either $u_\parallel=v_\parallel=0$, or $u_\perp=v_\perp=0$. Let us take $u_\perp=v_\perp=0$, so we have the first equation (9.149), which can be cast in the form

$$
(\omega^2-\omega_{ca}^2)(\omega^2-\omega_{cb}^2)=4\frac{\omega_{pa}^2\omega_{pb}^2}{(4\pi)^2}a^3b^3\frac{e^{2i(\lambda d+\varphi)}}{d^6} \ , \tag{9.150}
$$

where $\tan\varphi=\lambda d$. This equation implies $\lambda d+\varphi=\pi n$, where n is any integer, and

$$
(\omega^2-\omega_{ca}^2)(\omega^2-\omega_{cb}^2)=4\frac{\omega_{pa}^2\omega_{pb}^2}{(4\pi)^2}\frac{a^3b^3}{d^6} \ . \tag{9.151}
$$

It is easy to see that these two conditions cannot be satisfied simultaneously. The same result holds for the other dispersion equation, corresponding to $u_\parallel=v_\parallel=0$. We conclude that the dispersion equations have not solutions, and, therefore, the energy is not changed by the presence of the finite distance d, *i.e.* there is not a Casimir force.

In the non-retarded regime $\lambda=0$ and the dispersion equations (9.149) become

$$
(\omega^2-\omega_{ca}^2)(\omega^2-\omega_{cb}^2)=\frac{4\omega_{pa}^2\omega_{pb}^2}{(4\pi)^2}\frac{a^3b^3}{d^6} \tag{9.152}
$$

for the longitudinal components and

$$
(\omega^2-\omega_{ca}^2)(\omega^2-\omega_{cb}^2)=\frac{\omega_{pa}^2\omega_{pb}^2}{(4\pi)^2}\frac{a^3b^3}{d^6} \tag{9.153}
$$

for the transverse components. The two equations (9.152) and (9.153) differ by a factor 4. For realistic situations we may take $\omega_{ca}\simeq\omega_{cb}=\omega_c$, as for identical particles; we take also $\omega_{pa}=\omega_{pb}=\omega_p$. We may consider the rhs term of these equations as a small perturbation. The first equation (9.152) gives

$$\omega = \omega_c \left[1 \pm \frac{\omega_p^2}{4\pi\omega_c^2} \left(\frac{a^3 b^3}{d^6} \right)^{1/2} - \frac{\omega_p^4}{32\pi\omega_c^4} \frac{a^3 b^3}{d^6} + ... \right] \tag{9.154}$$

and a change in the zero-point energy

$$\Delta E = -\frac{\hbar\omega_p^4}{32\pi^2\omega_c^3} \cdot \frac{a^3 b^3}{d^6} \,. \tag{9.155}$$

This energy gives a $-1/d^7$-force, which is the classical van der Waals-London force acting between two pointlike particles. A similar result is obtained from the second dispersion equation, corresponding to the transverse modes. Statistically, the latter force appears with a weight factor 2 in comparison to the former, due to the two transverse polarizations of the displacement field. It is easy to see that the average energy contains a factor 1/2 which multiplies equation (9.155).

The difference between the result obtained here regarding the van der Waals-London force for δ-particles and the (null) result obtained by using the Fourier-transform approach is an extreme instance of the inadequacy of the Fourier-transform approach to problems concerning the matter polarization and the interaction of electromagnetically-coupled bodies. In general, the Fourier-transform approach to such problems is valid for bodies of a sufficient spatial extension, such that the surface effects can be neglected in comparison with the bulk contribution. However, for finite-size bodies, especially in the long-wavelength limit, this requirement is never achieved, due to the long-range character of the Coulomb interaction. Therefore, we expect deviations from our results presented here, whenever finite-size bodies are involved.

In conclusion, the polarization of matter can be represented by a displacement field of the mobile charges, within the well-known Lorentz-Drude model. By Maxwell's equations, the electromagnetic field generated by the polarization acquires a corresponding mechanical representation in terms of the displacement field. The collective motion of the polarization (reflected in the polarization electromagnetic field) is an example of emergent dynamics. In general, it consists of three infinite sets of harmonic oscillators, corresponding to longitudinal (plasmons, one set) and transverse (polaritons, two sets) degrees of freedom, which can be quantized. The electromagnetic field energy, Poynting vector, Lorentz force and charge-field interaction energy are represented in terms of these three types of eigenmodes. The polarizable matter stores a large amount of electromagnetic energy, arising from the zero-point (vacuum) fluctuations of the displacement (polarization) field.

Two electromagnetically-coupled bodies are treated within this framework, and their polarization eigenmodes are identified. The Casimir and van der Waals-London forces are derived, as arising from the zero-point fluctuations of the displacement (polarization) field. These forces incorporate the nature of the two bodies (by their individual longitudinal and transverse frequencies). The electromagnetic field (generated by the polarization fluctuations) brings its own contribution to the forces acting between two polarizable bodies. If a dielectric, at least, is present, the electromagnetic-field force goes like $\sim -1/d^2$, where d is the separation distance between the two bodies (it is a long-range force). If a conductor, at least, is present, the electromagnetic field adds its own contribution to the Casimir force ($\sim -1/d^4$). The magnitude and their range of action have been estimated here for all these forces.

The Fourier decomposition employed throughout as a general tool does not allow for including surface (or shape) effects. In general, the Fourier decomposition can be replaced by expansions in eigenmodes of the laplacian for specific, finite-size bodies (with outgoing-waves boundary condition at infinity), preserving the direct-space coordinate perpendicular to the surface. Such expansions may change, in general, the density of modes, which enters summations over states. In the long-wavelength limit, which gives usually the relevant contributions, this change is not significant; but the sharpness of the surfaces may bring important effects, through the depolarizing fields. Such effects are more evident for the van der Waals–London forces. The particular case of pointlike, δ-particles was analyzed in this context, as a limiting case of inadequacy of the Fourier-decomposition approach.

9.14 Coupled nanoplasmons

Enhanced and locally confined optical fields associated with nanoparticles, nanostructures and nanoaggregates enjoy an increasing interest in nanoscale manipulation of active devices, nanoantennas, surface enhanced Raman spectroscopy, light trapping, etc. The phenomenon is obviously related with dipolar fields, resonances and, as it was shown recently, with hybridization, charge building in narrow nano-gaps and quantum tunneling within nanodimers. All this dynamics is governed by coupled nano-plasmons.[17]

Plasmons is an old and fundamental concept in condensed matter physics: they are long-wavelength longitudinal oscillations of the charge density in matter. In a simple model, which is usually called the Drude-Lorentz model, matter can be represented as a plasma, consisting of identical mobile charges q of mass m and concentration (density) n (*e.g.*, electrons) moving uniformly and collectively against a quasi-rigid background of neutralizing charges $-q$ (*e.g.*, ions). The practical realization of long wavelength limit implies finite-size polarizable bodies, which entail, in turn, boundary conditions. Consequently, we may have many branches of plasmons: for instance, in a homogeneous conducting sphere the plasmon spectrum is given by $\Omega_l = \omega_p\sqrt{l/(2l+1)}$, where $\omega_p = \sqrt{4\pi nq^2/m}$ is called the plasma frequency and $l = 1, 2...$ is the azymuthal quantum number.[18] A much more convenient representation simplifies the things for point particles, which may be a reasonably useful model for the nowadays nano-plasmonics.

As it is well known, the Maxwell equations in matter imply four unknowns: **E** (electric field), **D** (electric displacement), **H** (magnetic field) and **B** (magnetic induction); and only two independent equations (Faraday and Maxwell-Ampere equations, which contain the *curl* and the time derivatives). In order to solve them, we introduce constitutive relations between these unknowns through the semi-phenomenological and quasi-empirical dielectric function ε and magnetic permeability μ. A large class of matter is quasi-non-magnetic,

---

[17] M. Apostol, S. Ilie, A. Petrut, M. Savu and S. Toba, "Coupled nano-plasmons", Appl. Phys. **A115** 387 (2014).

[18] M. Apostol and G. Vaman, "Plasmons and diffraction of an electromagnetic plane wave by a metallic sphere", Progr. Electrom. Res. PIER **98** 97 (2009).

such that we may equal $\mathbf{H}$ and $\mathbf{B}$ and put $\mu = 1$; still, we have three unknowns ($\mathbf{E}$, $\mathbf{D}$ and $\mathbf{H}$) and two equations.

On the other hand, the motion of the mobile charges in polarizable matter can be described by a displacement field $\mathbf{u}(t, \mathbf{r})$, which is a function of the time t and position $\mathbf{r}$. In the classical limit of small and slow variations (corresponding to classical electromagnetism), this displacement field generates a polarization charge density $\rho = -nqdiv\mathbf{u}$ and a corresponding current density $\mathbf{j} = nq\dot{\mathbf{u}}$. These charge and current densities generate in matter an electric field $\mathbf{E}$ and a magnetic field $\mathbf{H}$; but we still have two independent equations and three unknowns: $\mathbf{E}$, $\mathbf{H}$ and $\mathbf{u}$. However, the displacement field obeys an equation of motion, which, in this classical limit, is the Newton equation of motion

$$m\ddot{\mathbf{u}} = q(\mathbf{E} + \mathbf{E}_0) - m\omega_c^2\mathbf{u} - m\gamma\dot{\mathbf{u}} \; ; \qquad (9.156)$$

$\mathbf{E}$ is the internal (polarization) electric field, $\mathbf{E}_0$ is an external electric field, ω_c is a characteristic frequency and γ is a damping coefficient (much smaller than any relevant frequency). The magnetic part of the Lorentz force is absent in equation (9.156) because the velocities of the charges in matter are much smaller than the speed of light; the internal magnetic field is also absent, in accordance with our assumption of small $\mathbf{u}$ and non-magnetic matter. Equation (9.156) is the missing equation (the third equation), which helps solving the Maxwell equations.

Obviously, $\mathbf{P} = nq\mathbf{u}$ is the polarization (density of dipole moments); equation (9.156) leads immediately to the well-known Drude-Lorentz (plasma) dielectric function $\varepsilon(\omega) = (\omega^2 - \omega_c^2 - \omega_p^2)/(\omega^2 - \omega_c^2 + i\omega\gamma)$, where only the optical dispersion is included (through the dependence on the frequency ω). As it is well known, $\omega_c = 0$ corresponds to conductors, while $\omega_c \neq 0$ describes dielectrics. The model and equation (9.156) can be generalized in multiple ways. We limit ourselves here to use equation (9.156) in conjunction with Maxwell equations, in order to describe a simple situation regarding coupled nano-plasmons.

The longitudinal internal (polarization) electric field in Gauss equation $div\mathbf{E} = -4\pi nqdiv\mathbf{u}$ is given by $\mathbf{E} = -4\pi nq\mathbf{u}$ (*i.e.*, $\mathbf{E} = -4\pi\mathbf{P}$). In the long-wavelength limit, the finite size of the body is usually taken into account by a (de-) polarizing factor f, such as the field is

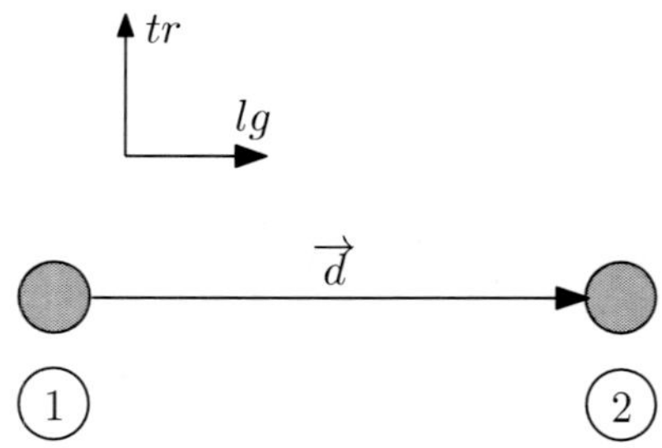

Figure 9.5: Two point dipoles.

given by $\mathbf{E} = -4\pi nqf\mathbf{u}$; for instance, for a sphere $f = 1/3$.[19] Introducing this polarization field in equation (9.156), taking the Fourier transform and leaving aside the coefficient γ, we get

$$(\omega^2 - \omega_c^2 - f\omega_p^2)\mathbf{u} = -\frac{q}{m}\mathbf{E}_0 \ ; \tag{9.157}$$

we can see that we have a plasmon resonance at frequency $\sqrt{\omega_c^2 + f\omega_p^2}$; for a conducting sphere with $\omega_c = 0$ and $f = 1/3$, we get the plasmon frequency $\omega_p/\sqrt{3}$, in accordance with the frequencies Ω_l given above for $l = 1$.

We consider two point particles, denoted by 1 and 2, each with its own plasmon frequency $\omega_{1,2}$, separated by the position vector $\mathbf{d}$, as shown in Fig. 9.5. We describe the motion of the mobile charges in each particle by a displacement vector $\mathbf{u}_{1,2}$; equation (9.157) becomes

$$(\omega^2 - \omega_{1,2}^2)\mathbf{u}_{1,2} = -\frac{q}{m}\mathbf{E}_{02,1} \ , \tag{9.158}$$

where $\mathbf{E}_{01,2}$ is the electric field generated by particle 1 (2) at the position of the particle 2 (1). In the long wavelength limit this is the field generated by a point dipole

$$\mathbf{E}_{01,2} = v_{1,2}n_{1,2}q\frac{3(\mathbf{u}_{1,2}\mathbf{d})\mathbf{d} - \mathbf{u}_{1,2}d^2}{d^5} \ , \tag{9.159}$$

[19] M. Apostol, *Essays in Electromagnetism and Matter*, Lambert, Saarbruecken (2013).

where $v_{1,2}$ are the volumes of the two particles and $n_{1,2}$ are the concentration of the mobile charges in the particles; equation (9.159) is valid in the near-field region $c/\omega \gg d$, where c is the speed of light. Since the particles are considered pointlike, we have also $v_{1,2}^{1/3} \ll d$. Introducing this field in equations (9.158) we get two coupled equations for the displacement vectors. It is convenient to use the projections of the displacement vectors on the vector **d** and on a direction perpendicular to the vector **d**; we call the former the longitudinal displacements and denote them by $u_{l1,2}$, while the latter, denoted by $\mathbf{u}_{t1,2}$, are called transverse displacements. The equations for the longitudinal displacements are decoupled from those corresponding to the transverse displacements; both sets of equations have the same structure. We limit ourselves here to the longitudinal displacements

$$(\omega^2 - \omega_{1,2}^2)u_{l1,2} = -\frac{\omega_{p2,1}^2 v_{2,1}}{2\pi d^3} u_{l2,1} \ . \tag{9.160}$$

The solution of these coupled-oscillators equations is straightforward. The factor $v_{1,2}/d^3$ plays the role of a weak-coupling constant. The eigenfrequencies of equations (9.160) are close to the plasmon frequencies $\omega_{1,2}$, which should satisfy the condition $c/\omega_{1,2} \gg d$; $c/\omega_{1,2}$ is usually called the plasma wavelength. For typical values $\omega_{1,2} \simeq 10^{15}s^{-1}$ we get a critical distance of the order $d \simeq 0.1\mu m$; the treatment given here holds for smaller distances, while for larger distances we need to take into account the retardation in estimating the polarization field.

An interesting situation occurs for two identical conducting particles $\omega_{c1} = \omega_{c2} = 0$, $\omega_{p1} = \omega_{p2} = \omega_p$ and $v_1 = v_2 = v$; in order to simplify the things we take also $\omega_1 = \omega_2 = \omega_p$. In this case the eigenfrequencies are given by

$$\Omega_{1,2} = \omega_p \left(1 \pm \frac{v}{2\pi d^3}\right)^{1/2} \simeq \omega_p \left(1 \pm \frac{v}{4\pi d^3}\right) \ . \tag{9.161}$$

The displacement vectors for the initial condition $u_{l2}(t=0) = 0$ are given by

$$\begin{aligned} u_{l1}(t) &= 2Ae^{i\omega_p t} \cos \tfrac{v}{4\pi d^3} t \ , \\ u_{l2}(t) &= -2iAe^{i\omega_p t} \sin \tfrac{v}{4\pi d^3} t \ ; \end{aligned} \tag{9.162}$$

we can see that the two coupled oscillations exhibit "beats", and the plasmons can be transferred periodically between the two particles, as expected. A similar situation holds for the transverse oscillations, with the factor 2π replaced by 4π in the above formulae.

A polarizable pointlike particle can be approximated by a dipole, with the current density $\mathbf{j} = vnq\dot{\mathbf{u}}\delta(\mathbf{r})$ and charge density $\rho = -vnq(\mathbf{u}grad)$ $\delta(\mathbf{r})$, where v is the volume of the particle placed at the origin. For these charge and current distributions we can compute easily the electromagnetic potentials (Fourier transforms):

$$\mathbf{A} = -i\lambda vnq\mathbf{u}\frac{e^{i\lambda r}}{r} \ , \ \Phi = -vnq\frac{\mathbf{ur}}{r}\frac{\partial}{\partial r}\frac{e^{i\lambda r}}{r} \ , \tag{9.163}$$

where $\lambda = \omega/c$. The polarization electric field is given by $\mathbf{E} = -(1/c)\partial\mathbf{A}/\partial t - grad\Phi$, so that we can include the retardation in the equation of motion (9.156). For the longitudinal oscillations of two identical conducting particles we get

$$(\omega^2 - \omega_p^2)u_{l1,2} = -\frac{\omega_p^2 v}{2\pi d^3}(1 - i\lambda d)e^{i\lambda d}u_{l2,1} \tag{9.164}$$

(and a similar set of equations for the transverse oscillations). We can see that in the non-retarded limit $\lambda d \ll 1$ equations (9.164) go into equations (9.160) derived above. We are interested now in the wave-zone limit $\lambda d \gg 1$. The eigenfrequencies of equations (9.164) are given by

$$(\omega^2 - \omega_p^2)^2 = \frac{\omega_p^4 v^2}{(2\pi d^3)^2}(1 - i\lambda d)^2 e^{2i\lambda d} \ , \tag{9.165}$$

or

$$\tan\lambda d = \lambda d \ , \ (\omega^2 - \omega_p^2)^2 = \frac{\omega_p^4 v^2}{(2\pi d^3)^2}(1 + \lambda^2 d^2) \ . \tag{9.166}$$

It is convenient to introduce the notations $g = v/2\pi d^3 \ll 1$ and $\omega_p d/c = A$; the solution can be found as a series of powers of g:

$$\Omega = \omega_p\left[1 \pm \frac{1}{2}g(1 + A^2)^{1/2} + \frac{1}{8}g^2(A^2 - 1) + ...\right] \ ; \tag{9.167}$$

it should satisfy the equation $\tan(\Omega d/c) = \Omega d/c$, which, in the limit $g \ll 1$ becomes $\tan A \simeq A$ ($\Omega \simeq \omega_p$); for large values of A we get

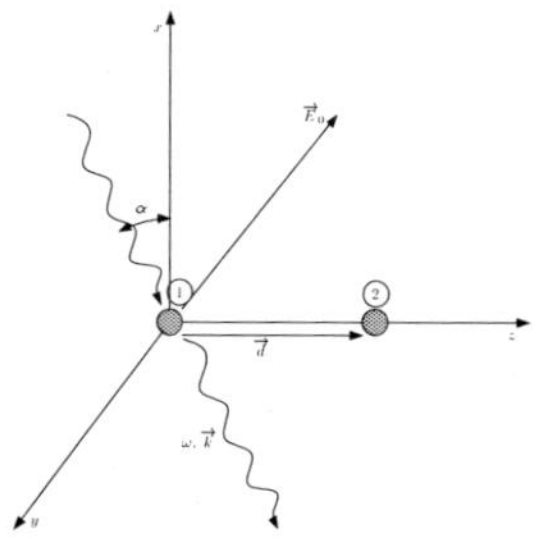

Figure 9.6: Two coupled point dipoles subjected to an external field.

$A = \omega_p d/c \simeq n\pi$. We can see that there are real solutions for the eigenfrequencies only for certain values of the distance $d_n \simeq n\pi c/\omega_p$, which are approximate multiples of the plasma wavelength (in this limit). The corresponding oscillations are usually called polaritons. For intermediate values of d the eigenfrequencies are complex, *i.e.* the coupling between the two particles is damped (the damping parameter γ in the equation of motion (9.156) should be retained in this case). We can equally well say that the two particles are not coupled in this case.

The zero-point (vacuum fluctuations) energy can be estimated as $E = \sum \hbar\Omega/2$, where the summation extends over all the eigenfrequencies. The motion of the transverse degrees of freedom leads to the eigenfrequencies equation

$$\begin{aligned} \tan\lambda d &= \tfrac{\lambda d}{1-\lambda^2 d^2} \ , \\ (\omega^2-\omega_p^2)^2 &= \tfrac{\omega_p^4 v^2}{(4\pi d^3)^2}(1+3\lambda^2 d^2+\lambda^4 d^4) \ . \end{aligned} \tag{9.168}$$

The solution is given by

$$\begin{aligned} \Omega = \omega_p\{1 \pm \tfrac{1}{4}g(1+3A^2+A^4)^{1/2}+ \\ +\tfrac{1}{32}g^2(3A^4+3A^2-1)+...\} \ . \end{aligned} \tag{9.169}$$

Now we can compute the zero-point energy (the transverse degrees of freedom have a double multiplicity):

$$E = \hbar\omega_p \left[3 + \frac{g^2}{16}(3A^4 + 5A^2 - 3) \right] \tag{9.170}$$

and the corresponding force

$$F = \frac{\hbar\omega_p v^2}{32\pi^2} \left(\frac{3\omega_p^4}{c^4 d^3} + \frac{10\omega_p^2}{c^2 d^5} - \frac{9}{d^7} \right) . \tag{9.171}$$

We can see that in the non-retarded limit ($\omega_p d/c \ll 1$) the force is attractive and goes like $-1/d^7$; this is the van der Waals-London force; it comes from the longitudinal degrees of freedom. In the opposite, retarded limit $\omega_p d/c \gg 1$ the force is repulsive and goes like $1/d^3$; this is the limit of the Casimir force, coming entirely from the transverse oscillations. The force changes sign around $\omega_p d/c \lesssim 1$ and has a maximum for $\omega_d d/c \gtrsim 1$. For intermediate distances the numerical coefficients in equation (9.171) are not reliable, since the transverse oscillations do not occur at the same distances d_n as the longitudinal ones; increasing the distance, the longitudinal and transverse oscillations contribute alternately to the repulsive force.

We consider two point dipoles acted by an external monochromatic plane wave with the electric field $\boldsymbol{E}_0(t, \boldsymbol{r}) = \boldsymbol{E}_0 \cos(\omega t - \boldsymbol{kr})$, $\omega = ck$ and $E_0 k = 0$ (Fig. 9.5). The two particles separated by the distance d lie along the z-axis and the external field (wavevector $\boldsymbol{k}$) makes an incidence angle α with the z-axis. For a p- (parallel) wave we have both a longitudinal (E_{0z}) and a transverse (E_{0x}) component of the external field, while for an s- (senkrecht) wave we have only a transverse component E_{0y}. We assume further the most interesting situation of two identical point dipoles with $v^{1/3} \ll d \ll c/\omega$ (optical excitation), as well as $d \ll c/\omega_p$. The equation of motion (9.156) gives

$$\begin{aligned} \ddot{u}_{l1,2} + \omega_p^2 u_{l1,2} - \frac{v\omega_p^2}{2\pi d^3} u_{l2,1} &= \frac{q}{m} E_{0l} \cos\omega t \ , \\ \ddot{u}_{t1,2} + \omega_p^2 u_{t1,2} + \frac{v\omega_p^2}{4\pi d^3} u_{t2,1} &= \frac{q}{m} E_{0t} \cos\omega t \ , \end{aligned} \tag{9.172}$$

where we have neglected the position dependence of the external field at the location of the two point dipoles. The solution of these equations is given by

$$\begin{aligned} u_{l1} &= u_{l2} = -\frac{q}{m}E_{0l}\frac{1}{\omega^2-\Omega_{2l}^2}\cos\omega t \ , \\ u_{t1} &= u_{t2} = -\frac{q}{m}E_{0t}\frac{1}{\omega^2-\Omega_{1t}^2}\cos\omega t \ , \end{aligned} \tag{9.173}$$

where

$$\Omega_{2l}^2 = \omega_p^2\left(1-\frac{v}{2\pi d^3}\right) \ , \ \Omega_{1t}^2 = \omega_p^2\left(1+\frac{v}{4\pi d^3}\right) \ . \tag{9.174}$$

Equations (9.173) give the polarizabilities of the two coupled point dipoles ($\boldsymbol{P} = nq\boldsymbol{u}$ is the polarization). It is worth noting that there exist two resonant frequencies Ω_{2l} (shifted to the red with respect to the characteristic frequency ω_p of the isolated dipole) and Ω_{1t} (shifted to the ultraviolet). At these resonances the light is trapped by the two coupled dipoles, and the field is enhanced appreciably.

Having known the displacements $\boldsymbol{u}_{1,2}$ we can use the potentials given by equations (9.163) to compute the field in the wave zone ($r \gg d$). The calculations are straightforward. We can use the results for estimating the cross-section defined as $d\sigma = \boldsymbol{Sr}d\Omega/S_0r^3$, where $\boldsymbol{S}$ is the Poynting vector of the scattered field and $\boldsymbol{S}_0$ is the Poynting vector of the incident field. We get the total cross-sections

$$\begin{aligned} \sigma_p &= \frac{2\omega_p^4v^2}{3\pi c^4}\omega^4\left[\frac{\sin^2\alpha}{\left(\omega^2-\Omega_{2l}^2\right)^2}+\frac{\cos^2\alpha}{\left(\omega^2-\Omega_{1t}^2\right)^2}\right] \ , \\ \sigma_s &= \frac{2\omega_p^4v^2}{3\pi c^4}\frac{\omega^4}{\left(\omega^2-\Omega_{1t}^2\right)^2} \end{aligned} \tag{9.175}$$

for the p- and the s-wave, respectively. We can see the resonant cross-section (especially the longitudinal resonance at lower frequencies), which indicates again the phenomenon of light trapping and field enhancement.

Under the action of an external field the particles get polarized; the displacement field given by equations (9.173) generate charge and current densities; for instance, the particle placed at $\boldsymbol{r} = 0$ acquires a charge density $\rho(\boldsymbol{r}) = -vnq(\boldsymbol{u}grad)\delta(\boldsymbol{r})$, where $\boldsymbol{u}$ denotes its displacement field. The external field acts upon this charge by a drift force given by

$$\boldsymbol{F}_d = \int d\boldsymbol{r}\rho(\boldsymbol{r})\boldsymbol{E}_0(t,\boldsymbol{r}) \ ; \tag{9.176}$$

the calculations are straightforward and we get

$$\boldsymbol{F}_d = vnq(\boldsymbol{k}\boldsymbol{u})\boldsymbol{E}_0 \sin\omega t \ . \tag{9.177}$$

A similar drift force acts upon the particle placed at $\boldsymbol{r} = \boldsymbol{d}$. We can see that an external s-wave, as well as an external plane wave acting upon an isolated particle brings no drift force. Making use of equations (9.173) for a p-wave (and identical particles), we get the drift force

$$\boldsymbol{F}_d = \frac{3v^2\omega_p^4\omega}{64\pi^2 cd^3}\frac{\sin 2\alpha \sin 2\omega t}{\left(\omega^2-\Omega_{2l}^2\right)\left(\omega^2-\Omega_{1t}^2\right)}E_0\boldsymbol{E}_0 \ . \tag{9.178}$$

We note the resonant character of this force for frequency ω approaching the frequencies Ω_{2l} or Ω_{1t}.

There is also an interaction force acting between the two particles. For instance we get the force acting upon the particle placed at $\boldsymbol{r} = 0$ on behalf of the particle placed at $\boldsymbol{r} = \boldsymbol{d}$ by using the dipole field given by equation (9.159) in equation (9.176). We get an interaction force

$$\begin{gathered} F_{12l} = \frac{3v^2\omega_p^4}{16\pi^2 d^4}\left[\frac{2E_{0l}^2}{\left(\omega^2-\Omega_{2l}^2\right)^2} - \frac{E_{0t}^2}{\left(\omega^2-\Omega_{1t}^2\right)^2}\right]\cos^2\omega t \ , \\ F_{12t} = -\frac{6v^2\omega_p^4}{16\pi^2 d^4}\frac{E_{0l}E_{0t}}{\left(\omega^2-\Omega_{2l}^2\right)\left(\omega^2-\Omega_{1t}^2\right)}\cos^2\omega t \ . \end{gathered} \tag{9.179}$$

We note again the resonance character of this force, as well as its non-central character, in general (for a p-wave it gives rise to a torque).

In estimating these forces we limited ourselves to the polarization (displacement field) generated by the external field. In general, we should also take into account the field generated by the particles in estimating the displacement field. The contribution of this "internal " field (which acts as an external field for each particle) is proportional to the volume v of the particles, and for a finite number of particles it may be neglected. However, for a large number of particles its contribution is important. It is worth noting also that under the action of such resonant interaction forces the nano-particles can get organized in super-structures.

In conclusion, we have analyzed here the electromagnetic coupling between two polarizable pointlike particles, modelled as point dipoles. This may be a reasonably useful model of coupled nano-plasmons and nano-polaritons. For small separation distances between the two particles (smaller than the plasma wavelength), where the non-retarded coupling regime dominates, the two particles exhibit coupled plasmons, which can be transferred from one particle to other. The zero-point fluctuations give the attractive van der Waals-London force in this case, acting between particles and behaving like $-1/d^7$, where d is the separation distance. For distances larger than the plasma wavelength the retardation comes into play and the coupling is realized through polaritons. This may happen only for certain discrete sets of separation distances (different for longitudinal and transverse oscillations of the charge density, with respect to the separation vector), in between the coupling being damped (non-coupling); it is realized, by turn, either by longitudinal or transverse oscillations of the charge density. Immediately after distances of the order of the plasma wavelength, the zero-point energy force acting between particles becomes repulsive, arising from transverse oscillations (polaritons) and going at infinity like $1/d^3$.

The polarizabilities of the two coupled nano-particles exhibit plasmonic resonances, which are also present in the cross-section of the scattered field as well as in the resonant force acting upon the particles both on behalf of the external field and of each other. The method used in getting the results presented here consists, basically, of the equation of motion of the electric polarization, related to the displacement field of the mobile charges. The method has also been employed previously for getting various other results.[20] In particular,

[20] M. Apostol and G. Vaman, "Electromagnetic field interacting with a semi-infinite plasma", J. Opt. Soc. Am. **A26** 1747 (2009); "Reflection and refraction of the electromagnetic field in a semi-infinite plasma", Opt. Commun. **282** 4329 (2009); "Plasmons and diffraction of an electromagnetic plane wave by a metallic sphere", Progr. Electrom. Res. PIER **98** 97 (2009); M. Apostol, S. Ilie, A. Petrut, M. Savu and S. Toba, "A generalization of the dipolar force", J. Appl. Phys. **112** 024905 (2012); M. Apostol, "On the molecular

the force acting between two pointlike polarizable dipoles when one of them is subjected to an external field was derived (generalized dipole force). It is a force generated by the mutual polarizations, in contrast to one of the forces derived here, when the external field acts upon, and polarizes, both particles.

9.15 Propagation of light in matter

In two classical papers[21] Sommerfeld and Brillouin examined the propagation velocity of the electromagnetic radiation in matter. Matter is known to be dispersive. This investigation was motivated by phase velocities which exceed the speed of light in vacuum c, in relation to the Theory of Relativity.

From ancient times it is known from experiment the Snell's law refraction law $\sin r/\sin i = 1/n$, where r is the refraction angle, i is the incidence angle and n is the refraction index. The law $\sin r/\sin i = v_\phi/c = \omega/ck$, where ω is the frequency and $\mathbf{k}$ is the wavevector of a monochromatic plane wave with the phase velocity $v_\phi = \omega/k$, is known from Huygens theory. From Maxwell equations it is also known $\varepsilon\omega^2 = c^2k^2$ (for non-magnetic matter), where ε is the dielectric constant (or $\varepsilon\mu\omega^2 = c^2k^2$ for magnetic matter, where μ is the magnetic permeability). The Snell's law of refraction reads

$$\frac{\sin r}{\sin i} = \frac{v_\phi}{c} = \frac{\omega}{ck} = \frac{1}{\sqrt{\varepsilon}} = \frac{1}{n} \,. \tag{9.180}$$

It is known from experiment that ε, v_ϕ, k, ε and n depend all on the frequency ω. For some frequencies, v_ϕ may trespass the speed of light c as, for instance, near a vanishing ε. The situation is further complicated by the anomalous dispersion region, where ε is negative (actually, the relation $\varepsilon\omega^2 = c^2k^2$ must be satisfied with $\varepsilon > 0$, so

forces acting between macroscopic bodies", Physica **B409** 57 (2013).

[21] A. Sommerfeld, "Ueber die Fortpflanzung des Lichtes in dispergierenden Medien", Ann. Phys. **44** 177 (1914); L. Brillouin, "Ueber die Fortpflanzung des Lichtes in dispergierenden Medien", Ann. Phys. **44** 203 (1914). Also, L. Brillouin, *Wave Propagation and Group Velocity*, Academic Press, NY (1960); A. Sommerfeld, *Vorlesungen ueber Theoretische Physik*, Band IV, (*Optik*), Akademische Verlag, Leipzig (1964).

the expression of the Snell's law with ε has a limited validity). The nature of the refraction law, as well as many other aspects related to the propagation of the electromagnetic waves in matter have been recently clarified to a large extent by the theory of polarizable matter and polaritonic eigenmodes. However, the problem of superluminal phase velocities persists.

In this context, it is worth recalling that the theory of polarizable matter and plasmonic and polaritonic eigenmodes is based on the Drude-Lorentz (plasma) model.[22] It is also known since long ago the so-called extinction theorem, which tells that free electromagnetic waves, *i.e.* electromagnetic waves propagating with the speed of light in vacuum c, cannot be propagated in matter.[23]

The propagation of the electromagnetic waves in matter implies both a separation surface of the body from the vacuum (or an interface between two bodies) and a signal of a finite duration (and finite spatial extension). The presence of the surface leads to refraction and the lack of translational symmetry, which complicates the analysis. A finite duration and a finite spatial extension imply a superposition of frequencies and wavevectors, related through the phase velocity. In a simplifying approach the analysis of Sommerfeld and Brillouin leads to an electromagnetic signal represented by

$$f(x,t) = \int dk e^{-i\Omega(k)t+ikx} \quad , \tag{9.181}$$

where $\Omega(k)$ is the polaritonic frequency. The body is assumed to be infinite, the extension of the wave along the transverse directions is supposed to be infinite (as for a plane wave, or a beam, ray in geometrical optics, in order to leave only one component of the wavevector). Typical polaritonic frequencies are given by $\Omega(k) = \sqrt{\omega_p^2 + c^2k^2}$ (*e.g.* for conductors) and $\Omega(k) = vk\omega_T/(vk + \omega_T)$ for dielectrics, where

[22] P. Drude, "Zur Elektronentheorie der Metalle", Ann. Phys. **306** 566 (1900); "2. Teile. Galvanomagnetische und thermomagnetische Effecte", **308** 369 (1900); H. A. Lorentz, *The Theory of Electrons*, Teubner, Leipzig (1916).

[23] P. P. Ewald, "Ueber die Grundlagen der Kristalloptik", Thesis, Munich (1912), Ann. Phys. **49** 1 (1916); C. W. Oseen, "Ueber die Wechselwirkung zwischen zwei elektrischen Dipolen und ueber die Drehung der Polarisationsebene in Kristallen und Fluessigkeiten", Ann. Phys. **48** 1 (1915); M. Born and E. Wolf, *Principles of Optics*, Pergamon, London (1959).

ω_T is the so-called transverse-mode frequency and $v = c\omega_T/\omega_L(< c)$, where $\omega_L$is the so-called frequency of the longitudianl modes ($\omega_L = \sqrt{\omega_p^2 + \omega_T^2}$).

The analysis of Brillouin and Sommerfeld proceeds as follows. We divide the whole domain of integration in equation (9.181) ($-\infty$ to $+\infty$) in some intervals Δ_i around some points k_i, such that within each interval Δ_i we may approximate the frequency by

$$\Omega(k) = \Omega(k_i) + v_i(k - k_i) + \frac{1}{2}\Omega^{''}(k_i)(k - k_i)^2 + ..., \tag{9.182}$$

where $v_i = \Omega^{'}(k_i) = \partial\Omega/\partial k|_{k=k_i}$. The integral in equation (9.181) becomes

$$f(x,t) \simeq \sum_i e^{-i\Omega_i t + ik_i x} \int_{\Delta_i} dq e^{i(x-v_i t)q} e^{-\frac{1}{2}i\Omega_i^{''} tq^2} \quad , \tag{9.183}$$

where $\Omega_i = \Omega(k_i)$, $\Omega_i^{''} = \Omega^{''}(k_i)$ and $q = k - k_i$ within each interval Δ_i. If we may neglect the second-order derivative and extend the integration over sufficiently large intervals, we can see that we get pulses of the form $\delta(x - v_i t)$ propagating with velocity v_i. This is called the group velocity. The second-order derivative leads to an (approximate) Fresnel integral and makes the pulses flat, with a spread increasing with increasing time, according to an imaginary Gauss function (as in an imaginary diffusion). Moreover, making use of the form of the polaritonic frequency $\Omega(k)$ we can see that the group velocity is always smaller than the speed of light in vacuum c, and for high frequency it approaches c, *i.e.* the front of the wave moves almost with the speed of light c in matter. Therefore, the wave in dispersive matter moves as a set of groups, each with a group velocity, getting flat in time; if there is one main group, there are others moving with higher velocity (but smaller than the speed of light in vacum), which we call precursors (or forerunners), beside others which lag behind the main group. This is the group-velocity picture based on the method of stationary phase (steepest descent or saddle point method).[24]

[24]P. Debye, "Naeherungsformeln fuer die Zylinderfunktionen fuer große Werte des Arguments und unbeschraenkt veraenderliche Werte des Index", Mathematische Annalen **67** 535 (1909).

Still, there is a discussion regarding superluminal velocities of electromagnetic signals.

Indeed, the picture described above is based on an approximation, concerning the extension of the integration interval Δ_i to, practically, infinity. Actually, the integration over a finite interval Δ_i leads to a wavepacket, of the form

$$\int_{\Delta_i} dq e^{i(x-v_i t)q} = \frac{2\sin(x-v_i t)\Delta_i/2}{x-v_i t} \ , \tag{9.184}$$

which, it is true, tends to $\pi\delta(x-v_i t)$ for $\Delta_i \to \infty$, but it extends to infinite $x - v_i t$ for a finite interval Δ_i. This means that each wave group extends instantaneously to infinity (of course, with a decreasing amplitude), which suggests indeed a velocity much higher than the speed of light in vacuum (practically an infinite one). This feature is due to the non-locality of the waves, and it implies in fact the "propagation" of no signal.

The non-local picture of the waves can be seen in fact more directly on the integral in equation (9.181). At $t = 0$ we have

$$f(x, t=0) = \int dk e^{-ikx} = 2\pi\delta(x) \ , \tag{9.185}$$

and the signal is highly localized (on $x = 0$); at the next infinitesimal moment of time Δt we have

$$f(x, \Delta t) = \int dk e^{-i\Omega(k)\Delta t + ikx} \ , \tag{9.186}$$

and the signal is already extended over the whole space; since $x/\Delta t \to \infty$ for any x and $\Delta t \to 0$ we may say that the signal has been propagated indeed with an infinite velocity.

10 Surface Electromagnetism

10.1 Polarization

With usual notations the Maxwell equations in matter read

$$div\mathbf{D} = 4\pi\rho_0 \ , \ div\mathbf{B} = 0 \ ,$$
$$curl\mathbf{E} = -\frac{1}{c}\frac{\partial\mathbf{B}}{\partial t} \ , \ curl\mathbf{H} = \frac{1}{c}\frac{\partial\mathbf{D}}{\partial t} + \frac{4\pi}{c}\mathbf{j}_0 \ , \tag{10.1}$$

where $\mathbf{E}$ is the electric field, $\mathbf{D}$ is the electric displacement, $\mathbf{B}$ is the magnetic induction and $\mathbf{H}$ is the magnetic field; ρ_0 is the external charge density and $\mathbf{j}_0$ is the external current density (obeying the continuity equation $\partial\rho_0/\partial t + div\mathbf{j}_0 = 0$). We have two independent Maxwell equations (10.1) (Faraday's and Maxwell-Ampere's equations) and four unknowns. In order to have a solution we introduce the quasi-phenomenological dielectric function ε and magnetic permeability μ, usually for the Fourier transforms. Apart from being unsatisfactory at the fundamental level, this procedure produces appreciable difficulties, especially with the finite size, inhomogeneous bodies.

Matter is polarizable, *i.e.* it consists of more-or-less mobile charges q, with mass m and concentration n (*e.g.*, electrons), which move against a neutralizing background of quasi-rigid charges $-q$ (*e.g.*, ions). A small displacement field $\mathbf{u}(t,\mathbf{r})$, which is a function of the time t and position $\mathbf{r}$, generates an imbalance $\delta n = -ndiv\mathbf{u}$ in the density of these charges, a charge density $\rho = -nqdiv\mathbf{u}$ and a corresponding current density $\mathbf{j} = nq\dot{\mathbf{u}}$. It is easy to see that the polarization is $\mathbf{P} = nq\mathbf{u}$ (density of the dipole moments) and the electric displacement is now represented as $\mathbf{D} = \mathbf{E} + 4\pi\mathbf{P}$. The displacement field obeys an equation of motion, which usually is Newton's equation

$$m\ddot{\mathbf{u}} = q\mathbf{E} - m\omega_c^2\mathbf{u} - m\gamma\dot{\mathbf{u}} \ ; \tag{10.2}$$

ω_c is a characteristic frequency (*e.g.*, for bound charges) and γ is a small damping coefficient. The magnetic term of the Lorentz force is usually absent in equation (10.2) (and the equation is non-relativistic), since the velocity of charges in matter is small, on one hand, and, on the other, the displacement $\mathbf{u}$ is sufficiently small to limit ourselves to linear terms only. This is the well-known Drude-Lorentz (plasma) model of polarizable matter.[1] The point is that the equation of motion (10.2) provides a third equation for the four unknowns: $\mathbf{E}$, $\mathbf{u}$, $\mathbf{B}$ and $\mathbf{H}$. It can be generalized in various ways, *e.g.* by including external fields.

Matter is also magnetizable. The continuity equation allows of a "magnetic" current $\mathbf{j} = c \cdot curl\mathbf{M}$; as it is well-known the magnetization $\mathbf{M}$ obeys the equation of motion of the angular momentum[2]

$$\frac{d\mathbf{M}}{dt} = \frac{q}{2mc}\mathbf{M} \times \mathbf{B} \ . \tag{10.3}$$

The magnetic induction is now represented as $\mathbf{B} = \mathbf{H} + 4\pi\mathbf{M}$. We have now four equations:

$$\begin{aligned} curl\mathbf{E} = -\tfrac{1}{c}\tfrac{\partial \mathbf{B}}{\partial t} \ , \ curl\mathbf{H} = curl(\mathbf{B} - 4\pi\mathbf{M}) = \\ = \tfrac{1}{c}\tfrac{\partial \mathbf{E}}{\partial t} + \tfrac{4\pi}{c}nq\dot{\mathbf{u}} + \tfrac{4\pi}{c}\mathbf{j}_0 \end{aligned} \tag{10.4}$$

and equations (10.2) and (10.3) and four unknowns: $\mathbf{E}$, $\mathbf{u}$, $\mathbf{B}$ and $\mathbf{M}$. These equations can provide the basis for treating the electromagnetic field in matter. Except for the important case of ferromagnetic (and related) matter, the usual matter is non-magnetic, so we may leave aside $\mathbf{M}$ and put $\mathbf{B} = \mathbf{H}$.

For the usual case of polarizable non-magnetic matter, we can find the plasmon and polariton eigenmodes, especially for infinite or semi-

[1] P. Drude, "Zur Elektronentheorie der Metalle", Ann. Phys. **306** 566 (1900); "2. Teile. Galvanomagnetische und thermomagnetische Effekte", Ann. Phys. **308** 369 (1900); H. A. Lorentz, *The Theory of Electrons*, Teubner, Leipzig (1916).

[2] L. Landau and E. Lifshitz, *Course of Theoretical Physics*, vol. 2 (*The Classical Theory of Fields*), Butterworth-Heinemann, Oxford (2003); J. Schwinger, L. L. deRaad, Jr., K. A. Milton and W. Tsai, *Classical Elctrodynamics*, ABP, Westview Press, Boulder, Co. (1998).

infinite (half-space) matter.[3] This way, we can describe the propagation of electromagnetic field in matter, as well as the interaction of the electromagnetic field with finite-size bodies, both in the near-field (sub-wavelength, quasi-static) regime and the wave (radiation) zone. This can be done in complex situations, where various fields are present for bodies with various shapes, a subject of high interest for nano-plasmonics.[4] The plasmon and polariton eigenmodes, respectively, are given by

$$\Omega_1 = \omega_L = \sqrt{\omega_c^2 + \omega_p^2} \ , \ \Omega_2(K) \simeq \sqrt{\omega_L^2 + c^2K^2} \ ,$$
$$\Omega_3(K) \simeq \omega_c cK/(\omega_L + cK) \ , \tag{10.5}$$

where $\omega_p = (4\pi nq^2/m)^{1/2}$ is the plasma wavevector and $\mathbf{K}$ is the wavevector. For a half-space we get a surface plasmon-polariton mode

$$\Omega^2 = \frac{2Ac^2k^2}{B+\sqrt{B^2-4Ac^2k^2}}$$
$$A = 2\omega_c^2 + \omega_p^2 \ , \ B = \omega_c^2 + \omega_p^2 + 2c^2k^2 \tag{10.6}$$

for $c^2k^2 > \omega_c^2$, where $\mathbf{k}$ is the wavevector parallel with the surface; it goes from ω_c ($ck = \omega_c$) to $\sqrt{\omega_c^2 + \omega_p^2/2}$ ($k \to \infty$). This mode is localized with respect to the direction perpendicular to the surface.

The zero-point energy (vacuum fluctuations) of the polarization eigenmodes leads to molecular forces like van der Waals-London-Casimir forces acting between macroscopic bodies.[5] The behaviour of the polarization eigenmodes in non-inertial motions may lead to interesting new effects.[6] The electromagnetic coupling between nano-structures can also be treated by this method, leading to plasmon transfer and

[3]M. Apostol and G. Vaman, "Electromagnetic field interacting with a semi-infinite plasma", J. Opt. Soc. Am. **A26** 1747 (2009); "Reflection and refraction of the electromagnetic field in a semi-infinite plasma", Opt. Commun. **282** 4329 (2009).

[4]M. Apostol and G. Vaman, "Plasmons and diffraction of an electromagentic plane wave by a metallic sphere", Progr. Electrom., Res. (PIER) **98** 97 (2009).

[5]M. Apostol, "On the molecular forces acting between macroscopic bodies", Physica **B409** 57 (2013).

[6]M. Apostol, "Non-inertial electromagnetic effects in matter. Gyromagnetic effect", Solid State Commun. **152** 1567 (2012).

resonances, or to electromagnetic forces with a resonant character.[7] The scattering of the electromagnetic waves by small particles or inhomogeneities, including the rough surface of a semi-infinite solid,[8] is also amenable to such a treatment.

10.2 Quasi-static fields

We write again the Maxwell equations in (non-magnetic, $\mathbf{H} = \mathbf{B}$) matter, with an explicit introduction of the polarization $\mathbf{P}$:

$$div\mathbf{E} = 4\pi\rho_0 - 4\pi div\mathbf{P} \ , \ div\mathbf{H} = 0 \ ,$$

$$curl\mathbf{E} = -\frac{1}{c}\frac{\partial \mathbf{H}}{\partial t} \ , \tag{10.7}$$

$$curl\mathbf{H} = \frac{1}{c}\frac{\partial \mathbf{E}}{\partial t} + \frac{4\pi}{c}\frac{\partial \mathbf{P}}{\partial t} + \frac{4\pi}{c}\mathbf{j}_0 \ ;$$

the fields satisfy wave equations with sources (including polarization charges $-div\mathbf{P}$ and currents $\partial\mathbf{P}/\partial t$); as it is well known, it is convenient to introduce the potentials $\mathbf{A}$ and Φ, such that $\mathbf{E} = -(1/c)\partial\mathbf{A}/\partial t - grad\Phi$, $\mathbf{H} = curl\mathbf{A}$; these potentials satisfy the wave equations with sources, providing the Lorenz gauge $div\mathbf{A}+(1/c)\partial\Phi/\partial t = 0$ is satisfied. We assume the quasi-static situation, *i.e.* we assume that the time variation is slow, such as to have $\omega/c \ll 1/l$, and $v/c \ll 1$, where ω is the frequency, l is the dimension of the bodies and v is the velocity of the charges. Then, it is easy to see that from all the Maxwell equations only Gauss's equation

$$div\mathbf{E} = 4\pi\rho_0 - 4\pi div\mathbf{P} \tag{10.8}$$

survives, and $curl\mathbf{E} = 0$; which means that we may neglect $\mathbf{A}$ and keep only Φ, such as $\mathbf{E} = -grad\Phi$. This can also be seen directly from Kirchhoff's solutions of the wave equations for potentials. This is the quasi-static approximation (or even static, when there is not a time dependence); it is also termed the sub-wavelength or near-field

[7] M. Apostol, S. Ilie, A. Petrut, M. Savu and S. Toba, "A generalization of the dipolar force", J. Appl. Phys. **112** 024905 (2012).

[8] B. F. Apostol, "Scattering of the electromagnetic waves from a rough surface" J. Mod. Phys. **59** 1607 (2012).

regime. It is relevant for external $\rho_0 \neq 0$ (or other external causes producing polarization charges). If this $\rho_0 = 0$ and no other external causes producing polarization charges are present, it is the next approximation which is relevant, corresponding to $\mathbf{j}_0$, with $div\mathbf{j}_0 = 0$ and, of course , $\mathbf{A} \neq 0$ and $\Phi = 0$. The former case is also called electrostatics (or quasi-electrostatics), the latter is magnetostatics, though, allowing for slow time dependence, they are both quasi-static.

It is convenient to work with an external (slowly varying in time) electric field $\mathbf{E}_0$ instead of the external charge density ρ_0 (a slow external magnetic field can also be present). With usual notations, the equation of motion (10.2) for the displacement $\mathbf{u}$ can be written as

$$m\ddot{\mathbf{u}} = q(\mathbf{E}_0 + \mathbf{E}) - m\omega_c^2\mathbf{u} - m\gamma\dot{\mathbf{u}} \ , \tag{10.9}$$

or, using the temporal Fourier transforms,

$$(\omega^2 - \omega_c^2 + i\omega\gamma)\mathbf{u} = -\frac{q}{m}(\mathbf{E}_0 + \mathbf{E}) \ , \tag{10.10}$$

where $\mathbf{E}$ is the internal (polarization) electric field, given by

$$div\mathbf{E} = 4\pi\rho = -4\pi div\mathbf{P} = -4\pi nq div\mathbf{u} \ ; \tag{10.11}$$

we note that $\rho = -div\mathbf{P} = -nq div\mathbf{u}$ denotes now the internal (polarization) charge density. This equation can easily be solved: $\mathbf{E} = -4\pi\mathbf{P} = -4\pi nq\mathbf{u}$ within this approximation (up to an irrelevant $curl$), so that the equation of motion (10.10) reads

$$(\omega^2 - \omega_c^2 - \omega_p^2 + i\omega\gamma)\mathbf{u} = -\frac{q}{m}\mathbf{E}_0 \ ; \tag{10.12}$$

this is the well-kown polarizability $\alpha(\omega)$,

$$\mathbf{u} = -\frac{q}{m}\mathbf{E}_0\frac{1}{\omega^2 - \omega_c^2 - \omega_p^2 + i\omega\gamma} = \frac{1}{nq}\alpha(\omega)\mathbf{E}_0 \ ; \tag{10.13}$$

we can see the presence of the plasmon eigenmode with frequency $\sqrt{\omega_c^2 + \omega_p^2}$.

It is worth noting that for a point external charge $div\mathbf{E}_0 = 4\pi q_0\delta(\mathbf{r})$, the field is $\mathbf{E}_0 = q_0\mathbf{r}/r^3$ and the displacement is given by

$$\mathbf{P} = nq\mathbf{u} = \alpha(\omega)q_0\frac{\mathbf{r}}{r^3} \ ; \tag{10.14}$$

the displacement acquires a maximum value near the external point charge and decreases like $1/r^2$. We note that, in principle, knowing the displacement (polarization) for an external point charge we can get the displacement (polarization) for an external charge of any shape.

The result obtained above is valid for an infinite piece of matter. As it is well known, for finite-size bodies, the relationship between $\mathbf{P}$ and $\mathbf{E}$ is $\mathbf{E} = -4\pi f\mathbf{P}$, where f is the (de-) polarizing factor; the polarizability is modified according to

$$\alpha(\omega) = -\frac{\omega_p^2}{4\pi}\frac{1}{\omega^2-\omega_c^2-\omega_p^2+i\omega\gamma} \rightarrow$$
$$\rightarrow \alpha(\omega) = -\frac{\omega_p^2}{4\pi}\frac{1}{\omega^2-f\omega_c^2-\omega_p^2+i\omega\gamma} \; . \tag{10.15}$$

Indeed, the presence of the surface of a finite-size body, as well as the boundary conditions (including, especially, the continuity conditions for the circular variables), lead to a charge density different from $\rho = -div\mathbf{P} = -nqdiv\mathbf{u}$. For instance, for a semi-infinite body with a plane surface we get $f = 1/2$, for the lowest mode of a sphere $f = 1/3$, etc.[9]

10.3 Surface plasmons

Let us suppose a half-space with a plane surface at $z = 0$, extending over the region $z > 0$.[10] Equation of motion (10.10) is preserved; it is convenient to write it as

$$(\omega^2 - \omega_c^2 + i\omega\gamma)(\mathbf{u}, u_z) = -\frac{q}{m}(\mathbf{E}_0 + \mathbf{E}) \; , \; z > 0 \; , \tag{10.16}$$

where $\mathbf{u}$ is the in-plane displacement vector (parallel to the surface) and u_z is the z-component of the displacement (perpendicular to the surface). The displacement field reads now $(\mathbf{u}, u_z)\theta(z)$, where $\theta(z) = 1$ for $z > 0$, $\theta(z) = 0$ for $z < 0$ is the step function; the charge density is $\rho = -nqdiv[(\mathbf{u}, u_z)\theta(z)]$, which can be written as

[9] M. Apostol, *Essays in Electromagnetism and Matter*, Lambert, Saarbruecken (2013).

[10] The half-spaces do not fulfill rigourously the conditions of quasi-static fields, because of their infinite extension.

$$\rho = -nq div[(\mathbf{u}, u_z)\theta(z)] = \\ = -nq[div(\mathbf{u}, u_z)]\theta(z) - nqu_z(z=0)\delta(z) \ , \tag{10.17}$$

where $\delta(z)$ is Dirac's delta function. We can see that there is a "volume" charge density $-nq[div(\mathbf{u}, u_z)]\theta(z)$, which generates the "volume" field $\mathbf{E}_v = -4\pi nq(\mathbf{u}, u_z)\theta(z)$ (equal to $-4\pi\mathbf{P}$, where $\mathbf{P} = nq(\mathbf{u}. u_z)\theta(z)$ is the polarization) and there is an additional "surface" charge density $\rho_s = -nqu_z(z=0)\delta(z)$, corresponding to a surface charge density $nqu_z(z=0)$. The "volume" charge corresponds to "volume" plasmons, with frequency $\sqrt{\omega_c^2 + \omega_p^2}$. For usual external fields the "volume" charge density does not build up inside the body, so we may leave it aside. We are left with the "surface" charge density and the equation for the internal (polarization) electric field

$$div\mathbf{E} = 4\pi\rho_s = -4\pi nqu_z(z=0)\delta(z) \ ; \tag{10.18}$$

the equation of motion reads

$$(\omega^2 - \omega_c^2 + i\omega\gamma)(\mathbf{u}, u_z) = -\frac{q}{m}(\mathbf{E}_0 + \mathbf{E}_s) \ , \ z > 0 \ ; \tag{10.19}$$

it remains to solve Gauss's equation (10.18) for the surface field.

This is a typical problem for finite-size bodies in the quasi-static approximation. Using $\mathbf{E}_s = -grad\Phi_s$, it amounts to solving the Poisson equation

$$\Delta\Phi_s = 4\pi nqu_z(z=0)\delta(z) \tag{10.20}$$

for the surface term. Equation (10.20) is also the Green function's equation for the laplacian (with a δ-source).[11] The standard way of tackling such equations consists in using a complete expansion in orthogonal eigenfunctions corresponding to the free coordinates. In the present case this is a Fourier expansion along the x, y-coordinates, with the wavevector $\mathbf{k}$. Leaving aside the irrelevant arguments, we get

$$\frac{d^2\Phi_s}{dz^2} - k^2\Phi_s = 4\pi nqu_z(z=0)\delta(z) \ , \tag{10.21}$$

[11] In time-dependent problems equation (10.20) is replaced by the Helmholtz equation for the Green function, of the form $\Delta\Phi + (\omega^2/c^2)\Phi = const\cdot\delta(\mathbf{r})$.

with the solution[12]

$$\Phi_s = -\frac{2\pi n q u_z(z=0)}{k} e^{-k|z|} \ . \tag{10.22}$$

We get the surface field

$$\mathbf{E}_s = 2\pi n q \left(\frac{i\mathbf{k}}{k}, -sgn(z) \right) u_z(z=0) e^{-k|z|} \ . \tag{10.23}$$

We denote the in-plane vectors by the subscript $\|$; the equation of motion (10.19) splits into two equations

$$\begin{gathered}
(\omega^2 - \omega_c^2 + i\omega\gamma)\mathbf{u}_{\|} = -\tfrac{q}{m}\mathbf{E}_{0\|} - \\
-\tfrac{i}{2}\omega_p^2 u_z(z=0)\tfrac{\mathbf{k}}{k}e^{-kz} \ , \ z > 0 \ , \\
(\omega^2 - \omega_c^2 + i\omega\gamma)u_z = -\tfrac{q}{m}E_{0z} + \\
+\tfrac{1}{2}\omega_p^2 u_z(z=0)e^{-kz} \ , \ z > 0 \ .
\end{gathered} \tag{10.24}$$

From the second equation we get

$$u_z(z=0) = -\frac{q}{m} \frac{E_{0z}(z=0)}{\omega^2 - \omega_c^2 - \frac{1}{2}\omega_p^2 + i\omega\gamma} \tag{10.25}$$

and

$$\begin{gathered}
u_z = -\tfrac{q}{m}\tfrac{1}{\omega^2 - \omega_c^2 + i\omega\gamma} \cdot \\
\cdot \left[E_{0z} + \tfrac{\omega_p^2}{2} \tfrac{E_{0z}(z=0)}{\omega^2 - \omega_c^2 - \frac{1}{2}\omega_p^2 + i\omega\gamma} e^{-kz} \right] \ ;
\end{gathered} \tag{10.26}$$

while from the first equation (10.24) we get

$$\begin{gathered}
\mathbf{u}_{\|} = -\tfrac{q}{m}\tfrac{1}{\omega^2 - \omega_c^2 + i\omega\gamma} \cdot \\
\cdot \left[\mathbf{E}_{0\|} - i\tfrac{\omega_p^2}{2} \tfrac{E_{0z}(z=0)}{\omega^2 - \omega_c^2 - \frac{1}{2}\omega_p^2 + i\omega\gamma} \tfrac{\mathbf{k}}{k} e^{-kz} \right] \ .
\end{gathered} \tag{10.27}$$

[12] M. Apostol, *Equations of Mathematical Physics*, Cambridge Scholars, Newcastle upon Tyne (2018).

It is convenient to introduce the "surface" polarizability

$$\alpha_s(\omega) = -\frac{\omega_p^2}{8\pi}\frac{1}{\omega^2 - \omega_c^2 - \frac{1}{2}\omega_p^2 + i\omega\gamma} \tag{10.28}$$

besides the "volume" polarizability

$$\alpha(\omega) = -\frac{\omega_p^2}{4\pi}\frac{1}{\omega^2 - \omega_p^2 - \omega_c^2 + i\omega\gamma} \tag{10.29}$$

(and recall the susceptibility

$$\chi(\omega) = -\frac{\omega_p^2}{4\pi}\frac{1}{\omega^2 - \omega_c^2}\ ) \ ; \tag{10.30}$$

the displacements can then be written as

$$\begin{aligned} u_z &= \tfrac{1}{nq}\chi(\omega)\cdot \\ &\cdot\left[E_{0z} - 4\pi\alpha_s(\omega)E_{0z}(z=0)e^{-kz}\right] \ , \\ \mathbf{u}_{\parallel} &= \tfrac{1}{nq}\chi(\omega)\cdot \\ &\cdot\left[\mathbf{E}_{0\parallel} + 4\pi i\alpha_s(\omega)E_{0z}(z=0)\tfrac{\mathbf{k}}{k}e^{-kz}\right] \ . \end{aligned} \tag{10.31}$$

First, we note the occurrence of an additional, surface eigenmode with frequency $\sqrt{\omega_c^2 + \omega_p^2/2}$; it is the surface plasmon. Next, we note the polarization $\mathbf{P} = nq(\mathbf{u},\, u_z)$ (for $z > 0$) and check the relationship $\mathbf{P} = \chi(\mathbf{E}_s + \mathbf{E}_0)$, by using equations (10.23), (10.26) and (10.27), as well as the continuity at the surface of the tangential components of the internal (polarization) electric field $\mathbf{E}_s$ and the transverse components of the electric displacement $\mathbf{D} = \mathbf{E}_s + 4\pi\mathbf{P}$. In addition for usual quasi-static external fields we have the conditions $div\mathbf{E}_0 = 0$ and $curl\mathbf{E}_0 = 0$, which lead, for Fourier transforms, to $\mathbf{E}_0 = \mathbf{E}_0(z=0)e^{-kz}$, as well as to $u_1 = -iu_z$. Consequently, from equations (10.26) and (10.27), we can also write the response as

$$(\mathbf{u},\, u_z) = -\frac{q}{m}\frac{1}{\omega^2 - \omega_c^2 - \frac{1}{2}\omega_p^2 + i\omega\gamma}\mathbf{E}_0 \ , \tag{10.32}$$

as expected.

By using the method described here we can treat a series of interesting situations within the quasi-static approximation, like a point charge or a point dipole placed in the vicinity of a half-space, or any other external field acting upon a half-space, or two half-spaces separated by a gap, a slab, etc. In all these situations we expect a resonance arising from the surface plasmons.

10.4 Half-space

The half-space can be given a full electromagnetic solution.[13] For a semi-infinite (homogeneous) body (half-space) extending over the region $z > d$ we take the polarization as

$$\mathbf{P} = nq(\mathbf{u},\, u_z)\theta(z-d) \ , \tag{10.33}$$

where $\theta(z) = 0$ for $z < 0$ and $\theta(z) = 1$ for $z > 0$ is the step function. The polarization charge and current densities are given by

$$\begin{gathered} \rho = -nq(div\mathbf{u} + \tfrac{\partial u_z}{\partial z})\theta(z-d) - nqu_z(d)\delta(z-d) \ , \\ \mathbf{j} = nq(\dot{\mathbf{u}},\, \dot{u}_z)\theta(z-d) \ . \end{gathered} \tag{10.34}$$

We use Fourier decompositions of the type

$$\mathbf{u}(\mathbf{r}, z; t) = \frac{1}{2\pi}\sum_{\mathbf{k}} \int d\omega \mathbf{u}(\mathbf{k}, z; \omega) e^{-i\omega t + i\mathbf{k}\mathbf{r}} \ , \tag{10.35}$$

where $\mathbf{R} = (\mathbf{r},\, z)$, and may omit ocassionally the arguments $\mathbf{k}$, ω, writing simply $\mathbf{u}(z)$, or $\mathbf{u}$. The electromagnetic potentials given by

$$\begin{gathered} \Phi(\mathbf{R}, t) = \int d\mathbf{R}' \frac{\rho(\mathbf{R}', t - \left|\mathbf{R}-\mathbf{R}'\right|/c)}{|\mathbf{R}-\mathbf{R}'|} \ , \\ \mathbf{A}(\mathbf{R}, t) = \tfrac{1}{c}\int d\mathbf{R}' \frac{\mathbf{j}(\mathbf{R}', t - \left|\mathbf{R}-\mathbf{R}'\right|/c)}{|\mathbf{R}-\mathbf{R}'|} \end{gathered} \tag{10.36}$$

[13] M. Apostol and G. Vaman, "Electromagnetic field interacting with a semi-infinite plasma", J. Opt. Soc. Am. **A26** 1747 (2009); "Reflection and refraction of the electromagnetic field in a semi-infinite plasma", Opt. Commun. **282** 4329 (2009); M. Apostol, "On the molecular forces acting between macroscopic bodies", Physica **B409** 57 (2013).

include the "retarded" Coulomb potential $e^{i\frac{\omega}{c}|\mathbf{R}-\mathbf{R}'|}/|\mathbf{R}-\mathbf{R}'|$, for which we use the well-known decomposition[14]

$$\frac{e^{i\lambda|\mathbf{R}-\mathbf{R}'|}}{|\mathbf{R}-\mathbf{R}'|}=\frac{i}{2\pi}\int d\mathbf{k}\frac{1}{\kappa}e^{i\mathbf{k}(\mathbf{r}-\mathbf{r}')}e^{i\kappa|z-z'|}\ , \tag{10.37}$$

where $\lambda=\omega/c$ and $\kappa=\sqrt{\lambda^2-k^2}$. It is more convenient to compute first the vector potential $\mathbf{A}$ and then derive the scalar potential Φ from the gauge equation $div\mathbf{A}-i\lambda\Phi=0$. The calculations are straightforward and we get the Fourier transforms of the potentials

$$\Phi(\mathbf{k},z;\omega)==\tfrac{2\pi}{\kappa}\int_d^\infty dz'\mathbf{ku}e^{i\kappa|z-z'|}-$$

$$-\tfrac{2\pi i}{\kappa}\tfrac{\partial}{\partial z}\int_d^\infty dz'u_z e^{i\kappa|z-z'|}\ , \tag{10.38}$$

$$\mathbf{A}(\mathbf{k},z;\omega)=\tfrac{2\pi\lambda}{\kappa}\int_d^\infty dz'(\mathbf{u},\,u_z)e^{i\kappa|z-z'|}$$

(where we have left aside the factor nq; it is restored in the final formulae). In order to compute the electric field ($\mathbf{E}=i\lambda\mathbf{A}-grad\Phi$) it is convenient to refer the in-plane vectors (*i.e.*, vectors parallel with the surface of the half-space) to the vectors $\mathbf{k}$ and $\mathbf{k}_\perp=\boldsymbol{e}_z\times\mathbf{k}$, where $\boldsymbol{e}_z$ is the unit vector along the z-direction; for instance, we write

$$\mathbf{u}=u_1\frac{\mathbf{k}}{k}+u_2\frac{\mathbf{k}_\perp}{k} \tag{10.39}$$

and a similar representation for the electric field parallel with the surface of the half-space. We get the electric field

$$E_1=2\pi i\kappa\int_d^\infty dz'u_1e^{i\kappa|z-z'|}-$$

$$-\tfrac{2\pi k}{\kappa}\tfrac{\partial}{\partial z}\int_d^\infty dz'u_z e^{i\kappa|z-z'|}\ ,$$

$$E_2=\tfrac{2\pi i\lambda^2}{\kappa}\int_d^\infty dz'u_2e^{i\kappa|z-z'|}\ , \tag{10.40}$$

$$E_z=-\tfrac{2\pi k}{\kappa}\tfrac{\partial}{\partial z}\int_d^\infty dz'u_1e^{i\kappa|z-z'|}+$$

$$+\tfrac{2\pi ik^2}{\kappa}\int_d^\infty dz'u_z e^{i\kappa|z-z'|}-4\pi u_z\theta(z-d)\ .$$

[14]M. Apostol, *Equations of Mathematical Physics*, Cambridge Scholars, Newcastle upon Tyne (2018).

Making use of equations (10.40), we can check easily the equalities

$$ikE_1 + \frac{\partial E_z}{\partial z} = -4\pi \left(iku_1 + \frac{\partial u_z}{\partial z}\right) \theta(z-d)$$
$$-4\pi u_z(z=d)\delta(z-d) \ , \qquad (10.41)$$

which is an expression of Gauss's law, and

$$k\frac{\partial E_1}{\partial z} + i\kappa^2 E_z = -4\pi i \lambda^2 u_z \theta(z-d) \ , \qquad (10.42)$$

which reflects Faraday's and Maxwell-Ampere's equations. From equation (10.41), we can check the transversality condition $div\mathbf{E} = 0$ for the electric field outside the half-space ($z < d$).

We use now the equation of motion (10.2) (with $\gamma = 0$) for E_2 given by equation (10.40) and for the combinations $iku_1 + \partial u_z/\partial z$ and $k\partial u_1/\partial z + i\kappa^2 u_z$ in the region $z > d$. Taking into account that $div\mathbf{E}_0 = 0$ and $k\partial E_{01}/\partial z + i\kappa^2 E_{0z} = 0$ (for a plane wave) we get

$$iku_1 + \frac{\partial u_z}{\partial z} = 0 \ , \ k\frac{\partial u_1}{\partial z} + i\kappa'^2 u_z = 0 \ , \qquad (10.43)$$

or

$$\frac{\partial^2 \mathbf{u}}{\partial z^2} + \kappa'^2 \mathbf{u} = 0 \ , \qquad (10.44)$$

where

$$\kappa'^2 = \kappa^2 - \frac{\lambda^2 \omega_p^2}{\omega^2 - \omega_c^2} \ . \qquad (10.45)$$

We emphasize that the *rhs* of equation (10.44) is vanishing only for a transverse external field. The components $u_{1,2}$ of the displacement field are given by $u_{1,2} = A_{1,2}e^{i\kappa' z}$, where $A_{1,2}$ are constants, while $u_z = -(k/\kappa')A_1 e^{i\kappa' z}$ (we restrict ourselves to outgoing waves, $\kappa' > 0$). The total electric field inside the half-space is given by the equation of motion (10.2):

$$\mathbf{E}_t = -\frac{m}{q}(\omega^2 - \omega_c^2)\mathbf{u} \qquad (10.46)$$

for $z > d$. We can see that the field propagates in the half-space with a modified wavevector κ', according to the Ewald-Oseen extinction theorem.[15] The modified wavevector κ' given by equation (10.45) can also be written as

$$\kappa'^2 = \varepsilon\frac{\omega^2}{c^2} - k^2 \ , \tag{10.47}$$

where

$$\varepsilon = 1 - \frac{\omega_p^2}{\omega^2 - \omega_c^2} \tag{10.48}$$

is the dielectric function (as given by equation (10.16), for instance). We can check the well-known polaritonic dispersion relation $\varepsilon\omega^2 = c^2K'^2$, where $\mathbf{K}' = (\mathbf{k}, \kappa')$ is the wavevector.

The amplitudes $A_{1,2}$ can be derived from the original equation (10.2) and the field equations (10.40) (for $z > d$). We get

$$\begin{aligned} &\tfrac{1}{2}A_1\omega_p^2\tfrac{\kappa\kappa'+k^2}{\kappa'(\kappa'-\kappa)}e^{i(\kappa'-\kappa)d}e^{i\kappa z} = \tfrac{q}{m}E_{01} \ , \\ &\tfrac{1}{2}A_2\omega_p^2\tfrac{\lambda^2}{\kappa(\kappa'-\kappa)}e^{i(\kappa'-\kappa)d}e^{i\kappa z} = \tfrac{q}{m}E_{02} \ . \end{aligned} \tag{10.49}$$

The (polarization) electric field, both inside and outside the half-space, can be computed from equations (10.40). We get

$$\begin{aligned} E_1 &= -4\pi nqA_1\tfrac{\omega^2-\omega_c^2}{\omega_p^2}e^{i\kappa' z} - \\ &-2\pi nqA_1\tfrac{\kappa\kappa'+k^2}{\kappa'(\kappa'-\kappa)}e^{i(\kappa'-\kappa)d}e^{i\kappa z} \ , \ z > d \ , \\ E_2 &= -4\pi nqA_2\tfrac{\omega^2-\omega_c^2}{\omega_p^2}e^{i\kappa' z} - \\ &-2\pi nqA_2\tfrac{\lambda^2}{\kappa(\kappa'-\kappa)}e^{i(\kappa'-\kappa)d}e^{i\kappa z} \ , \ z > d \ , \\ E_z &= 4\pi nqA_1\tfrac{k(\omega^2-\omega_c^2)}{\kappa'\omega_p^2}e^{i\kappa' z} + \\ &+2\pi nqA_1\tfrac{k(\kappa\kappa'+k^2)}{\kappa\kappa'(\kappa'-\kappa)}e^{i(\kappa'-\kappa)d}e^{i\kappa z} \ , \ z > d \ . \end{aligned} \tag{10.50}$$

[15] P. P. Ewald, "Ueber die Grundlagen der Kristalloptik", Thesis, Munich (1912), Ann. Phys. **49 (354)** 1 (1916); C. W. Oseen, "Ueber die Wechselwirkung zwischen zwei elektrischen Dipolen und ueber die Drehung der Polarisationsebene in Kristallen und Fluessigkeiten", Ann. Phys. **48 (353)** 1 (1915); M. Born and E. Wolf, *Principles of Optics*, Pergamon, London (1959).

for $z > d$. It is worth noting that the polarization electric field, as given by equations (10.50), includes both the external field $\sim e^{i\kappa z}$ (with opposite sign) and the displacement field $\mathbf{u} \sim e^{i\kappa' z}$. This can be checked easily by using equations (10.49) and (10.50). The (polarization) electric field outside the half-space (in the region $z < d$) is given by

$$\begin{aligned} E_1 &= -2\pi nqA_1 \tfrac{\kappa\kappa' - k^2}{\kappa'(\kappa'+\kappa)} \cdot \\ &\cdot e^{i(\kappa'+\kappa)d} e^{-i\kappa z} \ , \ z < d \ , \\ E_2 &= -2\pi nqA_2 \tfrac{\lambda^2}{\kappa(\kappa'+\kappa)} \cdot \\ &\cdot e^{i(\kappa'+\kappa)d} e^{-i\kappa z} \ , \ z < d \end{aligned} \tag{10.51}$$

and $E_z = (k/\kappa)E_1$ for $z < d$. We can see that it is the field reflected by the half-space ($\kappa \rightarrow -\kappa$). Making use of equations (10.49) and (10.51) we get also the total electric field $\mathbf{E}_t = \mathbf{E} + \mathbf{E}_0$ outside the half-space and the magnetic field $\mathbf{H} = curl\mathbf{A}$ both inside and outside the half-space. The amplitudes $A_{1,2}$ can be viewed either as being determined by the external field $\mathbf{E}_0$ (and $\mathbf{H}_0$) through equations (10.49), or as free parameters. In the latter case equations (10.49) are not valid anymore, but the (polarization) electric and magnetic fields (given, for instance, by equations (10.50), (10.51)) are valid. We can check also that all the fields are continuous at the surface $z = d$, except for E_z and E_{tz}, which exhibit a discontinuity ($E_{tz}(z = d^-) = \varepsilon E_{tz}(z = d^+)$), as expected. In particular, the surface plasmon-polariton mode given by equation (10.6) is obtained from the vanishing denominator $\kappa\kappa' + k^2$ of the response (equations (10.49)) for both κ and κ' purely imaginary, which means that both the incident and the refracted waves are localized at the surface and propagates only along the surface.

Having known the electromagnetic solution for a half-space we can derive the Fresnel formulae for reflection and refraction coefficients.[16] Similarly, we can treat two half-spaces, a slab (or a two dimensional sheet). In particular we can derive the van der Waals-London-Casimir

[16] M. Apostol and G. Vaman, "Electromagnetic field interacting with a semi-infinite plasma", J. Opt. Soc. Am. **A26** 1747 (2009); "Reflection and refraction of the electromagnetic field in a semi-infinite plasma", Opt. Commun. **282** 4329 (2009).

force for such electromagnetically coupled bodies,[17] or we can treat other interesting situations, like the presence of the inhomogeneities, rough surfaces,[18] etc. Here, we present a very interesting situation concerning a point-dipole and a half-space.

10.5 Point dipole and half-space

A point dipole placed at the origin has a charge density and a current density given by

$$\rho = -n_0 qv[(\mathbf{w}, w_z)grad]\delta(\mathbf{R}) \ ,$$
$$\mathbf{j} = n_0 qv(\dot{\mathbf{w}}, \dot{w}_z)\delta(\mathbf{R}) \ , \tag{10.52}$$

where n_0 is its density of charges q (with mass m), v is the (volume) of the dipole, $\mathbf{w}$ is the displacement field in the x, y-plane and w_z is the z-component of the displacement field; the displacement field depends only on the time. These charge and current densities introduced in equations (10.36) give the electromagnetic potentials; the vector potential is

$$\mathbf{A}_0 = -i\lambda n_0 qv(\mathbf{w}, w_z)\frac{e^{i\lambda R}}{R} \tag{10.53}$$

(temporal Fourier transform), where $\lambda = \omega/c$; the scalar potential is conveniently obtained from the Lorenz gauge $div\mathbf{A}_0 - i\lambda\Phi_0 = 0$:

$$\Phi_0 = -n_0 qv[(\mathbf{w}, w_0)grad]\frac{e^{i\lambda R}}{R} \ . \tag{10.54}$$

The electric field can be calculated from $\mathbf{E}_0 = i\lambda\mathbf{A}_0 - grad\Phi_0$. We represent the spherical wave by

$$\frac{e^{i\lambda R}}{R} = \frac{i}{2\pi}\int d\mathbf{k}\frac{1}{\kappa}e^{i\mathbf{kr}}e^{i\kappa|z|} \ , \tag{10.55}$$

[17] M. Apostol, "On the molecular forces acting between macroscopic bodies", Physica **B409** 57 (2013).

[18] B. F. Apostol, "Scattering of the electromagnetic waves from a rough surface" J. Mod. Phys. **59** 1607 (2012).

where $\mathbf{R} = (\mathbf{r}, z)$ and $\kappa = \sqrt{\lambda^2 - k^2}$; then, we select the $\mathbf{k}$-component by using $\sum_{\mathbf{k}} = [A/(2\pi)^2] \int d\mathbf{k}$, where A is the in-plane ((x, y)-plane) area; it is convenient to use the projections of the electric field on $\mathbf{k}$ (E_{01}) and $\mathbf{k}_\perp$ (E_{02}), where $\mathbf{k}_\perp$ is perpendicular to $\mathbf{k}$ and has the same magnitude k. We get

$$E_{01} = \frac{2\pi i n_0 qv}{A}[\kappa w_1 - k w_z sgn(z)]e^{i\kappa|z|} ,$$

$$E_{02} = \frac{2\pi i n_0 qv}{A} \cdot \frac{\lambda^2}{\kappa} w_2 e^{i\kappa|z|} , \tag{10.56}$$

$$E_{0z} = \frac{2\pi n_0 qv}{A} \cdot$$

$$[\frac{ik^2}{\kappa} w_z - ikw_1 sgn(z) - 2w_z\delta(z)]e^{i\kappa|z|} ;$$

we can check that this field is transverse ($div\mathbf{E}_0 = 0$).

From the calculations for a half-space described before we know the displacement field $u_{1,2} = A_{1,2}e^{i\kappa' z}$, $u_z = -(k/\kappa')A_1 e^{i\kappa' z}$, for $z > d$, where

$$\kappa' = \sqrt{\kappa^2 - \frac{\lambda^2\omega_p^2}{\omega^2 - \omega_c^2}} = \sqrt{\lambda^2\varepsilon - k^2} \tag{10.57}$$

and the coefficients $A_{1,2}$ are given by

$$\frac{1}{2}A_1\omega_p^2 \frac{\kappa\kappa' + k^2}{\kappa'(\kappa' - \kappa)} e^{i(\kappa'-\kappa)d} e^{i\kappa z} =$$

$$= \frac{q}{m}E_{01} ,$$

$$\frac{1}{2}A_2\omega_p^2 \frac{\lambda^2}{\kappa(\kappa' - \kappa)} e^{i(\kappa'-\kappa)d} e^{i\kappa z} = \tag{10.58}$$

$$= \frac{q}{m}E_{02} ,$$

ω_p being the plasma frequency for the half-space (equations (10.49)).

The field reflected by the half-space (equations (10.51)) is given by

$$E_{r1} = -2\pi nqA_1 \frac{\kappa\kappa' - k^2}{\kappa'(\kappa' + \kappa)} \cdot$$
$$\cdot e^{i(\kappa'+\kappa)d} e^{-i\kappa z} \ , \ z < d \ ,$$
$$E_{r2} = -2\pi nqA_2 \frac{\lambda^2}{\kappa(\kappa' + \kappa)} \cdot \tag{10.59}$$
$$\cdot e^{i(\kappa'+\kappa)d} e^{-i\kappa z} \ , \ z < d$$

and $E_{rz} = (k/\kappa)E_1$ for $z < d$, where n is the density of charges in the half-space. By using equations (10.58) we can express this reflected field by means of the field $\mathbf{E}_0$ and, by using equations (10.56), we can express it in terms of the displacement $\mathbf{W} = (\mathbf{w}, w_z)$. The reflected field plays the role of an external field in the equation of motion of the dipole displacement field

$$(\omega^2 - \omega_0^2)\mathbf{W} = -\frac{q}{m}\mathbf{E}_r(\mathbf{r} = 0, z = 0; \omega) \ , \tag{10.60}$$

where ω_0 is a characteristic frequency of the point dipole; we get

$$(\omega^2 - \omega_0^2)\mathbf{W} = \frac{i\omega_{p0}^2 v}{A} \cdot$$
$$\sum_{\mathbf{k}} \left\{ \frac{(\kappa' - \kappa)(\kappa\kappa' - k^2)}{(\kappa' + \kappa)(\kappa\kappa' + k^2)} (\kappa w_1 - k w_z)(\frac{\mathbf{k}}{k} + \frac{k}{\kappa}\mathbf{e}_z) + \right. \tag{10.61}$$
$$\left. + \frac{\lambda^2(\kappa' - \kappa)}{\kappa(\kappa' + \kappa)} w_2 \frac{\mathbf{k}_\perp}{k} \right\} e^{2i\kappa d} \ ,$$

where $\omega_{p0} = \sqrt{4\pi n_0 q^2/m}$. Here we use $\mathbf{k} = k(\cos\varphi, \sin\varphi)$ and $\mathbf{k}_\perp = k(-\sin\varphi, \cos\varphi)$ and integrate over the angle φ. We get the equations

$$(\omega^2 - \omega_0^2)w = \frac{i\omega_{p0}^2 v}{8\pi} \cdot$$
$$\cdot \int dk \cdot k \left[\frac{\kappa(\kappa' - \kappa)(\kappa\kappa' - k^2)}{(\kappa' + \kappa)(\kappa\kappa' + k^2)} + \frac{\lambda^2(\kappa' - \kappa)}{\kappa(\kappa' + \kappa)} \right] w e^{2i\kappa d} \ ,$$
$$(\omega^2 - \omega_0^2)w_z = -\frac{i\omega_{p0}^2 v}{4\pi} \cdot \tag{10.62}$$
$$\cdot \int dk \cdot k^3 \frac{(\kappa' - \kappa)(\kappa\kappa' - k^2)}{\kappa(\kappa' + \kappa)(\kappa\kappa' + k^2)} u_z e^{2i\kappa d} \ ;$$

w is an in-plane component of the displacement, the same equation holds for the other component. These equations can also be written as

$$(\omega^2-\omega_0^2)w=\frac{i\omega_{p0}^2 v}{8\pi}\cdot$$
$$\cdot\int dk\cdot k\left[\frac{\kappa(\kappa\varepsilon-\kappa')}{\kappa\varepsilon+\kappa'}+\frac{\lambda^2(\kappa'-\kappa)}{\kappa(\kappa'+\kappa)}\right]we^{2i\kappa d}\ , \tag{10.63}$$
$$(\omega^2-\omega_0^2)w_z=-\frac{i\omega_{p0}^2 v}{4\pi}\cdot$$
$$\cdot\int dk\cdot k^3\frac{\kappa\varepsilon-\kappa'}{\kappa(\kappa\varepsilon+\kappa')}w_z e^{2i\kappa d}\ ;$$

it is easy to see that these equations give the eigenfrequencies of the coupled point dipole and half-space.

We notice first that the dipole volume v should be much smaller than any relevant scale; this means that the solution ω should lie near ω_0 and the *rhs* of equations (10.63) should be treated as a small perturbation. Next, we notice that for large distances d the main contribution to the integrals come from small values of κ, *i.e.* from k near λ; this implies $d\lambda_0=d\omega_0/c\gg 1$ and small frequencies ω_0 in comparison with ω_c, ω_p; the main contribution comes in this case from small values of k; this is the retarded case; it is easy to see that equations (10.63) have not solutions in this case: the waves are damped, except for some discrete values of the distance d; in general, the two bodies do not couple to each other in this case. For small distances d ($d\omega_0/c\ll 1$), the main contribution to the integrals comes from large values of k, such as $\kappa\simeq\kappa'\simeq ik$; the first equation (10.63) can be written approximately

$$\omega^2-\omega_0^2\simeq\frac{i\omega_{p0}^2 v}{8\pi}\frac{\varepsilon-1}{\varepsilon+1}\int dk\cdot ik^2 e^{-2kd}\ ; \tag{10.64}$$

this is the non-retarded case. We get easily the solution

$$\Omega_{\parallel}\simeq\omega_0-\frac{\omega_{p0}^2 v}{64\pi\omega_0}\frac{\varepsilon_0-1}{\varepsilon_0+1}\cdot\frac{1}{d^3}\ , \tag{10.65}$$

where $\varepsilon_0=\varepsilon(\omega_0)$; hence, the zero-point energy $\hbar\Omega_{\parallel}/2$ leads to the van der Waals-London force (for a point dipole coupled to a half-space); we

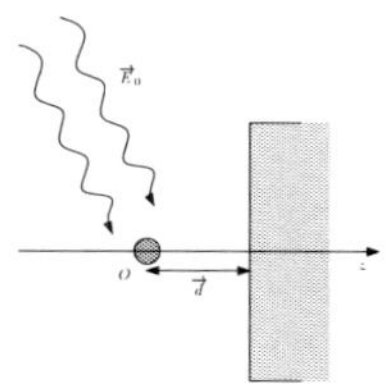

Figure 10.1: External field, point dipole and a half-space.

should take into account the multiplicity 2 for the transverse oscillations and the contribution of the z-component, whose eigenfrequency is given by

$$\Omega_z \simeq \omega_0 - \frac{\omega_{p0}^2 v}{32\pi\omega_0} \frac{\varepsilon_0 - 1}{\varepsilon_0 + 1} \cdot \frac{1}{d^3} . \qquad (10.66)$$

Finally, we get the force

$$F = -\frac{3\hbar\omega_{p0}^2 v}{32\pi\omega_0} \frac{\varepsilon_0 - 1}{\varepsilon_0 + 1} \cdot \frac{1}{d^4} . \qquad (10.67)$$

Making use of $\varepsilon = 1 - \omega_p^2/(\omega^2 - \omega_c^2)$, we note that one obtains an appreciable enhancement of this force for the characteristic frequency ω_0 of the point dipole approaching the frequency $\sqrt{\omega_c^2 + \omega_p^2/2}$ of the surface plasmon.

10.6 Surface enhanced Raman scattering

We consider a point dipole induced by an external field, placed in the vicinity of a half-space. The equation of motion (10.60) acquires now an additional term, arising from the contribution of the external field. We take a monochromatic plane wave of the form

$$\mathbf{E}_0(t, \mathbf{r}, z) = \mathbf{E}_0 e^{-i\Omega t} e^{i\mathbf{K}\mathbf{r} + K_z z} \qquad (10.68)$$

as an external (radiation) field. The calculations described above can easily be repeated in this case. We limit ourselves here to give the result (in the reference frame formed by the vectors $\mathbf{K}$, $\mathbf{K}_\perp$ and $\mathbf{K}_z$:

$$\begin{gathered}
(\omega^2-\omega_0^2)w_1=\frac{i\omega_{p0}^2 v}{8\pi}w_1\cdot\\
\cdot\int dk\cdot k\left[\kappa\frac{\varepsilon\kappa-\kappa'}{\varepsilon\kappa+\kappa'}+\frac{\lambda^2(\kappa'-\kappa)}{\kappa(\kappa'+\kappa)}\right]e^{2i\kappa d}+\\
+\frac{2\pi q}{m}\left[\frac{\varepsilon K_z-K_z'}{\varepsilon K_z+K_z'}e^{2iK_z d}-1\right]E_{01}\delta(\omega-\Omega)\ ,\\
(\omega^2-\omega_0^2)w_2=\frac{i\omega_{p0}^2 v}{8\pi}w_2\cdot\\
\cdot\int dk\cdot k\left[\kappa\frac{\varepsilon\kappa-\kappa'}{\varepsilon\kappa+\kappa'}+\frac{\lambda^2(\kappa'-\kappa)}{\kappa(\kappa'+\kappa)}\right]e^{2i\kappa d}+\\
+\frac{2\pi q}{m}\left[\frac{K_z'-K_z}{K_z'+K_z}e^{2iK_z d}-1\right]E_{02}\delta(\omega-\Omega)\ ,\\
(\omega^2-\omega_0^2)w_z=-\frac{i\omega_{p0}^2 v}{4\pi}w_z\int dk\cdot k^3\frac{\varepsilon\kappa-\kappa'}{\kappa(\varepsilon\kappa+\kappa')}e^{2i\kappa d}-\\
-\frac{2\pi q}{m}\left[\frac{\varepsilon K_z-K_z'}{\varepsilon K_z+K_z'}e^{2iK_z d}+1\right]E_{0z}\delta(\omega-\Omega)\ ,
\end{gathered}\tag{10.69}$$

where $K_z'=\sqrt{\varepsilon\Omega^2/c^2-K^2}$. As before, the main contributions to the integrals in equations (10.69) comes from large values of k; in addition, we are interested in small distances, such as $e^{2iK_z d}\simeq 1$. We get easily

$$\begin{gathered}
(\omega^2-\Omega_\parallel^2)w_1=-\frac{\cos r}{\sqrt{\varepsilon}\cos i+\cos r}\cdot\\
\cdot\frac{4\pi q}{m}E_{01}\delta(\omega-\Omega)\ ,\\
(\omega^2-\Omega_\parallel^2)w_2=-\frac{\cos i}{\cos i+\sqrt{\varepsilon}\cos r}\cdot\\
\cdot\frac{4\pi q}{m}E_{02}\delta(\omega-\Omega)\ ,\\
(\omega^2-\Omega_z^2)w_z=-\frac{\sqrt{\varepsilon}\cos i}{\sqrt{\varepsilon}\cos i+\cos r}\cdot\\
\cdot\frac{4\pi q}{m}E_{0z}\delta(\omega-\Omega)\ ,
\end{gathered}\tag{10.70}$$

where i is the incidence angle, r is the refraction angle ($\sin r/\sin i = 1/\sqrt{\varepsilon}$) and $\Omega_{\parallel}$, Ω_z are given by equations (10.65) and (10.66).

The displacement field $w_{1,2,z}$ given by equations (10.70) represents the response of the induced point dipole placed in the vicinity of a half-space to an external radiation field; multiplied by n_0qv it gives the induced dipole moment. Equations (10.70) answer the question regarding the effect of a semi-infinite solid placed in the vicinity of a small, neutral nano-particle which can acquire an induced dipole moment under the action of a radiation field. This is the typical problem in the surface enhanced Raman scattering.[19] We can see that the induced dipole may be enhanced appreciably for frequencies close to the resonance frequencies $\Omega_{\parallel,z}$.[20] We can compare this effect with the absence of the semi-infinite solid, which is obtained by putting $\varepsilon = 1$ (in this case $\Omega_{\parallel,z}$ reduce to ω_0).

In conclusion, the electromagnetic field in matter is described by a new method which includes explicitly the polarization and the magnetization degrees of freedom. The Maxwell equations are supplemented by the equations of motion of the polarization and magnetization, thus making possible a full solution. The method is applied to quasi-static fields and the plasmon response of a semi-infinite polarizable (non-magnetic) solid with a plane surface (half-space). The surface plasmons are highlighted and the surface polarizability is introduced. The electromagnetically coupled system of a neutral, polarizable nano-particle (induced point dipole) and a half-space is solved in the general situation, and the eigenfrequencies and eigenmodes of the zero-point (vacuum) fluctuations are obtained. The van der Waals-London-Casimir force is computed for this system, as well as its re-

[19] Z. Q. Tian, *Surface-Enhanced Raman Spectroscopy: It's Present Status* The Internet Journal of Vibrational Spectroscopy 4 ed. 2 (2000); P. Etchegoin, L. F. Cohen, H. Hartigan, R. J. C. Brown, M. J. T. Milton and J. C. Gallop, "Electromagnetic contribution to surface enhanced Raman scattering revisited", J. Chem. Phys. **119** 5281 (2003); M. Moskovits, *Surface-Enhanced Raman Spectroscopy: a Brief Perspective*, in *Surface-Enhanced Raman Scattering, Topics in Applied Physics*, vol. **103**, eds. K. Kneipp, M. Moskovits and H. Kneipp, Springer (2006), pp. 1–17; H. Xu, E. J. Bjerneld, M. Käll and L. Börjesson, "Spectroscopy of Single Hemoglobin Molecules by Surface Enhanced Raman Scattering", Phys. Rev. Lett. **83** 4357 (1999).

[20] M. Apostol, S. Ilie, A. Petrut, M. Savu and S. Toba, "Electromagnetic field in matter. Surface enhanced Raman scattering", Adv. Electrom. **2** 31 (2013).

sponse to an external radiation field, which is the typical problem for the surface enhanced Raman scattering.

10.7 Wave on surface

10.7.1 Cylindrical geometry

The Maxwell equations for electromagnetic waves in (non-magnetic) matter are

$$\begin{gathered} div\boldsymbol{D} = 0 \ , \ div\boldsymbol{H} = 0 \ , \\ curl\boldsymbol{E} = -\frac{1}{c}\frac{\partial \boldsymbol{H}}{\partial t} \ , \ curl\boldsymbol{H} = \frac{1}{c}\frac{\partial \boldsymbol{D}}{\partial t} \ , \end{gathered} \tag{10.71}$$

where $\boldsymbol{D}$ is the electric displacement, $\boldsymbol{E}$ is the electric field, $\boldsymbol{H}$ is the magnetic field, c denotes the speed of light in vacuum and t denotes the time. The electric displacement includes the internal field generated by the mobile charges in matter. We consider a homogeneous and isotropic body, such that the spatial dispersion may be neglected. For wavelengths longer than the dimensions of the body the Coulomb interaction is important; it gives rise to surface plasmons, which are additional plasmon branches besides the volume plasmons. We neglect these surface effects, and view the electric displacement as being related to the (total) electric field through $\boldsymbol{D} = \varepsilon\boldsymbol{E}$, where ε is the dielectric function. This function depends on the frequency ω, such that, for a monochromatic wave the Maxwell equations read

$$\begin{gathered} div\boldsymbol{E} = 0 \ , \ div\boldsymbol{H} = 0 \ , \\ curl\boldsymbol{E} = \frac{i\omega}{c}\boldsymbol{H} \ , \ curl\boldsymbol{H} = -\frac{i\varepsilon\omega}{c}\boldsymbol{E} \ . \end{gathered} \tag{10.72}$$

It is easy to see that in cylindrical geometry four field components are given by the other two; indeed, assuming a dependence $\sim e^{ikx}$ along the cylindrical axis, we get

$$\begin{gathered} \frac{i\omega}{c}H_r + ikE_\varphi = \frac{1}{r}\frac{\partial E_x}{\partial \varphi} \ , \ ikH_r + \frac{i\varepsilon\omega}{c}E_\varphi = \frac{\partial H_x}{\partial r} \ , \\ ikH_\varphi - \frac{i\varepsilon\omega}{c}E_r = \frac{1}{r}\frac{\partial H_x}{\partial \varphi} \ , \ \frac{i\omega}{c}H_\varphi - ikE_r = -\frac{\partial E_x}{\partial r} \ , \end{gathered} \tag{10.73}$$

or

$$i\kappa^2 H_r = \frac{\varepsilon\omega}{cr}\frac{\partial E_x}{\partial\varphi} - k\frac{\partial H_x}{\partial r} \; , \; i\kappa^2 E_\varphi = \frac{\omega}{c}\frac{\partial H_x}{\partial r} - \frac{k}{r}\frac{\partial E_x}{\partial\varphi} \; ,$$
$$i\kappa^2 H_\varphi = -\frac{\varepsilon\omega}{c}\frac{\partial E_x}{\partial r} - \frac{k}{r}\frac{\partial H_x}{\partial\varphi} \; , \; i\kappa^2 E_r = -\frac{\omega}{cr}\frac{\partial H_x}{\partial\varphi} - k\frac{\partial E_x}{\partial r} \; , \tag{10.74}$$

where $\kappa^2 = \varepsilon\omega^2/c^2 - k^2$. We can see that $E_{r,\varphi}$ and $H_{r,\varphi}$ are given by E_x and H_x. This is a typical eigenmode problem.[21] From equation (10.72) we get

$$\left(\Delta + \frac{\varepsilon\omega^2}{c^2}\right)\boldsymbol{E} = 0 \; , \; \left(\Delta + \frac{\varepsilon\omega^2}{c^2}\right)\boldsymbol{H} = 0 \; ; \tag{10.75}$$

these equations acquire a simple form

$$\left[\frac{1}{r}\frac{\partial}{\partial r}\left(r\frac{\partial}{\partial r}\right) + \frac{1}{r^2}\frac{\partial^2}{\partial\varphi^2} + \kappa^2\right](E_x, H_x) = 0 \tag{10.76}$$

for E_x, H_x.

We apply these equations to a long (and thin), straight cylindrical wire in vacuum. Since the field is continuous, k is the same inside and outside the wire. In vacuum $\varepsilon = 1$, and we denote $\kappa_0^2 = \omega^2/c^2 - k^2$. At the surface of the wire the tangential components of the field ($E_{x,\varphi}$, $H_{x,\varphi}$) should be continuous. Inside the wire the regular solutions of equation (10.76) are

$$E_x = AJ_m(\kappa r)e^{im\varphi}e^{ikx} \; , \; H_x = BJ_m(\kappa r)e^{im\varphi}e^{ikx} \; , \tag{10.77}$$

where A, B are constants and J_m are Bessel functions of m-th order. Outside the wire the solutions are

$$E_x = CH_m(\kappa_0 r)e^{im\varphi}e^{ikx} \; , \; H_x = DH_m(\kappa_0 r)e^{im\varphi}e^{ikx} \; , \tag{10.78}$$

where C, D are constants and H_m are Hankel functions of the first kind and m-th order (so-called Sommerfeld radiation condition[22]).

[21] A. Sommerfeld, "Ueber die Fortpflanzung elektrodynamischer Wellen laengs eines Drahtes", Ann. Phys. Chem. **67 (303)** 233 (1899); *Lectures on Theoretical Physics*, vol. 3, *Electrodynamics*, Academic Press, London (1964); D. Hondros, "Ueber elektromagnetische Drahtwellen", Ann. Physik **30 (335)** 905 (1909); D. Hondros und P. Debye, "Elektromagnetische Wellen an dielektrischen Draehten", Ann. Physik **32 (337)** 465 (1910).

[22] M. Apostol, *Equations of Mathematical Physics*, Cambridge Scholars, Newcastle upon Tyne (2018).

These functions go to infinity like

$$H_m(z) \sim \sqrt{\frac{2}{\pi z}} e^{i[z-(m+1/2)\pi/2]} \ , \ z \longrightarrow \infty \ ; \tag{10.79}$$

they are outgoing waves, and they are damped for a positive imaginary part of κ_0. The continuity of E_x, H_x at the boundary of the wire ($r = a$) reads

$$AJ_m = CH_m \ , \ BJ_m = DH_m \ ; \tag{10.80}$$

we omit the arguments κa, $\kappa_0 a$ because they can be seen easily from the context. The continuity of E_φ, H_φ at the surface of the wire leads to

$$\begin{aligned} &\tfrac{\varepsilon\omega}{c\kappa} AJ_m^{'} + \tfrac{imk}{a\kappa^2} BJ_m = \tfrac{\omega}{c\kappa_0} CH_m^{'} + \tfrac{imk}{a\kappa_0^2} DH_m \ , \\ &\tfrac{\omega}{c\kappa} BJ_m^{'} - \tfrac{imk}{a\kappa^2} AJ_m = \tfrac{\omega}{c\kappa_0} DH_m^{'} - \tfrac{imk}{a\kappa_0^2} CH_m \ . \end{aligned} \tag{10.81}$$

Equations (10.80) and (10.81) give the eigenmodes labelled by k (or κ, κ_0). From equations (10.80) we get

$$A = \frac{H_m}{J_m} C \ , \ B = \frac{H_m}{J_m} D \tag{10.82}$$

and equations (10.81) become

$$\begin{aligned} &\left(\tfrac{\varepsilon\omega}{c\kappa} \tfrac{J_m^{'}}{J_m} - \tfrac{\omega}{c\kappa_0} \tfrac{H_m^{'}}{H_m} \right) C + \left(\tfrac{imk}{a\kappa^2} - \tfrac{imk}{a\kappa_0^2} \right) D = 0 \ , \\ &\left(\tfrac{\omega}{c\kappa} \tfrac{J_m^{'}}{J_m} - \tfrac{\omega}{c\kappa_0} \tfrac{H_m^{'}}{H_m} \right) D + \left(\tfrac{imk}{a\kappa_0^2} - \tfrac{imk}{a\kappa^2} \right) C = 0 \end{aligned} \tag{10.83}$$

(the first equation (10.83) comes from the continuity of the electric field E_φ, the second equation (10.83) arises form the continuity of the magnetic field H_φ)

These equations raise an interesting problem of applied physics, pointed out for the first time by Sommerfeld.[23]

[23] A. Sommerfeld, "Ueber die Fortpflanzung elektrodynamischer Wellen laengs eines Drahtes", Ann. Phys. Chem. **67 (303)** 233 (1899); *Lectures on Theoretical Physics*, vol. 3, *Electrodynamics*, Academic Press, London (1964).

The problem arises from the dielectric function ε. In metals it reads

$$\varepsilon = 1 - \frac{\omega_0^2}{\omega^2 + i\omega\gamma} , \tag{10.84}$$

where ω_0 is the plasma frequency and γ is the attenuation coefficient. The latter is related to the conductivity

$$\sigma = \frac{1}{4\pi} \frac{i\omega_0^2}{\omega + i\gamma} , \tag{10.85}$$

which becomes the well-known static conductivity $\sigma_0 = \omega_0^2/4\pi\gamma$ for $\omega = 0$; $1/\gamma$ is the relaxation time. For typical metals ω_0 is of the order $10^{15}s^{-1}$ and γ (deduced from σ_0) is in the range $10^{13}s^{-1}$ – $10^{14}s^{-1}$. It follows that for electromagnetic waves with frequency ω up to, say, $10^{12}s^{-1}$ (terahertz) the dielectric function has a large positive imaginary part,

$$\varepsilon \simeq 1 + \frac{i\omega_0^2}{\omega\gamma} = 1 + \frac{4\pi i\sigma_0}{\omega} \tag{10.86}$$

(a relation which holds also for σ, $\varepsilon = 1 + 4\pi i\sigma/\omega$); metals have a large conductivity (and a reactive dielectric function). In typical metals the electromagnetic waves do not penetrate easily, since the mobile charges react and compensate the external field. Consequently, $\kappa^2 = \varepsilon\omega^2/c^2 - k^2$ may have a large positive imaginary part, which makes $J_m(\kappa r)$ irregular. It follows that k^2 (and k) in $\kappa^2 = \varepsilon\omega^2/c^2 - k^2$ should develop a large positive imaginary part, in order to compensate the large positive imaginary part of ε. This means that the electromagnetic waves are damped to a large extent ($\sim e^{ikx}$) and do not propagate, both in the wire and outside, providing they are determined by the system of equations (10.83). A large Joule heat is developed by these damped waves.

However, if $m = 0$ in equations (10.83) (symmetrical wave), the waves are not determined, and the solution is given by

$$\frac{\varepsilon\omega}{c\kappa} \frac{J_0'}{J_0} - \frac{\omega}{c\kappa_0} \frac{H_0'}{H_0} = 0 \tag{10.87}$$

(electric waves) and

$$\frac{\omega}{c\kappa} \frac{J_0'}{J_0} - \frac{\omega}{c\kappa_0} \frac{H_0'}{H_0} = 0 \tag{10.88}$$

(magnetic waves). In these equations κ has a large positive imaginary part. The asymptotic behaviour of J_0 at infinity is $J_0 \sim \cos(z - \pi/4)$ and, for $z = i\Gamma$, $\Gamma > 0$, we have $J_0 \sim e^{\Gamma}$ and $J_0^{'} \sim -ie^{\Gamma}$, such that we may replace $J_0^{'}/J_0$ by $-i$; equation (10.87) becomes

$$\frac{H_0}{H_0^{'}} = \frac{i\kappa}{\varepsilon\kappa_0} \ . \tag{10.89}$$

Let us examine the term on the right in equation (10.89). Since

$$\kappa^2 = \left(1 + \frac{i\omega_0^2}{\omega\gamma}\right)\frac{\omega^2}{c^2} - k^2 = \kappa_0^2 + \frac{i\omega_0^2\omega}{\gamma c^2} \tag{10.90}$$

and, assuming $\omega_0^2\omega/\gamma c^2 \gg \left|\kappa_0^2\right|$, we may write

$$\frac{i\kappa}{\varepsilon\kappa_0} \simeq \frac{\omega\sqrt{i\omega\gamma}}{\omega_0 c\kappa_0} \ ; \tag{10.91}$$

the term on the right of equation (10.91) is much smaller than unity in absolute value, *i.e.* we assume the inequalities

$$\frac{\omega^3\gamma}{\omega_0^2 c^2} \ll \left|\kappa_0^2\right| \ll \frac{\omega_0^2\omega}{\gamma c^2} \ ; \tag{10.92}$$

these inequalities imply $\omega\gamma \ll \omega_0^2$, which is a valid inequality. Consequently, we may use the asymptotic Hankel function $H_0 \simeq (2i/\pi)\cdot$ $\cdot\ln(Cz/2i)$ and $H_0^{'} \simeq (2i/\pi)/z$ for $\mid z \mid \ll 1$ in equation (10.89), where $\ln C = 0.577...$ is the Euler-Mascheroni constant ($C \simeq 1.78$). Actually, we may use $H_0 \simeq (2i/\pi)\ln z$ and equation (10.89) becomes

$$(\kappa_0 a)^2 \ln(\kappa_0 a)^2 = \frac{2\omega\sqrt{i\omega\gamma}}{\omega_0 c}a \tag{10.93}$$

for $\kappa_0 a \ll 1$. The solution of this equation is

$$(\kappa_0 a)^2 \simeq \frac{2\omega\sqrt{i\omega\gamma}}{\omega_0 c}a\frac{1}{\ln\left(\frac{2\omega\sqrt{\omega\gamma}}{\omega_0 c}a\right)} \tag{10.94}$$

for $\frac{\omega\sqrt{\omega\gamma}}{\omega_0 c}a \ll 1$, *i.e.*

$$\frac{\omega\gamma}{\omega_0^2} \ll \frac{c^2}{\omega^2 a^2} = \frac{\lambda^2}{4\pi^2 a^2} \ , \tag{10.95}$$

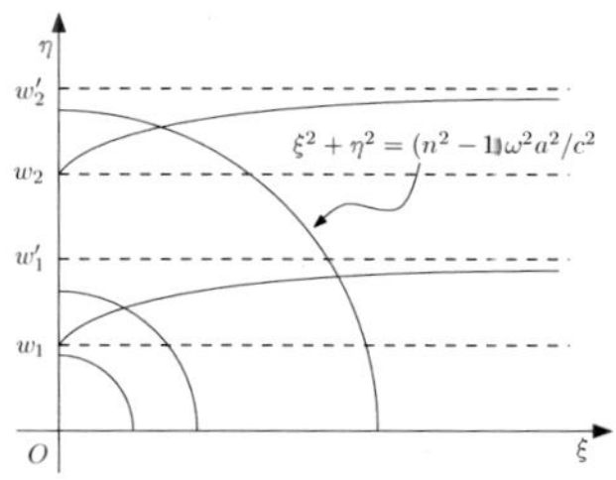

Figure 10.2: The roots of equation (10.97).

where $\lambda = c/\omega$ is the wavelength. This relation provides a constraint between ω and a. For instance, for $\omega = 10^{-2}s^{-1}$ this condition leads to $a^2 \ll 1cm^2 - 10cm^2$ ($\gamma = 10^{13}s^{-1} - 10^{14}s^{-1}$). Noteworthy, $Re(\kappa_0 a)^2 <$ 0 and $Im(\kappa_0 a) > 0$.

We can see that the electric component of the wave propagates, practically, undamped ouside the wire ($\omega/c \simeq k$), with a long radial wavelength (small $\kappa_0 a$). This is not so for the magnetic component (equation (10.88)), which gives a damped H_x; therefore $B = D = 0$, from equations (10.77) and (10.82). Inside the wire we are left with $E_x \sim AJ_0(\kappa r)$, $E_r \sim A(ik/\kappa)J_0'(\kappa r)$ and $H_\varphi \sim A(i\varepsilon\omega/c\kappa)J_0'(\kappa r)$ (equations (10.77) and (10.74)), where κ (and ε) has a large positive imaginary part (which gives a small penetration depth of a skin effect). Consequently, $A = 0$ and the waves do not propagate inside the wire. Outside, from equations (10.83) the constant C is free, such that we may have an electric wave outside the wire, which propagates practically undamped with $\kappa_0 = \sqrt{\omega^2/c^2 - k^2} \simeq 0$; its phase velocity is slightly smaller than c. The components of this wave are E_x, E_r and H_φ, where E_r and H_φ are much larger than E_x. This is the Sommerfeld (electric) surface wave.

For dielectrics the dielectric function is real and greater than unity; it is the square of the refractive index (for non-magnetic dielectrics), $\varepsilon = n^2$. The electromagnetic waves propagate in dielectrics, so k is real. However, the wavenumber $\kappa_0 = \sqrt{\omega^2/c^2 - k^2}$ must be purely imaginary, $\kappa_0 a = i\xi$, since the phase velocity ω/k must be smaller than c. Therefore, the waves are damped outside the wire and guided inside (waveguide), with the wavevector k and the real wavenumber

$\kappa = \sqrt{\varepsilon\omega^2/c^2 - k^2} = \eta/a$; we have the relationship

$$\xi^2 + \eta^2 = \left(n^2 - 1\right)\frac{4\pi^2 a^2}{\lambda^2} \ , \qquad (10.96)$$

where $\lambda = 2\pi c/\omega$ is the wavelength. Equations (10.83) (boundary conditions) have, in general, solutions. Let us examine the symmetric electric wave (main, principal electric mode) given by equation (10.87),

$$i\xi\frac{H_0(i\xi)}{H_0'(i\xi)} = \frac{\eta}{n^2}\frac{J_0(\eta)}{J_0'(\eta)} \ . \qquad (10.97)$$

Making use of the asymptotic behaviour of H_0 we can see that the term on the left in equation (10.97) varies between 0 ($-\xi^2 \ln\xi$) and ∞ (ξ) for ξ in the range $\xi = 0$ and $\xi \longrightarrow \infty$, *i.e.* it goes, more or less, like ξ. In this interval η varies between the roots w_i of $J_0(\eta) = 0$ and the roots w_i' of $J_0'(\eta) = 0$, which are alternating. For $\eta \longrightarrow 0$, $J_0'(\eta) \sim \eta$. It follows that the roots of the equation (10.97) are given by the intersection of these curves with the circle given by equation (10.96), as shown in Fig. 10.2. We can see that the first root appears for $\left(n^2-1\right)4\pi^2a^2/\lambda^2 = w_1^2$, where $w_1 = 2.4$ is the first root of $J_0(w) = 0$; it is $\xi = 0$ ($\kappa_0 = i0^+$) and $\eta = \kappa a = w_1$. It follows that the maximum wavelength admissible is $\lambda_{max} = 2\pi a(n^2-1)^{1/2}/w_1$ (for water in the high-frequency range $\varepsilon \simeq 9$). Longer wavelengths cannot be propagated. For shorter wavelengths, many other modes may appear. The asymptotic behaviour of $H_0(\kappa_0 r)$ for large values of the argument indicates a damped wave $\sim e^{-\xi r/a}$ outside the wire, where ξ are the roots of equation (10.97). This behaviour corresponds to an inverse skin effect (a damping outside the wire), with the penetration length of the order a/ξ.

A similar analysis holds for the magnetic-type waves, corresponding to equation (10.88).

10.7.2 Half-space

The above calculations can be transposed immediately to a half-space. We consider a half-space in the region $z < 0$ and a field dependence along the x-axis of the form $\sim e^{ikx}$. The four field components are given by E_x and H_x according to equations

$$i\kappa^2 H_y = \frac{\varepsilon\omega}{c}\frac{\partial E_x}{\partial z} - k\frac{\partial H_x}{\partial y} \ , \ i\kappa^2 H_z = -k\frac{\partial H_x}{\partial z} - \frac{\varepsilon\omega}{c}\frac{\partial E_x}{\partial y} \ ,$$
$$i\kappa^2 E_y = -k\frac{\partial E_x}{\partial y} - \frac{\omega}{c}\frac{\partial H_x}{\partial z} \ , \ i\kappa^2 E_z = \frac{\omega}{c}\frac{\partial H_x}{\partial y} - k\frac{\partial E_x}{\partial z} \ , \tag{10.98}$$

where

$$\left(\Delta_{y,z} + \kappa^2\right)(E_x, H_x) = 0 \ ; \tag{10.99}$$

as above, $\kappa^2 = \frac{\varepsilon\omega^2}{c^2} - k^2$ inside the body (half-space) and $\kappa_0^2 = \frac{\omega^2}{c^2} - k^2$ outside. It is easy to see that the y-dependence leads to damped modes along the x-direction (corresponding to asymmetric modes above). The principal mode is uniform along the y-direction. Then, the electric-type modes are decoupled from the magnetic modes in equations (10.98). The solution of equation (10.99) is

$$E_x^i = Ae^{i\kappa z}e^{ikx} \ , \ E_x^o = Ae^{i\kappa_0 z}e^{ikx} \tag{10.100}$$

for the inside (i) and the outside (o) electric component E_x(electric-type mode); this component is continuous at $z = 0$. From equations (10.98) we get

$$H_y^i = \frac{\varepsilon\omega}{c\kappa}Ae^{i\kappa z}e^{ikx} \ , \ H_y^o = \frac{\omega}{c\kappa_0}Ae^{i\kappa_0 z}e^{ikx} \ ,$$
$$E_z^i = -\frac{k}{\kappa}Ae^{i\kappa z}e^{ikx} \ , \ E_z^o = -\frac{k}{\kappa_0}Ae^{i\kappa_0 z}e^{ikx} \ ; \tag{10.101}$$

the continuity of the tangential field H_y at $z = 0$ leads to $\varepsilon\kappa_0 = \kappa$, or

$$-\frac{\omega_0^4}{\omega^2\gamma^2}\kappa_0^2 = i\frac{\omega_0^2\omega}{c^2\gamma} \tag{10.102}$$

for the large imaginary part of $\varepsilon \simeq 1 - \frac{\omega_0^2}{i\omega\gamma}$. We get

$$\kappa_0 = -\frac{\omega\sqrt{\omega\gamma}}{c\omega_0}\sqrt{-i} \ , \tag{10.103}$$

which shows that the field is practically undamped outside the half-space ($|\ \kappa_0\ | \ll \omega/c$), with large components H_y^o and E_z^o and damped inside ($\kappa \simeq -\frac{\omega}{c}\frac{\omega_0}{\sqrt{\omega\gamma}}\sqrt{i}$, $|\ \kappa\ | \gg \omega/c$); the inverse of the imaginary part of κ gives the penetration depth of a skin effect. A similar analysis shows that the magnetic mode does not exist. For dielectrics, the longitudinal components E_x, H_x are zero, $\kappa = 0$ ($k = \sqrt{\varepsilon}\omega/c > \omega/c$) and κ_0 is purely imaginary; the waves penetrate the body and are damped outside.

10.8 Surface plasmon-polaritons. Superficial modes

10.8.1 Internal field. Polarization

Let us assume the presence of a free electromagnetic field $\boldsymbol{E}_0$ (electric field) and $\boldsymbol{H}_0$ (magnetic field). It obeys the Maxwell equations

$$\begin{gathered} div\boldsymbol{E}_0 = 0 \ , \ div\boldsymbol{H}_0 = 0 \ , \\ curl\boldsymbol{E}_0 = -\frac{1}{c}\frac{\partial \boldsymbol{H}_0}{\partial t} \ , \ curl\boldsymbol{H}_0 = \frac{1}{c}\frac{\partial \boldsymbol{E}_0}{\partial t} \ , \end{gathered} \tag{10.104}$$

where c is the speed of light in vacuum.

In matter, the mobile charges q, with concentration n, under the action of an external field, are displaced by a displacement field $\boldsymbol{u}$. The displacement field generates a charge density imbalance $-nqdiv\boldsymbol{u}$ and a current density $nq\frac{\partial \boldsymbol{u}}{\partial t}$. In addition, in magnetic matter a magnetic current may appear, given by $c \cdot curl\boldsymbol{M}$, where $\boldsymbol{M}$ is called magnetization. These charge and current densities should be added to the right side of the first and fourth Maxwell equations (Gauss and Maxwell-Ampere equations), and an additional internal field should also be added.[24] By historical tradition, in matter, $\boldsymbol{E}_0$ is denoted by $\boldsymbol{D}$ (electric displacement) and the total magnetic field $\boldsymbol{H}$ is denoted by $\boldsymbol{B}$ (magnetic induction). Let us assume a non-magnetic matter ($\boldsymbol{M} = 0$) and preserve the notation $\boldsymbol{H}$ for the total magnetic field (magnetic induction). In addition, we denote by $\boldsymbol{E} = \boldsymbol{E}_0 + \boldsymbol{E}_i$ the total electric field, $\boldsymbol{E}_i$ being the internal electric field, and use $\boldsymbol{E}_0$ (or

[24] M. Apostol, *Essays in Electromagnetism and Matter*, Lambert, Saarbruecken (2013).

$\boldsymbol{D}$) for the external electric field. The Maxwell equations become

$$\begin{gathered} div\boldsymbol{E} == -4\pi nq div\boldsymbol{u} \ , \ div\boldsymbol{H} = 0 \ , \\ curl\boldsymbol{E} = -\frac{1}{c}\frac{\partial \boldsymbol{H}}{\partial t} \ , \ curl\boldsymbol{H} = \frac{1}{c}\frac{\partial \boldsymbol{E}}{\partial t} + \frac{4\pi}{c}nq\frac{\partial \boldsymbol{u}}{\partial t} \ , \end{gathered} \tag{10.105}$$

where

$$div\boldsymbol{E}_i = -4\pi nq div\boldsymbol{u} \ . \tag{10.106}$$

From equation (10.106) we get $\boldsymbol{E}_i = -4\pi nq\boldsymbol{u}$. This solution is valid for infinite matter. $\boldsymbol{P} = nq\boldsymbol{u}$ is called polarization; the first equation (10.105) becomes $div(\boldsymbol{E} + 4\pi\boldsymbol{P}) = 0$, whence $\boldsymbol{E} + 4\pi\boldsymbol{P} = \boldsymbol{D} = \boldsymbol{E}_0$, $div\boldsymbol{D} = 0$; the fourth equation (10.105) becomes $curl\boldsymbol{H} = \frac{1}{c}\frac{\partial \boldsymbol{D}}{\partial t}$. The equation of motion of an elementary mobile charge with mass m is

$$m\ddot{\boldsymbol{u}} + m\gamma\dot{\boldsymbol{u}} = q\boldsymbol{E}_0 + q\boldsymbol{E}_i \ , \tag{10.107}$$

where γ is the attenuation coefficient. In equation (10.107) we neglect internal forces and view the charge velocity as being sufficiently small, such that we may neglect also the Lorentz force. In finite bodies Coulomb interaction brings additional normal modes for the internal force.[25] Equation (10.107) can be written as

$$\ddot{\boldsymbol{u}} + \omega_0^2\boldsymbol{u} + \gamma\dot{\boldsymbol{u}} = \frac{q}{m}\boldsymbol{E}_0 \ , \tag{10.108}$$

where ω_0 given by $\omega_0^2 = 4\pi nq^2/m$ is the plasma frequency. Equation (10.108) is suitable for metals; for dielectrics an additional important internal (elastic) force appears. The solution of equation (10.108) for a monochromatic wave with frequency ω is

$$\boldsymbol{u} = -\frac{q\boldsymbol{E}_0}{m}\frac{1}{\omega^2 - \omega_0^2 + i\omega\gamma} \ ; \tag{10.109}$$

we get

$$\begin{gathered} \boldsymbol{E}_i = \frac{\omega_0^2}{\omega^2 - \omega_0^2 + i\omega\gamma}\boldsymbol{E}_0 \ , \ \boldsymbol{P} = -\frac{1}{4\pi}\frac{\omega_0^2}{\omega^2 - \omega_0^2 + i\omega\gamma}\boldsymbol{E}_0 \ , \\ \boldsymbol{E} = \boldsymbol{E}_0 + \boldsymbol{E}_i = \frac{\omega^2 + i\omega\gamma}{\omega^2 - \omega_0^2 + i\omega\gamma}\boldsymbol{E}_0 \end{gathered} \tag{10.110}$$

[25] M. Apostol, *Essays in Electromagnetism and Matter*, Lambert, Saarbrucken (2013); "Resonant coupling of the electron cloud to the nucleus in heavy atoms", Roum. Reps. Phys. **71** 210 (2019).

and, from $\boldsymbol{D} = \boldsymbol{E}_0 = \varepsilon \boldsymbol{E}$, $\boldsymbol{P} = \chi \boldsymbol{E}$, $\varepsilon = 1 + 4\pi\chi$,

$$\varepsilon = 1 - \frac{\omega_0^2}{\omega^2 + i\omega\gamma} \ , \ \chi = -\frac{1}{4\pi}\frac{\omega_0^2}{\omega^2 + i\omega\gamma} \ , \tag{10.111}$$

where ε is the dielectric function and χ is the electric susceptibility. In addition, since the current density $nq\frac{\partial \boldsymbol{u}}{\partial t}$ is $\boldsymbol{j} = -i\omega nq\boldsymbol{u} = \sigma \boldsymbol{E}$, we get the conductivity

$$\sigma = \frac{1}{4\pi}\frac{i\omega_0^2}{\omega + i\gamma} \ , \ \varepsilon = 1 + \frac{4\pi i\sigma}{\omega} \ . \tag{10.112}$$

We can see that the internal field has a resonance at the plasma frequency ω_0 (the plasmon).

By equation (10.107) the first Maxwell equation (10.105) (Gauss law) is solved. In the fourth Maxwell equation (10.105)

$$\begin{aligned} curl\boldsymbol{H} &= \tfrac{1}{c}\tfrac{\partial \boldsymbol{E}}{\partial t} + \tfrac{4\pi}{c} nq \tfrac{\partial \boldsymbol{u}}{\partial t} = \\ &= -\tfrac{i\omega}{c}\left(\boldsymbol{E}_0 + \boldsymbol{E}_i + 4\pi nq\boldsymbol{u}\right) \end{aligned} \tag{10.113}$$

(Maxwell-Ampere equation) $\boldsymbol{E}_i + 4\pi nq\boldsymbol{u} = 0$, such that this equation reduces to

$$curl\boldsymbol{H} = -\frac{i\omega}{c}\boldsymbol{E}_0 \ ' \tag{10.114}$$

therefore, $\boldsymbol{H} = \boldsymbol{H}_0$, as expected for non-magnetic matter. The second Maxwell equation (magnetic Gauss law) is satisfied. The third Maxwell equation (Faraday law) can be written as

$$curl\boldsymbol{E} = curl\boldsymbol{E}_0 + curl\boldsymbol{E}_i = \tfrac{i\omega}{c}\boldsymbol{H}_0 \ , \tag{10.115}$$

which shows that the internal field $\boldsymbol{E}_i = -4\pi nq\boldsymbol{u}$ is given by a gradient, as expected when charges are present. It is worth noting that the internal field and the total electric field are complex, due to the presence of the attenuation coefficient γ. Also, it is worth noting that these fields propagate in matter with the velocity $c/\sqrt{\varepsilon}$, where $\sqrt{\varepsilon}$ is the (complex) refractive index; usually, its large positive imaginary part gives a high dissipation in metals.

10.8.2 Surface plasmon-polaritons

We consider a straight wire with a circular cross-section with radius a in cylindrical coordinates (r,φ,x). We introduce the notation $\boldsymbol{u}=\boldsymbol{v}\theta(a-r)$. We recall that the function $\theta(a-r)$ is not defined for $r=a$. In the charge density $-nqdiv\boldsymbol{u}$ we have

$$div\boldsymbol{u}=div\boldsymbol{v}\cdot\theta(a-r)-v_r^s\delta(a-r)\ , \tag{10.116}$$

where $v_r^s=v_r(a,\varphi,x)$ is the r-th component of the displacement $\boldsymbol{v}$. This equation is valid for $r\leq a$, according to the definition of the charge density. Therefore, the δ-function in equation (10.116) should be viewed as $\frac{1}{2}\delta$. Due to this circumstance v_r^s should be viewed as a distinct (surface) displacement denoted by $w(r,\varphi)$. This introduces a discontinuity. The Gauss law for the internal electric field reads

$$div\boldsymbol{E}_i=-4\pi nqdiv\boldsymbol{v}\cdot\theta(a-r)+2\pi nqw\delta(a-r)\ . \tag{10.117}$$

the (discontinuous) solution of this equation is $E_i^v=-4\pi nq\boldsymbol{v}$ for $r<a$ and

$$E_{ir}^s=-2\pi nqw\ . \tag{10.118}$$

The term E_i^v is the bulk contribution. The Maxwell equations are solved as in the preceding section for the bulk contribution. The term E_{ir}^s is the surface contribution, arising from the surface charge density $-\frac{1}{2}nqw$. This contribution is decoupled from the bulk term. We write the Maxwell equations for the surface fields, denoted by $\boldsymbol{e}$ and $\boldsymbol{h}$; for instance, $E_{ir}^s=e_{ir}$. Wherever we encounter the derivative $\partial/\partial r$ we integrate it out over the thickness of the surface; in $curl's$ only tangential fields appear, associated with this derivative, which are continuous, while h_r in the magnetic Gauss law is also continuous. We get the two-dimensional Maxwell equations

$$\begin{gathered}(curl\boldsymbol{e})_\varphi=\frac{\partial e_r}{\partial x}=\frac{i\omega}{c}h_\varphi\ ,\ (curl\boldsymbol{e})_x=-\frac{1}{a}\frac{\partial e_r}{\partial\varphi}=\frac{i\omega}{c}h_x\ ,\\ (curl\boldsymbol{h})_r=\frac{1}{a}\frac{\partial h_x}{\partial\varphi}-\frac{\partial h_\varphi}{\partial x}=-\frac{i\omega}{c}e_{0r}+\frac{i\omega}{c}\cdot 2\pi nqw\end{gathered} \tag{10.119}$$

and

$$(curl\boldsymbol{h})_{\varphi,x}=-\frac{i\omega}{c}e_{0\varphi,x}\ ,\ (div\boldsymbol{h})_{\varphi,x}=0\ . \tag{10.120}$$

We note the occurence of the so-called uni-polar current density $\frac{1}{2}nq\dot{w}$ on the right of the radial Maxwell-Ampere equation (10.119); the fields denoted $\boldsymbol{e}_0$ and $\boldsymbol{h}_0$ are the restrictions of the corresponding external fields to the surface of the cylinder.

The equation of motion for the radial surface displacement w is

$$m\ddot{w} + m\gamma\dot{w} = qe_{0r} + qe_{ir} \ , \tag{10.121}$$

or

$$\ddot{w} + \frac{1}{2}\omega_0^2 w + \gamma\dot{w} = \frac{q}{m}e_{0r} \ ; \tag{10.122}$$

the other components of the displacement are those of the bulk displacement. From equation (10.122) we get

$$w = -\frac{qe_{0r}}{m}\frac{1}{\omega^2 - \frac{1}{2}\omega_0^2 + i\omega\gamma} \ . \tag{10.123}$$

We can see that the surface electric field has a resonance at $\omega_0/\sqrt{2}$ (surface plasmon). If the Coulomb interaction eigenmodes are included, the plasmon (bulk and surface) branches become dispersive and the plasmons pass over gradually into polaritons.[26] A similar result ($\omega_0/\sqrt{2}$) is well known for a half-space.

Making use of equation (10.123), the system of equations (10.119) and (10.120) can be solved. We assume a dependence $\sim e^{ikx}e^{il\varphi}$ for the external field $\boldsymbol{e}_0$, $\boldsymbol{h}_0$, where l is an integer, and get

$$\begin{aligned} e_r &= e_{0r} + e_{ir} = e_{0r} - 2\pi nqw = \\ &= \tfrac{\omega^2 + i\omega\gamma}{\omega^2 - \frac{1}{2}\omega_0^2 + i\omega\gamma}e_{0r} = \tfrac{1}{\varepsilon_s}e_{0r} \ , \end{aligned} \tag{10.124}$$

where

$$\varepsilon_s = 1 + \frac{\frac{1}{2}\omega_0^2}{\omega^2 - \frac{1}{2}\omega_0^2 + i\omega\gamma} \tag{10.125}$$

is the surface (radial) dielectric function, and

$$\begin{gathered} h_\varphi = \tfrac{ck}{\omega\varepsilon_s}e_{0r} \ , \ h_x = -\tfrac{lc}{a\omega\varepsilon_s}e_{0r} \ , \\ h_r = h_{0r} \ . \end{gathered} \tag{10.126}$$

[26] M. Apostol and G. Vaman, "Plasmons, polaritons and diffraction of the electromagnetic field for an infinite circular cylinder", Roum. J. Phys. **55** 483 (2010).

From the second equation (10.119) we get the dispersion relation

$$\omega^2 = c^2\left(\frac{l^2}{a^2} + k^2\right) \; ; \tag{10.127}$$

it defines the surface polaritons, associated with the magnetic field $h_{\varphi,x}$; since this field is resonant for the surface plasmon frequency $\omega = \omega_0/\sqrt{2}$, they are called surface plasmon-polaritons. We can see that there are many branches of plasmon-polaritons, and all are dispersive for $l \neq 0$ (asymmetric modes). There exists only one non-dispersive branch $l = 0$ (symmetric mode), with $h_x = 0$ and $h_\varphi \neq 0$. The symmetric surface plasmon-polariton mode may guide the electromagnetic field along the wire, non-dispersively and with the speed of light in vacuum.[27] Noteworthy, it is a magnetic-type mode, excited by a transverse external field (e_{0r}). It is different from Sommerfeld "surface wave", which is a bulk, electric-type mode, excited by a longitudinal electric field. A completely analogous situation occurs for a half-space.

10.8.3 Superficial modes

The solution of equation (10.116) given here is a discontinuous solution. This equation admits also a continuous, differentiable solution. For instance, we may assume that the internal electric field arises from a scalar potential Φ, such that equation (10.116) becomes a Poisson equation with bulk and surface sources. As long as the retardation is neglected the Coulomb interaction may be decomposed in the normal modes of the finite-size body, so we get plasmons, including surface plasmons, with dispersion, associated with surface electromagnetic waves. More generally, the retardation may be included and the eigenmodes we get are the polaritons.[28]

[27] C. Miziumski, "Utilization of a cylindrical geometry to promote radiative interaction with slow surface excitations", Phys. Lett **A40** 187 (1972); C. A. Pfeiffer, E. N. Economou and K. L. Ngai, "Surface polaritons in a circularly cylindrical interface: surface plasmons", Phys. Rev. **B10** 3038 (1974); J. C. Ashley and L. C. Emerson, "Dispersion relations for non-radiative surface plasmons on cylinders", Surf. Sci. **41** 615 (1974); H. Raether, *Surface Plasmons on Smooth and Rough Surfaces and on Gratings*, Springer (1988).

[28] M. Apostol, *Essays in Electromagnetism and Matter*, Lambert, Saarbruecken (2013); "Resonant coupling of the electron cloud to the nucleus in heavy

In the treatment described above the finite size of the body is accounted by the θ function, with respect to the surface of the body. There exists a more direct, and simplified, treatment of the surface modes.[29] The θ function accounts for the geometric surface discontinuity of a continuous macroscopic body. However, in finite-size bodies there exists another, physical discontinuity at their surface, which arises from the formulation of the macroscopic Electromagnetism.

Indeed, the macroscopic Electromagnetism is based on a spatial (and implicitly temporal) averaging of the highly-varying atomic electromagnetic fields. The macroscopic fields associated with each point in the macroscopic Maxwell equations are these average fields. The conceptual procedure of this averaging is to take spheres about each point, with various radii, and average the fields over these spheres. Obviously, the boundary of these spheres should be such that the contribution of the boundary atomic fields is not too large in comparison with the bulk contribution; on the contrary, the boundary contribution should be much smaller, in fact. It is obvious, therefore, that there should be a lower limit of the size of these spheres. An estimate of their minimal radius is give by the ratio of the interatomic volume $v \simeq a^3$ to the atomic volume $v_0 \simeq a_0^3$, where a is the mean inter-atomic distance and a_0 is the mean atomic dimension. For instance, if we take $a \simeq 4$ Å, as in typical metals, and $a_0 \simeq 1$Å, we get $v/v_0 \simeq 64$, which yields a minimal radius of the averaging spheres $r_0 = 64a \simeq 256$Å. An order of magnitude 10^2Å is estimated, by other arguments.[30] Now let assume that we have a homogeneous and isotropic finite body. The (mean) density (concentration) of the mobile charges is n everywhere inside the body, but in a superficial layer of thickness $2r_0$ this concentration decreases from n to zero. In accord with the procedure of the macroscopic Electromagnetism we should attribute to the surface a concentration $\frac{1}{2}n$. Obviously, inside this su-

---

atoms", Roum. Reps. Phys. **71** 210 (2019); M. Apostol and G. Vaman, "Plasmons, polaritons and diffraction of the electromagnetic field for an infinite circular cylinder", Roum. J. Phys. **55** 483 (2010); "Electromagnetic field interacting with a semi-infinite plasma", J. Opt. Soc. Am. **A26** 1747 (2009); "Plasmons and diffraction of an electromagnetic plane wave by a metallic sphere", Progr, Electrom. Res. (PIER) **98** 97 (2009).

[29] Bo E. Sernelius, *Surface Modes in Physics*, Wiley-VCH, Berlin (2001).

[30] J. D. Jackson, *Classical Electrodynamics*, Wiley, NY (1975) and References therein.

perficial layer there exists a discontinuity of the density of the mobile charges and electromagnetic fields, which is distinct from the geometric discontinuity. The new feature brought about by this discontinuity is associated with the displacement of the superficial charges normal to the surface and the corresponding electromagnetic fields are limited to the superficial layer; they are fields in two dimensions. It is worth noting that the superficial displacement (change in concentration) and the fields produced by it are macroscopic and two-dimensional, in contrast to the atomic-size double layer which occurs at the surfaces, as a result of the cohesion of the condensed matter.[31] This particularity of the macroscopic Electromagnetism is the origin of the superficial, discontinuous elctromagnetic modes, like surface plasmon-polaritons, which are distinct from the bulk and the surface modes.[32]

The above calculations (equations (10.116) to (10.127)) can be applied to a half-space $z < 0$ (with $\theta(-z)$). The transverse superficial displacement is $v_z^s = w$ and the superficial internal field is $E_{iz}^s = -2\pi nqw = e_{iz}$. The solution of equations (10.119) and (10.120) is

$$\begin{gathered} e_z = \frac{1}{\varepsilon_s} e_{0z} \ , \ h_x = \frac{ck_y}{\omega\varepsilon_s} e_{0z} \ , \\ h_y = -\frac{ck_x}{\omega\varepsilon_s} e_{0z} \ , \ h_z = h_{0z} \ , \end{gathered} \tag{10.128}$$

where ε_s is given by equation (10.125); the second equation (10.119) yields the photon dispersion relation $\omega^2 = c^2k^2$ ($k^2 = k_x^2 + k_y^2$), as expected.

Let us apply these calculations to a sphere with radius a (with $\theta(a - r)$), where the transverse superficial displacement is $v_r^s = w$. It is easy to see that the integration of $\delta(a - r)$ in equation (10.117) gives $E_{ir}^s = -\frac{4\pi}{3} nqw = e_{ir}$. Indeed, this result can be obtained by comparing the volume of the sphere $\frac{4\pi}{3}a^3$ with the volume $4\pi a^3$ of a cylinder, or by integrating the projection of the spherical surface on the tangent plane (for both the spherical coordinates θ and φ), which gives

[31] M. Apostol, *Physical Kinetics*, Cambridge Scholars, Newcastle upon Tyne (2020).

[32] M. Apostol, "Surface plasmon-polaritons. Superficial modes", Roum. J. Phys. **65** 604 (2020).

$\frac{1}{4\pi}\int d\varphi d\theta \sin\theta \cdot \cos^2\theta = \frac{1}{3}$. It follows

$$w = -\frac{qe_{0r}}{m}\frac{1}{\omega^2 - \frac{1}{3}\omega_0^2 + i\omega\gamma} \; , \; \varepsilon_s = 1 + \frac{\frac{1}{3}\omega_0^2}{\omega^2 - \frac{1}{3}\omega_0^2 + i\omega\gamma} \; , \quad (10.129)$$

where $\omega_0/\sqrt{3}$ is the frequency of the spherical plasmon. We write equations (10.119) and (10.120) in spherical coordinates. Making use of the first two equations (10.119) we can see immediately that the second equation (10.120) is satisfied; from the first equation (10.120) we get $h_r = h_{0r}$. Making use again of the first two equations (10.119) the second equation (10.119) becomes

$$\frac{c^2}{a^2\omega^2}\Delta_{\theta,\varphi}e_r = -\frac{1}{\varepsilon_s}e_{0r} \; , \quad (10.130)$$

where $e_r = \frac{1}{\varepsilon_s}e_{0r}$. It follows the dispersion relation

$$\omega^2 = \frac{c^2}{a^2}l(l+1) \quad (10.131)$$

for the spherical plasmon-polaritons, where l is the index of the spherical functions Y_{lm} (the external field e_{0r} is decomposed in terms of these functions). Making use of $e_r = \frac{1}{\varepsilon_s}e_{0r}$ we get the tangential fields $h_{\theta,\varphi}$ from equations (10.119) (they are components of the Hansen vectors (vector surface harmonics)).

In matter there exist bulk electromagnetic fields. The polarization gives rise to bulk plasmons and polaritons. The usual boundary conditions for the Maxwell equations account for the geometric discontinuity of the finite-size bodies. This discontinuity generates a polarization surface charge, whose motion gives the surface plasmons. The surface charge occurs in a surface (superficial) layer (sheath) with undetermined thickness, which is predicted by the macroscopic Electromagnetism. Consequently, in this superficial layer appear discontinuous electromagnetic fields (distinct from the bulk fields associated with the geometric discontinuity), whose coupling with the surface charge produces surface (superficial) plasmon-polaritons.

10.9 Topological conductors

10.9.1 Topological conductors

Let us consider an ensemble of many fermions, leaving aside their interaction and the question of their stability. This is a very idealized picture, where we may expect surprising things. One of such things is related to the dimensionality of the space the fermions move in. Indeed, in three dimensions the density dn of fermionic states is proportional to $\sqrt{\varepsilon}d\varepsilon$, where ε is the energy of a fermionic state, while in two dimensions $dn \sim d\varepsilon$ and in one dimension $dn \sim d\varepsilon/\sqrt{\varepsilon}$. We can see that at the Fermi level there exists a large state density in three dimensions, which indicates the possibility of a normal Fermi liquid picture, when interaction is present; in two dimensions the state density does not depend on the Fermi level, so we may expect no difference between the thermodynamics of a two-dimensional gas of fermions and the thermodynamics of a two-dimensional gas of bosons. Indeed, this indication is verified by a direct calculation.[33] In one dimension the fermionic state density is very small at the Fermi level, such that we expect a dynamics governed entirely by bosonic density degrees of freedom, and no Fermi surface (points). We say that the one-dimensional models of fermions are bosonized.[34]

More surprises are claimed to appear in two-dimensional ensembles of many fermions. For instance, since in two dimensions there exist rotations only about one axis, it is suggested that the spin does not exist anymore for fermions in two dimensions. Moreover, an electron state around a localized magnetic flux may have any phase; in particular, for an integral number of magnetic flux quanta the electric charge of the electron would be fractional, because the electric charge multiplies the magnetic flux in the phase. So, the idea of composite electrons appeared, *i.e.* electrons with an associated number of quanta of magnetic flux, which would have a fractional occupation number and a fractional statistics (any fraction, hence their denomination of

---

[33] M. Apostol, "Dimensionality effects in the ideal Bose and Fermi gases", Phys. Rev. **E56** 4854 (1997), and References therein.

[34] D. C. Mattis, ed., *The Many-Body Problem, An Encyclopedia of Exactly Solved Models in One Dimension*, World Scientific Singapore (1993).

anyons).[35]

The fermion ensembles in two dimensions received a new impetus with the discovery of the graphene.[36] Since graphene sheets occur with a small size, the edge electronic states have been brought into discussion.[37] The electric field preserves a Kramers degeneracy of the electronic spin states, related to the time reversal symmetry; but a magnetic field removes this degeneracy, and transforms the graphene bulk in an insulator. This is not so for the edge states, which have a completely different spectrum than the bulk ones. The graphene edges remain conducting the electrical current. It is said that the edge states are protected by the time reversal symmetry, and they define a conductor surrounding an insulator; the latter is called a topological one, because its surface is a conductor.[38]

However, the surface is the deed of the devil. It seems that Pauli said "God made the bulk; the surface was invented by the devil". We show here that quasi-free surface electrons are ubiquituous.

Let us consider a solid consisting of N atoms (ions), placed at positions $\boldsymbol{r}_i$, $i = 1, 2, ...N$, and N electrons. The wavefunction of a free electron is $\frac{1}{\sqrt{V}}e^{i\boldsymbol{kr}}$,where $\boldsymbol{k}$ is the wavevector and V is the volume. An electron can be trapped in an atom (ion), where its wavefunction is $\frac{1}{\sqrt{v}}\chi(\boldsymbol{r}-\boldsymbol{r}_i)$; χ is a function localized over the volume v, which is of the order of the volume assigned in the solid to an atom placed at $\boldsymbol{r}_i$, such that $\frac{1}{v}\int d\boldsymbol{r} \mid \chi \mid^2 = 1$. The wavefunction of an electron trapped in any atom in solid is

$$\frac{1}{\sqrt{N}}\sum_i \frac{1}{\sqrt{v}}\chi(\boldsymbol{r}-\boldsymbol{r}_i) \ . \tag{10.132}$$

[35]F. Wilczek, *Fractional Statistics and Anyon Superconductivity*, Word Scientific, Singapore (1990).

[36]A. H. Castro Neto, F. Guinea, N. M. R. Peres, K. S. Novoselov and A. K. Geim, "The electronic properties of graphene", Revs. Mod. Phys. **81** 109 (2009); "The electronic properties of graphene", Revs. Mod. Phys. **81** 109 (2009).

[37]Oana-Ancuta Dobrescu and M. Apostol, "Tight-binding approximation for bulk and edge electronic states in graphene", Can. J. Phys. **93** 580 (2015).

[38]X.-L. Qi and S.-C. Zhang, "Topological insulators and superconductors", Revs. Mod. Phys. **83** 1057 (2011).

The amplitude of probability for an electron to be trapped in solid is

$$\frac{1}{\sqrt{N}}\sum_i \frac{1}{\sqrt{vV}}\int d\boldsymbol{r}\chi^*(\boldsymbol{r}-\boldsymbol{r}_i)e^{i\boldsymbol{kr}} = \frac{1}{\sqrt{N}}\sum_i \sqrt{\frac{v}{V}}e^{i\boldsymbol{kr}_i} \ , \quad (10.133)$$

such that the probability for the electron to be trapped is

$$P = \frac{v}{NV}\sum_{ij} e^{i\boldsymbol{k}(\boldsymbol{r}_i-\boldsymbol{r}_j)} \ . \quad (10.134)$$

If the atoms are distributed randomly, the summation in equation (10.134) gives $N^2\delta_{\boldsymbol{k}0}$, such that we get $P = Nv/V$. Since $Nv = V$, this probability is equal to unity. It follows that in an amorphous solid all the electrons can be trapped, and the solid is an insulator, as it is well known. If the solid is crystalline, the summation in equation (10.134) gives $N^2\delta_{\boldsymbol{kg}}$, where $\boldsymbol{g}$ is a vector of the reciprocal lattice. It follows that the electron states lying on the walls of the Brillouin zone are trapped. The corresponding energy bands are blocked by the exclusion principle. Therefore, again the solid is an insulator (actually, there are $2N$ electrons in each band, as a consequence of their spin states). If the band is not completed, the crystalline solid is a conductor, as it is well known.

The situation is different in a two-dimensional solid, *i.e.* a solid with the atoms arranged on a plane surface. In this case the electrons are delocalized over a transverse distance d, where $d > a$, $v = a^3$. The above probability reads $P = (N\sigma/S)\frac{a}{d}$, where S is the area of the solid, $V = Sd$, and $\sigma = a^2$ is the area of the cross-section of the volume assigned to an atom. Now, for an amorphous two-dimensional solid $N\sigma/S = 1$, such that $P = a/d < 1$. It follows that an amorphous two-dimensional solid is always a conductor. A similar conclusion holds for a crystalline two-dimensional solid, and for a one-dimensional solid.

An estimation of the distance d can be obtained from simple geometrical arguments. An electron moving along a single plane layer of atoms has a transverse distance $d = 2a$ at its disposition, while it has only the distance $d = a$ if the layer of atoms is sandwiched between two other atomic layers. If the layer of atoms is deposited on a solid surface, the distance d becomes equal to a, but the extent of the atomic volume in the transverse direction is reduced to $a/2$. In both cases the ratio a/d is equal to $1/2$.

It is well known that the mean inter-atomic separation distance is modified at the surface of a solid, in comparison to the bulk, albeit to a very small extent. For instance, in an anharmonic solid the lattice parameter is slightly increased at the surface, or has an oscillatory behaviour.[39] This makes the electrons at the surface to have a small additional room, measured by the transverse distance $d > a$. On the other hand, the electrons and the ions at the surface form a charge inversion layer, where the electrons spill over the surface, leaving behind a positive ionic charge; the characteristic dimension of this layer is of the order of the mean inter-atomic distance.[40] Therefore, we may expect that a small amount of electrons at the surface of an insulator are quasi-free. Their motion is completely decoupled from the bulk electrons. We may say that this circumstance defines a topological conductor. A special discussion requires metals consisting of micro-crystallites, or metallic glasses, alloys, etc, where the electron band structure varies slightly locally (in a quasi-classical description). This variation is abrupt at the surface, where the energy bands are split and the surface quasi-free electrons occupy an incomplete band.

We give here an estimation of the thickness of the distorted superficial layer of a solid. As it is well known, at the atomic level the physical quantities vary abruptly and indeterminately in space and time. Usually, we average this motion over small spatial regions and small durations. Thereby, we get homogeneous domains, which microscopically may be sufficiently large, but macroscopically they are very small. At the surface this homogeneity is lost, such that we may estimate the thickness of the superficial layer as the dimension of a domain. The variation of the volume due to the atomic extent of the particles is $\delta\Omega/\Omega = a_0^3/a^3$, where a_0 is the atomic dimension and a is the mean inter-atomic separation distance; it follows $\delta\Omega = (a_0/a)^3 a^3 N_d$, where N_d is the number of atoms in a domain. On the other hand, the vol-

[39] M. Apostol, "A semi-infinite solid", J. Theor. Phys. **24** (2001).

[40] J. R. Smith, "Self-consistent many-electron theory of electron work functions and surface potential characteristics for selected metals, Phys. Rev. **181** 522 (1969); W. Jones and N. H. March, *Theoretical Solid-State Physics*, Wiley-Interscience, London (1973); L. C. Cune and M. Apostol, *Theory of Atomic Clusters; Metals Deposited on Surfaces*, in *Low-Dimensional Systems: Theory, Preparation and some Applications*, eds. L. M. Liz-Marzan and M. Giersig, Kluwer, (2003).

ume variation of a domain is $\delta\Omega = Aa = N_d^{2/3}a^3$, where A is the area of the surface which encloses the domain. These two relations lead to $N_d^{1/3} = (a/a_0)^3$, which indicates a dimension $D = N_d^{1/3}a = (a/a_0)^3 a$ for a domain. For instance, making use of typical values $a_0 = 1\text{Å}$ and $a = 4\text{Å}$, we get $D = 64a = 256\text{Å}$ and $N_d = (R/a)^3 \simeq 10^5$. This is an estimation of the size of the domains and the thickness of the superficial layer. An order of magnitude 10^2Å is estimated, by other arguments.[41] Averaging over fractions of domains at the surface leads to surface electromagnetic fields distinct from the bulk fields, which sustain surface plasmon-polaritons.[42] In addition, the change in the electronic structure at the surface highlights surface electrons decoupled from the bulk.

It is expected that the density n of the surface electrons is much reduced in comparison to their bulk density n_b; we write $n = \beta n_b$ (the typical electron density in a solid is $n_b = 10^{22}cm^{-3}$). Indeed, the maximum deviation from homogeneity of a single domain described above is one atom in 64 atoms, which indicates a factor $\beta = 1/64$ of the order $\beta = 10^{-2}$. On the other hand, the relative variation of the lattice parameter at the surface of an anharmonic solid is of the order 10^{-6}.[43] We may take an average order of magnitude 10^{-4} for the parameter β.

10.9.2 Collective motion

As it is well known, the existence of a (quasi-) plane surface leads to the existence of a surface plasmonic mode with frequency $\omega_p/\sqrt{2}$, where $\omega_p = (4\pi n_b e^2/m)^{1/2}$ is the bulk plasmon frequency; $-e$ and m are the electron charge and the electron mass.[44] This mode corresponds to a motion perpendicular to the surface and exists only in conjunction with the bulk plasmon. On the other hand, polaritonic modes may exist in the superficial layer, due to electromagnetic

[41] J. D. Jackson, *Classical Electrodynamics*, Wiley, NY (1975) and References therein.

[42] M. Apostol, "Surface plasmon-polaritons. Superficial modes", Roum. J. Phys. **65** 604 (2020).

[43] M. Apostol, "A semi-infinite solid", J. Theor. Phys. **24** (2001).

[44] Bo E. Sernelius, *Surface Modes in Physics*, Wiley-VCH, Berlin (2001).

waves propagating along the surface. These modes have been identified recently in a wire with a circular cross-section;[45] they may guide electromagnetic waves along the surface, propagating without dispersion with the speed of light in vacuum. These modes are called surface plasmon-polariton modes, since they have a resonant behaviour at the surface plasmon frequency.

A different plasmon mode may appear along the finite dimension of a superficial layer. In general, the bulk plasmon is generated by the local internal (polarization) field. Indeed, if the charges q ($q = -e$) with concentration n_b suffer a local displacement $\boldsymbol{u}$, then a local charge density imbalance $-n_b q div\boldsymbol{u}$ occurs, which generates an internal field given by $div\boldsymbol{E} = -4\pi n_b q div\boldsymbol{u}$. This equation holds locally, with a spatial variation in all the three directions, in general. Consequently, the internal field is given by $\boldsymbol{E} = -4\pi n_b q\boldsymbol{u}$. In the equation of motion for a charge

$$m\ddot{\boldsymbol{u}} + m\omega_c^2\boldsymbol{u} + m\gamma\dot{\boldsymbol{u}} = q\boldsymbol{E}_0 + q\boldsymbol{E} \ , \tag{10.135}$$

where ω_c is a characteristic (internal) frequency of the charge, γ is a small dissipation coefficient and $\boldsymbol{E}_0$ is the external electric field, the internal field generates the term $-4\pi n_b q^2\boldsymbol{u}$, which leads to the plasma frequency $\omega_p = (4\pi n_b q^2/m)^{1/2}$. For the motion of the superficial charges along a finite dimension L the div does not hold locally anymore; the smallest charge density imbalance is given approximately by $\delta n \simeq nu/L$, where $n = \beta n_b$ ($n_b = 1/a^3$) is the surface electron density, u is the displacement and L is of the order of the linear dimension of the surface, for a body with a relatively smooth surface. The electric potential of the surface charges is of the order is $\varphi \simeq \beta^{1/3}q/a$. The corresponding interaction energy for a charge is $q\varphi(u/L)^2 = (\beta^{1/3}q^2/a)(u/L)^2$, which indicates that a charge has an eigenfrequency ω_0 given by $(\beta^{1/3}q^2/a)(u/L)^2 = m\omega_0^2u^2$, *i.e.*

$$\omega_0 = \frac{1}{L}\sqrt{\beta^{1/3}\frac{q^2}{ma}} = \beta^{1/6}\frac{a}{L}\sqrt{\frac{q^2}{ma^3}} = \beta^{1/6}\omega_p\frac{a}{L} \ . \tag{10.136}$$

This eigenfrequency of the motion of the surface electrons is much lower than the bulk plasma frequency; indeed, for $L = 10cm$ we get

[45] M. Apostol, "Surface plasmon-polaritons. Superficial modes", Roum. J. Phys. **65** 604 (2020).

$\omega_0 \simeq 10^{-8}\omega_p$ (for typical distances $a = 4$Å. we may neglect the factor $\beta^{1/6}$). The equation of motion for these superficial charges $m\ddot{\boldsymbol{u}} + m\omega_c^2\boldsymbol{u} + m\omega_0^2\boldsymbol{u} + m\gamma\dot{\boldsymbol{u}} = q\boldsymbol{E}_0$ gives an internal field $\boldsymbol{E}_i = -m\omega_0^2\boldsymbol{u}/q = \boldsymbol{E}_0\omega_0^2/(\omega^2 - \omega_c^2 - \omega_0^2 + i\omega\gamma)$, a total field $\boldsymbol{E}_t = \boldsymbol{E}_0 + \boldsymbol{E}_i = \boldsymbol{E}_0(\omega^2 - \omega_c^2 + i\omega\gamma)/(\omega^2 - \omega_c^2 - \omega_0^2 + i\omega\gamma)$, a surface dielectric function $\varepsilon = 1 - \omega_0^2/(\omega^2 - \omega_c^2 + i\omega\gamma)$ and a conductivity $\sigma = (nq^2/m)i\omega/((\omega^2 - \omega_c^2 + i\omega\gamma)$ $(\boldsymbol{j} = nq\dot{\boldsymbol{u}} = \sigma\boldsymbol{E}_t)$. The resonance exhibited by the polarizability $\alpha = \omega_0^2/((\omega^2 - \omega_c^2 - \omega_0^2 + i\omega\gamma)$ at $\omega^2 = \omega_c^2 + \omega_0^2$ may be tested experimentally $(\boldsymbol{E}_i = \alpha\boldsymbol{E}_0)$. For the surface quasi-free electrons $\omega_c = 0$. We note that the surface conductivity is smaller by the factor β than the bulk conductivity. $\omega_{ps} = (4\pi\beta n_b q^2/m)^{1/2}$ may be viewed as the plasma frequency of the surface electrons.

We note that the plasma frequency given above can be written as $\omega_0 = v/R$, where $v = \sqrt{q^2/ma}$ is a characteristic velocity (we leave aside the factor $\beta^{1/6}$); for electrons and typical values of a this velocity is of the order $10^8 cm/s$; for $L = 10cm$ we get $\omega_0 = 10Mhz$ (and $\omega_p = 10^{15}s^{-1}$). A similar set of frequencies cl/R, where c is the speed of light in vacuum and $l = 1, 2, , , ,$, was obtained for polaritons moving along the circumference of a wire with radius R.

10.9.3 Single-particle properties. Thermodynamics

We consider here the surface electrons of a long and thin wire, with length L and radius R of the circular circumference, $L \gg R$. We consider both L and R much larger than the mean inter-atomic separation distance a and omit the motion across the small thickness D of the surface layer. As it is well known, the thermodynamic properties of a two-dimensional gas of electrons are equivalent with the thermodynamic properties of a gas of bosons.[46] (In two dimensions there is no superfluid transition). The single-particle wavefunctions of the surface electrons are $\frac{1}{\sqrt{2\pi}}e^{il\varphi}\frac{1}{\sqrt{L}}e^{ikx}$, where φ is the angle on the circumference, l is any integer and k is the continuous wavevector along the length of the wire (direction x). The corresponding single-electron

[46] M. Apostol, "Dimensionality effects in the ideal Bose and Fermi gases", Phys. Rev. **E56** 4854 (1997), and References therein.

energies are given by

$$\varepsilon = \frac{\hbar^2}{2mR^2}l^2 + \frac{\hbar^2}{2m}k^2 \ . \tag{10.137}$$

These energies define a continuous two-dimensional Fermi sea, which is a circular disk in the variables $\sqrt{\hbar^2/2mR^2}l$ and $\sqrt{\hbar^2/2m}k$. The superficial density of electrons is given by

$$n_s = \frac{m}{\pi\hbar^2}\int_0^\infty d\varepsilon \frac{1}{e^{(\varepsilon-\mu)/T}+1} \ , \tag{10.138}$$

an equation which defines the chemical potential μ (spin included). We can see that the thermal wavelength $\lambda = (\pi\hbar^2/2mT)^{1/2}$, where T is the temperature, is much smaller than the (macroscopic) dimensions R, L and the thickness of the layer of the superficial electrons, over a wide range of temperatures, such that the thermodynamics is well defined and the gas is degenerate. From equation (10.138) we get $\mu = \pi\hbar^2 n_s/m$. Therefore, the chemical potential of the surface electrons is smaller by the factor $\beta^{2/3}$ than the chemical potential of the bulk electrons. The grand-partition potential is given by

$$\Omega = -\frac{mA}{\pi\hbar^2}\int_0^\infty d\varepsilon \frac{\varepsilon}{e^{(\varepsilon-\mu)/T}+1} = -E = -pA \ , \tag{10.139}$$

where A is the area of the cylindrical surface of the wire, E is the energy and p is the "pressure", *i.e.* the force per unit length. We get

$$E = \frac{mA}{2\pi\hbar^2}\mu^2\left(1 + \frac{\pi^2}{3}T^2/\mu^2 + ...\right) \ , \tag{10.140}$$

whence, by means of $E = -T^2\partial(F/T)/\partial T$, we may get the free energy and all the other thermodynamic properties. The contribution of the surface electrons to the specific heat is $C_s = \pi mAT/3\hbar^2$.

10.9.4 Transport properties

The two-dimensional Fermi sea defined by the chemical potential μ computed above has a Fermi wavevector of the order $\beta^{1/3}1/a$, where a denotes the mean separation distance between the (bulk) electrons, in both directions, *i.e.* along the wire and along the circumference of the wire. In typical metals the Fermi velocity $v_F = \hbar/ma$ is of the order $10^7 - 10^8 cm/s$ ($n = 10^{22} cm^{-3}$). For the surface electrons it is diminished by the factor $\beta^{1/3}$. The single-particle elementary excitations are quasi-particles, with a lifetime governed by the thermal uncertainty, phonons, impurities, etc. The static electrical conductivity computed above reads $\sigma = nq^2/m\gamma = \beta\omega_p^2/4\pi\gamma$. This equation may be used to get the dissipation coefficient γ. In typical metals at room temperature $\gamma = 10^{13} - 10^{14} s^{-1}$ (and $\omega_p = 10^{15} s^{-1}$) for bulk electrons; we may assume that the same order of magnitude is preserved for the surface electrons. The parameter γ defines a lifetime $\tau_{lf} = 1/\gamma$ and a mean freepath $\Lambda = v_F \tau_{lf}$ of the order $0.1 - 1\mu m$ for bulk electrons at room temperature. For the surface electrons the mean freepath is $\Lambda_s = \beta^{1/3}\Lambda$. Similarly, the quasi-particles lifetime is of the order $\tau_{lf} = \hbar\mu/T^2$, which gives $\gamma = 10^{12} s^{-1}$ for bulk electrons ($\mu = 1eV$) at room temperature. For the surface electrons we get $\tau_{lf} = \beta^{2/3}\tau_{lfbulk}$, $\gamma_s = \beta^{-2/3}\gamma$ and $\Lambda_s = \beta\Lambda$. We can see that the mean freepath of the surface electrons may be appreciably smaller than the mean freepath of the bulk electrons. The transport properties of the surface electrons are much poorer than those of the bulk electrons, as a consequence of their much lower density. The dissipation coefficient is diminished in oscillations ($\omega \neq 0$). We note that the surface current along the wire is smaller by a factor D/R than the bulk current, where D is the thickness of the surface layer (estimated above as 256Å).

10.9.5 Solenoidal effect

Let us assume a uniform harmonic displacement

$$u_0 e^{-i\omega t} D\delta(r - b) \tag{10.141}$$

along the circumference of the wire with radius b; D is the tickness of the superficial region where this displacement is confined. It gives rise to a current density

$$\boldsymbol{j} = -i\omega n q u_0 (0, -\sin\Phi, \cos\Phi) e^{-i\omega t} D\delta(r - b) \ , \tag{10.142}$$

were Φ is the angle in the (y,z)-plane. The vector potential is given by

$$\boldsymbol{A}=\frac{1}{c}\int d\boldsymbol{R}'\frac{\boldsymbol{j}(\boldsymbol{R}',t-\mid\boldsymbol{R}-\boldsymbol{R}'\mid/c)}{\mid\boldsymbol{R}-\boldsymbol{R}'\mid}\ , \tag{10.143}$$

where $\boldsymbol{R}=(x,r\cos\varphi,r\sin\varphi)$ and $\boldsymbol{R}'=(x',r'\cos\Phi,r'\sin\Phi)$. We introduce the angle $\varphi'=\varphi-\Phi$ and notice that the contribution to the integral comes from even functions of $\cos\varphi'$. The integral over x' gives the Hankel function $H_0^{(1)}$, such that we get[47]

$$\begin{gathered}\boldsymbol{A}=\tfrac{2\pi nq\omega}{c}u_0bD\left(0,-\tfrac{z}{r},\tfrac{y}{r}\right)\cdot\\ \cdot\int_0^{\pi}d\varphi'\cos\varphi' H_0^{(1)}\left(\tfrac{\omega}{c}\sqrt{r^2+b^2-2rb\cos\varphi'}\right)e^{-i\omega t}\ .\end{gathered} \tag{10.144}$$

In this expression we can use the asymptotic behaviour of the Hankel functions. We limit ourselves to quasi-static fields, *i.e.* fields, which satisfy the condition $\omega r/c$, $\omega b/c\ll 1$ (wavelengths much larger than the dimensions of the wire). The Hankel function goes like $H_0^{(1)}(z)\simeq\frac{2i}{\pi}\ln z$ in this limit, such that we get

$$\boldsymbol{A}=-\frac{2\pi inq\omega}{c}u_0b^2D\left(0,-\frac{z}{r^2+b^2},\frac{y}{r^2+b^2}\right)e^{-i\omega t} \tag{10.145}$$

and the magnetic field

$$\begin{gathered}\boldsymbol{H}=curlA=-\tfrac{4\pi inq\omega}{c}u_0b^2De^{-i\omega t}\left(\tfrac{b^2}{(r^2+b^2)^2},0,0\right)=\\ =\tfrac{4\pi}{c}Dj\left(\tfrac{b^4}{(r^2+b^2)^2},0,0\right)\ ;\end{gathered} \tag{10.146}$$

this formula coincides with the well-known magnetic field along the axis of an infinite solenoid, $H_x=4\pi Dj/c\ (r\longrightarrow 0)$. The electric field is smaller by a factor $\omega b/c$ than the magnetic field.

Let us consider a uniform, quasi-static longitudinal magnetic field H applied along the axis of the wire. According to the Faraday equation $curl\boldsymbol{E}=-\frac{1}{c}\frac{\partial\boldsymbol{H}}{\partial t}$ it generates an electric field E along the circumference, given by $E=i\omega rH/2c$. If electrical charges are present along

[47] M. Apostol, *Essays in Electromagnetism and Matter*, Lambert, Saarbruecken (2013).

this circumference, they get a displacement $u = -iqrH/2mc(\omega + i\gamma)$ and a current density $j = -nq^2\omega rH/2mc(\omega + i\gamma) = \sigma E$. For reasonable values of these parameters the displacement is very large. We can estimate the induced magnetic field H_i from the Maxwell-Ampere equation $curl\boldsymbol{H}_i = \frac{1}{c}\frac{\partial \boldsymbol{E}}{\partial t} + \frac{4\pi}{c}\boldsymbol{j}$, which leads to

$$H_i = \left(\frac{\omega^2}{2c^2} - \frac{\omega_p^2}{2c^2}\frac{\omega}{\omega + i\gamma}\right)\frac{r^2}{3}H = \frac{1}{2c^2}\left(\omega^2 + 4\pi i\sigma\omega\right)\frac{r^2}{3}H \ . \tag{10.147}$$

This field is uniform and directed along the axis of the wire. For reasonable values of the parameters (even if the factor β is included for the surface electrons), this induced magnetic field is much larger than the applied field and directed in the opposite direction. Therefore, we may expect that the applied magnetic field is expelled from the wire. This result is valid for metals. For insulators with surface electrons the conductivity is small and real, and the external magnetic field may penetrate.

The above estimation does not include the mutual dependence of the external and induced fields. This dependence is included by $curl\boldsymbol{H} = \frac{1}{c}\frac{\partial \boldsymbol{E}}{\partial t} + \frac{4\pi}{c}\sigma\boldsymbol{E}$, or $\Delta\boldsymbol{H} - \frac{1}{c^2}\frac{\partial^2 \boldsymbol{H}}{\partial t^2} - \frac{4\pi\sigma}{c^2}\frac{\partial \boldsymbol{H}}{\partial t} = 0$; the second-order time derivative may be neglected, such that, the solution of this equation in cylindrical coordinates is a Bessel function of imaginary argument, which is damped over distances of the order $c/\sqrt{\mid\sigma\mid\omega} \simeq c/\omega_p$, *i.e.* the plasma wavelength. This is the well-known skin effect. For metals this wavelength is very small, for insulators it is large, irrespective of the presence of the surface electrons.

10.9.6 Quantum-mechanical effects

Let us focus now on the surface electrons in a wire with radius R; their single-particle wavefunctions are $\frac{1}{\sqrt{2\pi}}e^{il\varphi}\frac{1}{\sqrt{L}}e^{ikx}$ and their energy is given by equation (10.137). We assume an external uniform quasi-static electromagnetic field with the magnetic component $\boldsymbol{H}$ directed along the axis of the wire, $\boldsymbol{H} = (H, 0, 0)$, where $H = H_0\cos\omega t$. The vector potential is $\boldsymbol{A} = \frac{1}{2}H(0, -z, y)$; in cylindrical coordinates it has only the angular component $A_\varphi = \frac{1}{2}Hr$, where r is the radius of the cross-section. For the moment we allow also a motion along the small thickness D of the surface layer (coordinate r) and write the hamiltonian of these free electrons as

$$\mathcal{H} = \frac{1}{2m}\left(\boldsymbol{p_r} - \frac{q}{c}\boldsymbol{A}\right)^2 + \frac{1}{2m}p_x^2 \ . \tag{10.148}$$

The canonical transformation $\psi = e^{\frac{iq}{c\hbar}\int^r \boldsymbol{A}d\boldsymbol{l}}\chi$ leads to the Schroedinger equation

$$i\hbar\frac{\partial\chi}{\partial t} = \frac{1}{2m}\left(p_{\boldsymbol{r}}^2 + p_x^2\right)\chi - q\int^{\boldsymbol{r}} \boldsymbol{E}d\boldsymbol{l}\chi \ ; \tag{10.149}$$

we can see that this canonical transformation removes the vector potential and introduces the work done by the electric field $\boldsymbol{E}$ upon the electrons. This electric field is given by $\boldsymbol{E} = -\frac{1}{c}\frac{\partial \boldsymbol{A}}{\partial t}$ and has only the φ-component (acts along the circumference); it is proportional to the vector potential, dephased in time. Now, we can neglect the transverse motion and writes the Schroedinger equation as

$$i\hbar\frac{\partial\chi}{\partial t} = -\frac{\hbar^2}{2m}\left(\frac{\partial^2}{R^2\partial\varphi^2} + \frac{\partial^2}{\partial x^2}\right)\chi - \frac{q\omega H_0 R^2}{2c}\varphi\sin\omega t \ . \tag{10.150}$$

In this equation we may view the H_0-term as a perturbation, denoted by $V\sin\omega t$. The amplitude of transition from the initial state i to the final state f is given by

$$a_{fi} = -\frac{i}{\hbar}\int_0^t dt' V_{fi}\sin\omega t' e^{\frac{i}{\hbar}(\varepsilon_f - \varepsilon_i)t'} \ . \tag{10.151}$$

For absorption we retain only the term $e^{-i\omega t}$ in $\sin\omega t$. The transition does not imply the motion along the x-coordinate; it is a transition from the state l to the state l'. For a long time we get the number of transitions per unit time

$$\mid a_{fi}\mid^2 /t = \frac{\pi^2 C^2}{\hbar}\frac{1}{(l'-l)^2}\delta\left(\frac{\hbar^2}{2mR^2}(l'^2 - l^2) - \hbar\omega\right) \ , \tag{10.152}$$

where $C = -q\omega H_0 R^2/2c$. We sum (integrate) over the final states l' and get

$$\mid a\mid^2 /t = \frac{\pi^2 C^2}{\hbar}\frac{2mR^2}{\hbar^2}F(l) \ , \tag{10.153}$$

where

$$F(l) = \frac{16l^3}{B^2}\frac{2l^2 + B}{(B + 4l^2)^2} \tag{10.154}$$

and $B = 2mR^2\omega/\hbar$. In order to get the total rate of absorption we should sum (integrate) over all the states l up to l_{max} given by $\mu = \frac{\hbar^2}{2mR^2}l_{max}^2$, which amounts to $B = \frac{\hbar\omega}{\mu}l_{max}^2$. It is easy to see that in equation (10.154) $l/l_{max} < 1$ is much larger than $\sqrt{B}/l_{max}$. Finally, we multiply by $\hbar\omega$ to get the absorbed power given by

$$P = \frac{\pi^2 q^2\omega^2 H_0^2 R^4}{4c^2\hbar} \; . \tag{10.155}$$

The electrons near the Fermi surface are quasi-classical, except those with small k (and large l) and those with small l (and large k). The states with small values of l are quantum-mechanical. If we neglect the transverse motion, their wavefunction can be written as

$$\psi = e^{\frac{iq}{2c\hbar}HR^2\varphi}\chi \; , \tag{10.156}$$

where $\chi = \frac{1}{\sqrt{2\pi}}e^{il\varphi}\frac{1}{\sqrt{L}}e^{ikx}$. The current density associated with the hamiltonian given by equation (10.148) is

$$\boldsymbol{j} = \frac{iq\hbar}{2m}\left(grad\psi^* \cdot \psi - \psi^* grad\psi\right) - \frac{q^2}{mc}\boldsymbol{A}\psi^*\psi \tag{10.157}$$

(leaving aside the spin contribution). If we neglect the perturbation and use the wavefunctions given by equation (10.156), the diamagnetic contribution to the current (the $\boldsymbol{A}$-term) cancels out exactly the paramagnetic contribution, such that we get

$$\boldsymbol{j} = \frac{1}{2\pi L}\left(\frac{q\hbar k}{m}, \frac{q\hbar l}{mR}, 0\right) \tag{10.158}$$

(in cylindrical coordinates x, φ, r; a proper normalization introduces a factor $1/RD$). Since there exist two values $\pm k$ and two values $\pm l$, the total current is zero.

If $qHR^2/2c\hbar = qH\pi R^2/ch = integer$, *i.e.* the flux $\Phi = \frac{ch}{q} \times integer$, for a constant magnetic field, then the wavefunctions ψ are the same as the wavefunctions χ; ch/q is the quantum of magnetic flux.

10 Surface Electromagnetism

Quasi-free electrons are identified on the surface of solids, irrespective of the structure and the state, which define topological conductors. The surface electrons are decoupled from the bulk electrons. Their occurrence is caused by the surface discontinuity, where the solid properties vary abruptly. The surface electrons form a quasi-two-dimensional electron gas, with poor transport properties. They exhibit specific plasmonic modes, sustain surface plasmon-polariton modes and are active in absorption of quasi-static magnetic fields.

Index